MEMS Accelerometers

MEMS Accelerometers

Special Issue Editors

Mahmoud Rasras
Ibrahim (Abe) M. Elfadel
Ha Duong Ngo

MDPI • Basel • Beijing • Wuhan • Barcelona • Belgrade

Special Issue Editors

Mahmoud Rasras
New York University Abu Dhabi
UAE

Ibrahim (Abe) M. Elfadel
Khalifa University
Abu Dhabi, UAE

Ha Duong Ngo
University of Applied Sciences Berlin
Fraunhofer Institute for Reliability and Microintegration IZM
Germany

Editorial Office
MDPI
St. Alban-Anlage 66
4052 Basel, Switzerland

This is a reprint of articles from the Special Issue published online in the open access journal *Micromachines* (ISSN 2072-666X) from 2018 to 2019 (available at: https://www.mdpi.com/journal/micromachines/special_issues/MEMS_Accelerometers)

For citation purposes, cite each article independently as indicated on the article page online and as indicated below:

LastName, A.A.; LastName, B.B.; LastName, C.C. Article Title. *Journal Name* **Year**, *Article Number*, Page Range.

ISBN 978-3-03897-414-7 (Pbk)
ISBN 978-3-03897-415-4 (PDF)

Contents

About the Special Issue Editors

Mahmoud Rasras is an Associate Professor of the Electrical and Computer Engineering at New York University Abu Dhabi (NYUAD). He received a PhD degree in physics from the Catholic University of Leuven, Belgium. Dr. Rasras has more than 11 years of industrial research experience as a Member of Technical Staff at Bell Labs, Alcatel-Lucent, NJ, USA. Prior to joining NYUAD. Dr. Rasras was a faculty member and former Director of the SRC/GF Center-for-Excellence for Integrated Photonics at Masdar Institute (part of Khalifa University). He authored and co-authored more than 120 journal and conference papers and holds 33 US patents. Dr. Rasras is an Associate Editor of Optics Express, Guest Editor – MDPI, and a Senior IEEE Member.

Ibrahim (Abe) M. Elfadel has been a professor of electrical and computer engineering at Khalifa University, Abu Dhabi, UAE, since 2011. From May 2014 to April 2018, he served as the program manager of Mubadala's TwinLab MEMS, a joint collaboration with the Institute of Microelectronics and GLOBALFOUNDRIES, Singapore, on next-generation MEMS platforms. Prior to his current position, Dr. Elfadel had a 15-year career with the corporate CAD organizations at IBM, Yorktown Heights, NY. His current MEMS research interests include IMU, piezoelectric energy harvesting, piezoresistive force sensing, and CAD methodologies for MEMS-CMOS co-design. Dr. Elfadel is the inventor of 55 US patents, the author of more than 140 archival articles, and the editor of 3 books. He is the recipient of six Invention Achievement Awards, one Outstanding Technical Achievement Award, and one Research Division Award, all from IBM. In 2014, he was the recipient of the Donald O. Pederson Best Paper Award from the IEEE Transactions on CAD. In 2018, he received the Board of Directors Recognition Award from the US-based Semiconductor Research Corporation for "Pioneering Semiconductor Research in Abu Dhabi". Dr. Elfadel has served on the technical program committees of several major conferences, including the Symposium on Design, Test, Integration, and Packaging of MEMS/MOEMS (DTIP). He is an associate editor of the IEEE Transactions on VLSI and was a general co-chair of the IFIP/IEEE 25th International Conference on Very Large Scale Integration (VLSI-SoC 2017). Dr. Elfadel received his PhD from MIT in 1993.

Ha Duong Ngo studied electrical engineering in UdSSR, in Ukraine and microsystem technologies in Chemnitz, Germany. He received a German diploma in 1998 from Technical University Chemnitz. He joined MAT (Microsensors and Actuators Technology Center) in 1998 and worked on MEMS sensors and actuators. From 2004 to 2006 he worked with Schott AG, where his research and development work focused on CMOS image sensors and wafer-level packaging technologies. He received a PhD on MOEMS from Technical University Berlin in 2006. By the end of 2006, he joined the Electrical Faculty and the Research Center for Microperipheric Technologies. He was head of the Microsensors and Actuator Technology Center at Technical University. He is now a professor at the University of Applied Sciences Berlin and a group leader of microsensors technology and high-density integration at Fraunhofer Institute IZM. His present research interests include silicon, SOI and silicon carbide technology, microsensors and actuators, AeroMEMS, printed MEMS, and sensors packaging.

 *micromachines*

Editorial

Editorial for the Special Issue on MEMS Accelerometers

Mahmoud Rasras [1,*], Ibrahim (Abe) M. Elfadel [2,*] and Ha Duong Ngo [3,4,*]

[1] Electrical and Computer Engineering, Engineering Division, New York University Abu Dhabi, Abu Dhabi, UAE

[2] Department of Electrical and Computer Engineering, Khalifa University, Abu Dhabi, UAE

[3] Hochschule für Technik und Wirtschaft Berlin, University of Applied Sciences, Treskowallee 8, 10318 Berlin, Germany

[4] Fraunhofer Institute for Reliability and Microintegration IZM, Department Wafer Level Integration, Group Leader Microsensors Technology, Gustav-Meyer-Allee 25, 13355 Berlin, Germany

* Correspondence: mr5098@nyu.edu (M.R.); ibrahim.elfadel@ku.ac.ae (I.M.E.); HaDuong.Ngo@HTW-Berlin.de (H.D.N.)

Received: 26 April 2019; Accepted: 26 April 2019; Published: 29 April 2019

Micro-Electro-Mechanical Systems (MEMS) devices are widely used for motion, pressure, light, and ultrasound sensing applications. They are also used as micro switches and micro actuators in control applications. Research on integrated MEMS technology has undergone extensive development driven by the requirements of compact footprint, low cost, and increased functionality. Accelerometers are among the most widely used sensors implemented in MEMS technology. MEMS Accelerometers are showing a growing presence in almost all industries, ranging from consumer electronics to transportation and from games and entertainment to healthcare. Their MEMS embodiment has evolved from single, stand-alone devices to the integrated, 6-axis and 9-axis inertial motion units that are available on the market today. A traditional MEMS accelerometer employs a proof mass suspended to springs, which displaces in response to an external acceleration. A single proof mass can be used for one- or multi-axis sensing. A variety of transduction mechanisms have been used to detect the displacement. They include—capacitive, piezoelectric, piezoresistive, thermal, tunneling, and optical. Capacitive accelerometers are widely used due to their DC measurement interface, thermal stability, reliability, and low-cost. However, they are sensitive to electromagnetic field interferences and have poor performance for high-end applications (e.g., precise attitude control for satellites). Over the past three decades, steady progress has been made in the area of optical accelerometers for high-performance and high-sensitivity applications but several challenges are still to be tackled by researchers and engineers to fully realize Opto-Mechanical Accelerometers, such as chip-scale integration, scaling, low bandwidth, etc. Currently, optical technologies are still used in navigation systems and tactical guidance. New applications have been enabled by low-cost MEMS sensors, and significant progress has been made in the past few years in terms of their reliability. MEMS accelerometers are now accepted in high-reliability environments, and are even starting to replace optical and other established technologies.

This Special Issue on "MEMS Accelerometers" includes research papers, short communications, and review articles. There are 16 papers published covering the design, fabrication, modeling and applications of MEMS accelerometers. Half of the papers discuss accelerometer integration [1,2], piezoresistive sensing [3,4] multi-axis accelerometers, and review current technologies [4–6]. Three papers investigate MEMS accelerometer multi-physics modeling [7–10]. The rest of the papers are focused on the application domains, including environmental monitoring [11] and WiFi positioning [12]. Healthcare monitoring, positioning and daily activity monitoring are discussed in [13,14], while wearable body sensors for patients with gait impairments and the classification of horse gaits for self-coaching are covered in [15,16].

On the device design and integration, H. Liu et al. [1] demonstrate a hybrid-integrated, high-precision, vacuum accelerometer based on field emission. It shows a sensitivity of 3.081 V/g, the non-linearity is 0.84% in the acceleration range of −1 g to 1 g, while the average noise spectrum density value is 36.7 μV/Hz in the frequency range of 0–200 Hz. H. Liu et al. [2] develop a differential capacitive accelerometer based on low-temperature co-fired ceramic (LTCC) technology for harsh-environment applications. The device has a full-scale range of 10 g with a sensitivity of 30.27 mV/g. X. Hu et al. [3] report on a family of silicon-on-insulator (SOI)–based high-g MEMS piezoresistive sensors for the measurement of accelerations up to 60,000 g. In this device, four piezoresistors are connected in a Wheatstone bridge to measure acceleration. X. Zhao et al. [4] also develop a silicon-on-insulator (SOI) piezoresistive, three-axis acceleration sensor with demonstrated sensitivities along x-axis, y-axis, and z-axis of 0.255 mV/g, 0.131 mV/g, and 0.404 mV/g, respectively. A thermal convection-based accelerometer is fabricated and characterized by J. Kim et al. [5]. They investigate the impact of cavity volume, gas medium density and viscosity with a focus on the Z-axis response. Z. Mohammed et al. [6] provide an in-depth review of monolithic multi-axis capacitive MEMS accelerometers, including a detailed analysis of recent advancements aimed at addressing various challenges such as size, noise floor, cross-axis sensitivity, and process aware modeling.

As for multi-physics modeling, X. Dong et al. [7] develop an experimental method for measuring the parasitic capacitance mismatch in a MEMS accelerometer. This result is helpful for improving bias performance and the scale factor. F. Wang et al. [8] report on the design, modeling, and fabrication of an elastic-beam delay element. Chen D. et al. [9] propose using a fifth-order $\Sigma\Delta$ closed-loop interface for a capacitive MEMS accelerometer that includes a digital built-in self-testing feature. By a single-bit $\Sigma\Delta$-modulation, the noise and linearity of excitation is effectively improved, and a higher detection level for distortion is achieved. Yang Z. et al. [10] show that the angular-rate sensing based on mode splitting offers good suppression of Kerr noise. They demonstrate that at an angular rate of $5 \times 106°/s$, a Kerr noise of 1.913×10^{-5} Hz is measured which corresponds to an angular rate deviation of $9.26 \times 10^{-9}°/s$.

As for MEMS accelerometer applications, Tian B. et al. [11] design a probe for marine environmental monitoring to estimate the ocean turbulent kinetic energy dissipation rate. They achieve a sensitivity of 3.91×10^{-4} (Vms²)/kg over a measurement range of 10^{-8}–10^{-4} W/kg. Lai M. et al. [12] study a large amount of raw data measured by a MEMS accelerometer-based wrist-worn device. This device is used to monitor different levels of physical activities (PAs) for subjects wearing it continuously 24 h a day. Lin W. et al. [13] develop a method using multi-mounted devices to construct a lightweight site-survey radio map (LSS-RM) for WiFi positioning. Their experimental results show that their method can reduce the time required to construct a WiFi-received signal strength index (RSSI) radio map from 54 min to 7.6 min. Yuan C. et al. [14] propose a novel framework for fault-tolerant visual-inertial odometry (VIO) navigation and positioning. Qiu. S. et al. [15] show promising results for a low-cost, intelligent and lightweight wearable gait analysis platform based on body IMU sensor networks. They have assembled the IMU from accelerometers/gyroscopes chipsets. A multi-sensor fusion algorithm is used to estimate the gait parameters. The method has great potential as an auxiliary for medical rehabilitation assessment. Lee J. et al. [16] investigate the classification of horse gaits using MEMS inertial sensor technology with the goal of developing a horse-gait self-coaching platform based on machine learning methods. In the experimental setup, the authors employ a camera-less 3D human motion measurement system based on state-of-the-art MEMS inertial sensors, biomechanical models, and sensor fusion algorithms.

We would like to thank all the authors for submitting their original papers to this special issue. Special thanks are also due to all the reviewers for their dedicated efforts in helping to improve the quality of the submitted papers. Finally, we are grateful to Ms. Mandy Zhang and the MDPI team for all their editorial assistance.

Conflicts of Interest: The authors declare no conflict of interest.

References

1. Liu, H.; Wei, K.; Li, Z.; Huang, W.; Xu, Y.; Cui, W. A Novel, Hybrid-Integrated, High-Precision, Vacuum Microelectronic Accelerometer with Nano-Field Emission Tips. *Micromachines* **2018**, *9*, 481. [CrossRef] [PubMed]

2. Liu, H.; Fang, R.; Miao, M.; Zhang, Y.; Yan, Y.; Tang, X.; Lu, H.; Jin, Y. Design, Fabrication, and Performance Characterization of LTCC-Based Capacitive Accelerometers. *Micromachines* **2018**, *9*, 120. [CrossRef] [PubMed]

3. Hu, X.; Mackowiak, P.; Bäuscher, M.; Ehrmann, O.; Lang, K.; Schneider-Ramelow, M.; Linke, S.; Ngo, H. Design and Application of a High-G Piezoresistive Acceleration Sensor for High-Impact Application. *Micromachines* **2018**, *9*, 266. [CrossRef] [PubMed]

4. Zhao, X.; Wang, Y.; Wen, D. Fabrication and Characteristics of a SOI Three-Axis Acceleration Sensor Based on MEMS Technology. *Micromachines* **2019**, *10*, 238. [CrossRef] [PubMed]

5. Kim, J.; Han, M.; Kang, S.; Kong, S.; Jung, D. Multi-axis Response of a Thermal Convection-based Accelerometer. *Micromachines* **2018**, *9*, 329. [CrossRef] [PubMed]

6. Mohammed, Z.; Elfadel, I.; Rasras, M. Monolithic Multi Degree of Freedom (MDoF) Capacitive MEMS Accelerometers. *Micromachines* **2018**, *9*, 602. [CrossRef] [PubMed]

7. Dong, X.; Yang, S.; Zhu, J.; En, Y.; Huang, Q. Method of Measuring the Mismatch of Parasitic Capacitance in MEMS Accelerometer Based on Regulating Electrostatic Stiffness. *Micromachines* **2018**, *9*, 128. [CrossRef] [PubMed]

8. Wang, F.; Zhang, L.; Li, L.; Qiao, Z.; Cao, Q. Design and Analysis of the Elastic-Beam Delaying Mechanism in a Micro-Electro-Mechanical Systems Device. *Micromachines* **2018**, *9*, 567. [CrossRef] [PubMed]

9. Chen, D.; Liu, X.; Yin, L.; Wang, Y.; Shi, Z.; Zhang, G. A ΣΔ Closed-Loop Interface for a MEMS Accelerometer with Digital Built-In Self-Test Function. *Micromachines* **2018**, *9*, 444. [CrossRef] [PubMed]

10. Yang, Z.; Li, D.; Sun, Y. Analysis of Kerr Noise in Angular-Rate Sensing Based on Mode Splitting in a Whispering-Gallery-Mode Microresonator. *Micromachines* **2019**, *10*, 150. [CrossRef] [PubMed]

11. Tian, B.; Li, H.; Yang, H.; Zhao, Y.; Chen, P.; Song, D. Design and Performance Test of an Ocean Turbulent Kinetic Energy Dissipation Rate Measurement Probe. *Micromachines* **2018**, *9*, 311. [CrossRef] [PubMed]

12. Yang, W.; Xiu, C.; Ye, J.; Lin, Z.; Wei, H.; Yan, D.; Yang, D. LSS-RM: Using Multi-Mounted Devices to Construct a Lightweight Site-Survey Radio Map for WiFi Positioning. *Micromachines* **2018**, *9*, 458. [CrossRef] [PubMed]

13. Lin, W.; Verma, V.; Lee, M.; Lai, C. Activity Monitoring with a Wrist-Worn, Accelerometer-Based Device. *Micromachines* **2018**, *9*, 450. [CrossRef] [PubMed]

14. Yuan, C.; Lai, J.; Lyu, P.; Shi, P.; Zhao, W.; Huang, K. A Novel Fault-Tolerant Navigation and Positioning Method with Stereo-Camera/Micro Electro Mechanical Systems Inertial Measurement Unit (MEMS-IMU) in Hostile Environment. *Micromachines* **2018**, *9*, 626. [CrossRef] [PubMed]

15. Qiu, S.; Liu, L.; Zhao, H.; Wang, Z.; Jiang, Y. MEMS Inertial Sensors Based Gait Analysis for Rehabilitation Assessment via Multi-Sensor Fusion. *Micromachines* **2018**, *9*, 442. [CrossRef] [PubMed]

16. Lee, J.; Byeon, Y.; Kwak, K. Design of Ensemble Stacked Auto-Encoder for Classification of Horse Gaits with MEMS Inertial Sensor Technology. *Micromachines* **2018**, *9*, 411. [CrossRef] [PubMed]

Article

Fabrication and Characteristics of a SOI Three-Axis Acceleration Sensor Based on MEMS Technology

Xiaofeng Zhao *, Ying Wang and Dianzhong Wen

The Key Laboratory of Electronics Engineering, College of Heilongjiang Province, Heilongjiang University, Harbin 150080, China; 2181212@s.hlju.edu.cn (Y.W.); wendianzhong@hlju.edu.cn (D.W.)
* Correspondence: zhaoxiaofeng@hlju.edu.cn; Tel.: +86-451-8660-8457

Received: 31 January 2019; Accepted: 7 April 2019; Published: 9 April 2019

Abstract: A silicon-on-insulator (SOI) piezoresistive three-axis acceleration sensor, consisting of four L-shaped beams, two intermediate double beams, two masses, and twelve piezoresistors, was presented in this work. To detect the acceleration vector (a_x, a_y, and a_z) along three directions, twelve piezoresistors were designed on four L-shaped beams and two intermediate beams to form three detecting Wheatstone bridges. A sensitive element simulation model was built using ANSYS finite element simulation software to investigate the cross-interference of sensitivity for the proposed sensor. Based on that, the sensor chip was fabricated on a SOI wafer by using microelectromechanical system (MEMS) technology and packaged on a printed circuit board (PCB). At room temperature and V_{DD} = 5.0 V, the sensitivities of the sensor along x-axis, y-axis, and z-axis were 0.255 mV/g, 0.131 mV/g, and 0.404 mV/g, respectively. The experimental results show that the proposed sensor can realize the detection of acceleration along three directions.

Keywords: three-axis acceleration sensor; MEMS technology; sensitivity; L-shaped beam

1. Introduction

Accelerometers have been used in many different fields, such as automotive industry, aviation and national security, aerospace engineering, biological engineering, etc. [1]. The main sensing mechanisms to convert acceleration into electrical signals include piezoresistive, capacitive, piezoelectric and resonant types, etc. Nevertheless, piezoresistive technique among of them has been attracted more attention due to its simple structures design and read out circuits, good direct current (DC) response, high sensitivity, linearity, and reliability as well as low cost. In 1979, Roylance et al. proposed a piezoresistive microsilicon accelerometer for the first time [2]. In addition, with the development of microelectromechanical system (MEMS) technology, acceleration sensors have been widely used in the field of inertial systems to test the acceleration of moving object [3–5]. Up to date, the three-axis acceleration sensor has realized the measurements of the velocity and posture for moving objects including unmanned aerial vehicle, gravity gradiometer, wearable acceleration sensor for monitoring human movement behavior, etc. [6,7]. Due to the extensive applications in many different fields, increasing demands for detection has triggered a particular research attention to improve the properties of three-axis acceleration sensor, such as miniaturization, high sensitivity, good consistency and low cross-interference of sensitivity, etc. For example, in 2011, Hsieh et al. designed a three-axis piezoresistive accelerometer with a stress isolation guard-ring structure, a low disturbance of environment and a big sensitivity range of 0.127 to 0.177 mV/(g·V) [8]. In 2016, Xu et al. fabricated a novel piezoresistive accelerometer with axially stressed sensing beams, not only improving the sensitivity and the resonant frequency at a supply voltage of 3.0 V, but also reducing the cross-axis sensitivity along x-axis and z-axis by less than 4.875×10^{-6} mV/g and 4.425×10^{-6} mV/g, respectively [9]. Thereafter, in 2017, Jung et al. proposed a monolithic piezoresistive high-g (20000 g) three-axis accelerometer with a single proof mass suspended using thin eight beams, achieving sensitivities of 0.243 mV/g, 0.131 mV/g, and 0.307 mV/g along the *x*-axis, *y*-axis and *z*-axis at a supply voltage of 5.0 V, respectively [10]. Meanwhile, Wang et al. presented a high-performance piezoresistive micro-accelerometer based on slot etching in an eight-beam structure to detect the vibration of a high speed spindle, improve the sensitivity and the natural frequency, as well as realize an average sensitivity of 0.785 mV/g at a supply voltage of 5.0 V [11]. In 2018, Marco et al. proposed a piezoresistive accelerometer based on a progressive moment of inertia (MMI) increment of the sensor proof mass in three-axis head injuries monitoring, obviously enhancing the sensitivity of the optimized structure along the *z*-axis up to 0.22 mV/g and obtaining low cross-interference less than 1% F.S. [12]. Meanwhile, Han et al. proposed a low cross-axis sensitivity piezoresistive accelerometer based on masked–maskless wet etching, which consisted of a proof mass, eight supporting beams, and four sensing beams, and achieved cross-axis sensitivities along *x*-axis and *y*-axis of 1.67% and 0.82%, respectively [13]. As the characteristics of the sensor are closely related to the sensing structure and the sensitive element of sensor, currently available methods have been adopted to improve the sensitivity and reduce the cross-interference, including modifying structure and selecting novel sensitive materials.

In this paper, a silicon-on-insulator (SOI) three-axis acceleration sensor with four L-shaped beams, intermediate double beams, and two masses was presented. To detect the acceleration vector (a_x, a_y, and a_z) along three directions and reduce the size of the chip, a basic structure of sensor was designed by using MEMS technology, and the corresponding working principle was investigated. Meanwhile, in order to reduce the cross-interference of sensitivity, how the sensitive element influences the cross-interference of sensitivity was analyzed by using ANSYS finite element simulation software. Based on that, the chip was fabricated on the SOI wafer by using MEMS technology and the thicknesses of cantilever beams can be effectively controlled, avoiding the effects of beams' thickness on the sensitive characteristics. The study on the proposed sensor provides a new strategy for fabricating three-axis acceleration sensor to detect the acceleration vector.

2. Basic Structure and Sensing Principle

2.1. Basic Structure

To easily release the structure of beams and better control the thickness of the beams by using the self-stop technology of inductively-coupled plasma (ICP) etching technology, a SOI wafer was utilized as a substrate of the proposed three-axis acceleration sensor. Figure 1a,b show the top and back views of the SOI three-axis acceleration sensor, respectively. The chip is composed of an elastic structure and a piezo-sensitive element as shown in Figure 1a, where the elastic structure includes four L-shaped beams (L_1, L_2, L_3, and L_4) and an intermediate double beam (L_5 and L_6). l_1 (1200 μm) and w_1 (200 μm) are the length and the width of the L-shaped beams for the proposed sensor, respectively. l_3 (300 μm) and w_3 (150 μm) are the length and the width of the double beams, respectively. d (100 μm) is the thicknesses of the L-shaped beams (L_1, L_2, L_3, L_4, L_5, and L_6), named d_{L1}, d_{L2}, d_{L3}, d_{L4}, d_{L5}, and d_{L6}, respectively. l_2 (2600 μm) and w_2 (850 μm) are the length and the width of the two masses. Twelve piezoresistors are exploited as the sensitive elements, where the four piezoresistors (R_{x1}, R_{x2}, R_{x3}, and R_{x4}) far away from the mass were fabricated at the roots of L-shape beams (L_1, L_2, L_3, and L_4) to form the first Wheatstone bridge (W_x). Meanwhile, the four piezoresistors (R_{y1}, R_{y2}, R_{y3}, and R_{y4}) close to the mass were fabricated at the roots of L-shaped beams (L_1, L_2, L_3, and L_4) to construct the second Wheatstone bridge (W_y), and the other piezoresistors (R_{z1}, R_{z2}, R_{z3}, and R_{z4}) at the roots of the double beams (L_5, L_6) form the third Wheatstone bridge (W_z) in response. W_x, W_y, and W_z are used to measure the acceleration along *x*-axis, *y*-axis, and *z*-axis (a_x, a_y, and a_z), respectively. Based on that, through analyzing the effect of conduction type and doping concentration on piezoresistive coefficient, the piezoresistors were selected as p-Si, and its resistivity was designed in the range of 0.01 to 0.1 Ω·cm.

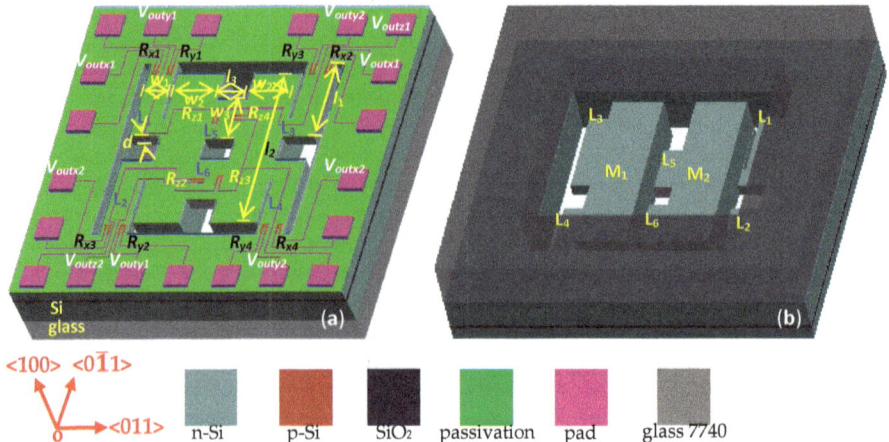

Figure 1. Basic structure of the silicon-on-insulator (SOI) three-axis acceleration sensor: (**a**) top view and (**b**) back view.

To realize a free movement of the middle double masses in the space, the back side of the chip was bonded with a glass sheet with a hole in the middle by using bonding technology, as shown in Figure 1b.

2.2. Theoretical Analysis of Sensing Principle

To study the sensing principle of the chip under different accelerations, theoretical analysis was presented based on piezoresistive effect. In the condition of stress, the relative variation of the silicon piezoresistor along the same crystal orientation can be expressed as [14]

$$\frac{\Delta R}{R_0} = \pi_\parallel \sigma_\parallel + \pi_\perp \sigma_\perp, \tag{1}$$

where ΔR is the variation of the piezoresistor. R_0 is the value of the piezoresistor without stress. $\pi_\parallel$ and $\pi_\perp$ are the longitudinal and the lateral piezoresistive coefficients, respectively. $\sigma_\parallel$ and $\sigma_\perp$ are the longitudinal and the lateral stresses, respectively.

From Equation (1), we can see that the main factors to influence ΔR include piezoresistive coefficients ($\pi_\parallel$ and $\pi_\perp$) and stresses ($\sigma_\parallel$ and $\sigma_\perp$). Due to the silicon belongs to the cubic crystal system, the piezoresistive coefficient along any crystal orientation can be expressed as [14,15]

$$\begin{cases} \pi_\parallel = \pi_{11} - 2(\pi_{11} - \pi_{12} - \pi_{44})(l_1^2 m_1^2 + m_1^2 n_1^2 + n_1^2 l_1^2) \\ \pi_\perp = \pi_{12} + (\pi_{11} - \pi_{12} - \pi_{44})(l_1^2 l_2^2 + m_1^2 m_2^2 + n_1^2 n_2^2) \end{cases}, \tag{2}$$

where π_{11} and π_{12} are the longitudinal and the lateral piezoresistive coefficients along the crystal axis orientation, respectively. π_{44} is the shear piezoresistive coefficient. l_1, m_1 and n_1 are the cosine of the piezoresistor's longitudinal orientation. l_2, m_2 and n_2 are the cosine of the piezoresistor's lateral orientation.

As shown in Equation (2), it can be found that the piezoresistive coefficients of silicon along different orientations are different from each other. As is well-known, the piezoresistive coefficient of p-Si is better than that of n-Si. According to theoretical analysis, the $\pi_\parallel$ and $\pi_\perp$ on the (100) plane of p-Si are positive and negative, respectively, i.e., $\pi_\parallel$ along [011] is $\pi_{44}/2$ and $\pi_\perp$ along [0$\bar{1}$1] is $-\pi_{44}/2$. Thus, it is possible to obtain a maximum piezoresistive coefficient. Based on the above analysis, the piezoresistors were designed along [011] and [0$\bar{1}$1] orientations.

To analyze the working principle of the chip, a simulation model was built by using ANSYS finite element software. Based on this model, the effects of acceleration on the deformations of the L-shaped beams and the middle double beams were investigated. Figure 2 shows the deformation diagrams of the beams in the condition of $a = 0$ g, $a = a_x$, $a = a_y$, and $a = a_z$, respectively. To further analyzing the sensing characteristics of the acceleration sensor, twelve piezoresistors on the beams are equivalent to three Wheatstone bridge circuits, with an equivalent circuit diagram under the action of $a = 0$ g, $a = a_x$, $a = a_y$, and $a = a_z$, respectively, as shown in Figure 3.

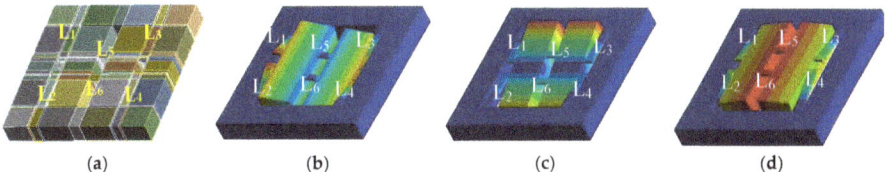

Figure 2. The deformation diagram of the chip under the acceleration along three-axis directions: (**a**) $a_x = a_y = a_z = 0$ g; (**b**) $a = a_x$; (**c**) $a = a_y$; and (**d**) $a = a_z$.

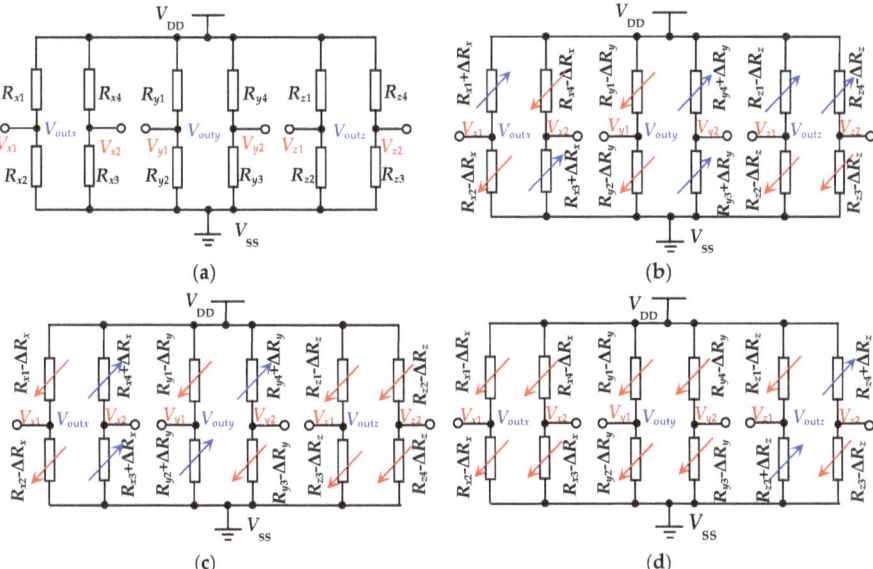

Figure 3. Equivalent circuit of the SOI three-axis acceleration sensor: (**a**) $a_x = a_y = a_z = 0$ g; (**b**) $a = a_x$; (**c**) $a = a_y$; and (**d**) $a = a_z$.

In an ideal case, no deformation exhibits in the structure of the proposed sensor under no acceleration along x-axis, y-axis, and z-axis, leading to an equal resistance value of the twelve piezoresistors and no output of V_{outx}, V_{outy}, and V_{outz} for the three Wheatstone bridges, as shown in Figure 3a. According to Newton's second law, the two masses would cause a displacement along x-axis under the action of a_x when applying acceleration along x-axis as shown in Figure 3b. As a result, the two L-shaped beams (L_1 and L_2) were squeezed and the other two L-shaped beams (L_3 and L_4) were stretched, as shown in Figure 2b. The deformation of L-shaped beams causes a different stress distribution at the roots of beams, resulting in the increase of R_{x1}, R_{x3}, R_{y3}, and R_{y4} and the decrease of R_{x2}, R_{x4}, R_{y1}, R_{y2}, R_{z1}, R_{z2}, R_{z3}, and R_{z4} based on the elastic theory and piezoresistive effect [16,17], as shown in Figure 3b. In view of the change of V_{outx} with the external acceleration, a_x can be measured.

In response, L_1 and L_3 were squeezed and L_2 and L_4 were stretched under the action of a_y for the chip shown in Figure 2c; the different stress distributions at the roots of beams would cause the increase of R_{x3}, R_{x4}, R_{y2}, and R_{y4} and the decrease of R_{x1}, R_{x2}, R_{y1}, R_{y3}, R_{z1}, R_{z2}, R_{z3}, and R_{z4}, as shown in Figure 3c. From the change of V_{outy} with the external acceleration, a_y can be measured. When a_z is applied, the middle double masses would form a displacement along z-axis under the action of a_z, as shown in Figure 2d, resulting in the four L-shaped beams (L_1, L_2, L_3, and L_4) to be squeezed or stretched at the same time, and intermediate double beams (L_5 and L_6) to be bent under an external force. Since the combined actions of R_{z1} and R_{z3} acting as longitudinal resistances, and R_{z2} and R_{z4} acting as lateral resistances, it is inevitable that a reduction in R_{z1} and R_{z3} as well as increase in R_{z2} and R_{z4} will occur, as shown in Figure 3d. According to the V_{outz} changes with the external acceleration, which depends on the stress distribution on double beams (L_5 and L_6) and the resistance changes of z-axis's piezoresistors caused by the middle double beam deformations, it is possible to achieve the measurement of a_z.

Based on the piezoresistive effect and the above equivalent circuit analysis, the relationship between output voltage (V_{outx}, V_{outy} and V_{outz}) and relative variation of piezoresistors can be expressed as Equation (3):

$$\begin{cases} V_{outx} = V_{x1} - V_{x2} = \frac{\Delta R_x}{R_0} \cdot V_{DD} \\ V_{outy} = V_{y1} - V_{y2} = \frac{\Delta R_y}{R_0} \cdot V_{DD} \\ V_{outz} = V_{z1} - V_{z2} = \frac{\Delta R_z}{R_0} \cdot V_{DD} \end{cases} \tag{3}$$

where V_{outx}, V_{outy}, and V_{outz} are the output voltages of the three Wheatstone bridges along the x-axis, y-axis, and z-axis, respectively. V_{DD} is the supply voltage and R_0 is the resistance value of the piezoresistor under no external acceleration. In an ideal case, ΔR_x, ΔR_y, and ΔR_z are the variations of piezoresistors along x-axis (R_{x1}, R_{x2}, R_{x3}, and R_{x4}), y-axis (R_{y1}, R_{y2}, R_{y3}, and R_{y4}) and z-axis (R_{z1}, R_{z2}, R_{z3}, and R_{z4}), respectively.

Under no accelerations along x-axis, y-axis, and z-axis, ΔR_x, ΔR_y, and ΔR_z are equal to zero, resulting in no output of V_{outx}, V_{outy}, and V_{outz}. Nevertheless, the absolute values of ΔR_x, ΔR_y, and ΔR_z are approximately equal under the action of acceleration along x-axis, y-axis, or z-axis, ideally contributing to the same values of V_{outx}, V_{outy}, and V_{outz} for the proposed sensor.

Based on the above theoretical analysis, it is possible to realize the measurement of accelerations along x-axis, y-axis, and z-axis by using the proposed sensor. According to the definition of sensitivity and Equation (4), when applying acceleration to the sensor, the output voltages can be expressed as

$$\begin{bmatrix} V_{outx} \\ V_{outy} \\ V_{outz} \end{bmatrix} = \begin{bmatrix} S_{xx} & S_{xy} & S_{xz} \\ S_{yx} & S_{yy} & S_{yz} \\ S_{zx} & S_{zy} & S_{zz} \end{bmatrix} \begin{bmatrix} a_x \\ a_y \\ a_z \end{bmatrix} \tag{4}$$

where a_x, a_y, and a_z are the components of acceleration along x-axis, y-axis, and z-axis, respectively. S_{xx}, S_{yy}, and S_{zz} are the sensitivities along x-axis, y-axis, and z-axis, respectively. S_{xy} and S_{xz} are the x-axis cross-axis sensitivity under the actions of a_y and a_z, respectively. S_{yx} and S_{yz} are the y-axis cross-axis sensitivity under the actions of a_x and a_z, respectively. S_{zx} and S_{zy} are the cross-axis sensitivity of z-axis under a_x and a_z, respectively.

2.3. Simulation Analysis of Sensing Principle

To analyze the sensitive characteristics and the cross-interference of sensitivity, a sensitive element simulation model of the proposed sensor was installed by using ANSYS finite element simulation software. According to Equation (1), i.e., the relative variation of piezoresistors is proportional to the stress acted, every acceleration in x, y, and z directions has similar behaviors associated with measurement principle, and only one of them in the three directions is needed for detailed measurement. In order to investigate the stress distributions of the four piezoresistors (R_{y1}, R_{y2}, R_{y3}, and R_{y4}) at the roots of L-shaped beams, the y-axis acceleration taken as example was tested by using the resulted

Besides noise, linearity is another key performance parameter which often imposes an upper limit on the range of signal amplitude that can be precisely represented. As mentioned in Equation (6), in our system, the main source of non-linearity comes from the fact that the feedback electrostatic force has a nonlinear displacement modulation effect, since the square-law effect of it is linearized by the use of ΣΔ modulation. For a closed-loop operation, the most straightforward way to suppress the residue displacement is to increase the loop gain seen by it. However, note that, if the loop is breaking at different node, the effective loop gain is different. As a result, although the loop gain seen by quantization noise is sufficiently high, this is not the case for residue displacement. When breaking at the displacement node, the inside 3-order modulator should be considered as a whole, thus, the loop gain seen by displacement x can be expressed as

$$G_x(z) = \frac{F_0 K_{x/C} K_{C/V} H_{ms}(z) H_C(z) STF_{3-order}(z)}{m}.$$

(16)

As evident, the 3-order integrating effect, which is the major force in gain enhancement, is missing in this expression. As a result, the loop gain seen by residue displacement is much less than that seen by quantization noise. In order to make up for this loss, an additional gain stage is needed. However, if it is inserted in the feed-forward path, there will be more energy appearing in front of the quantizer, which will result in an earlier saturation. An alternative way is to scale the coefficient in the feedback path L_1, which can easily be realized by scaling the feedback voltage. As shown in Equation (11), the decrease of L_1 will give rise to $STF_{3-order}$. The distribution plot of the input voltage in front of quantizer is shown in Figure 6. As shown, the scaling of L_1 will also help in reducing the signal amplitude in the feed-forward path, which will make the behavior of the electrical circuit more ideal. However, we must admit that the decreasing of L_1 will cause a reduction in the quantization noise suppressing ability due to the reduced loop gain. Besides that, there will need more continuous logic levels to draw the input back, which means more susceptible to instability. In our system, L_1 is scaled as one-tenth of L_0, to make a compromise to those trade-offs.

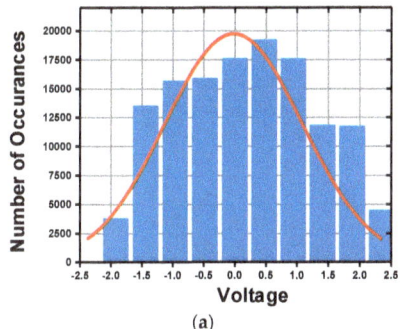
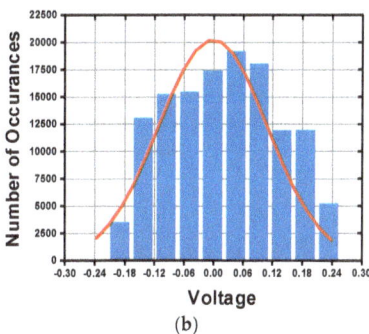

(a) (b)

Figure 6. Distribution plot of the amplitude in front of quantizer: (**a**) loop gain enhanced by reinforcing the feed-forward path; (**b**) loop-gain enhanced by scaling of the feedback path L_1.

2.3.2. Stability

Next, we come to the stability problem. In order to maintain stability, a phase compensator is inserted in the feed-forward path, whose expression is given by

$$H_C(z) = \frac{z - \alpha}{z},$$

(17)

where α is a compensation factor whose value is in the range of $(0, 1)$. The distribution plot of poles and zeros is shown in Figure 7a. The frequency response of $H_C(z)$ can be found by graphically considering

sensitive element. Figure 4 shows the relationship curves between the max stress values along the analysis path of four piezoresistors and a_x, a_y, and a_z, respectively.

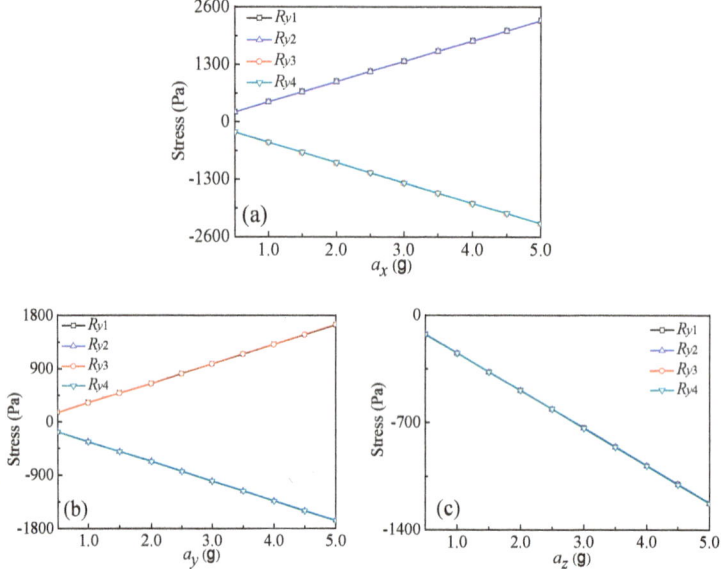

Figure 4. Relationship curves between the stress of L-shaped beams' root along y-axis and applied acceleration: (a) $a = a_x$; (b) $a = a_y$; and (c) $a = a_z$.

The analysis results indicate that when applying a_y to the chip, displacements of the two masses would be caused based on Newton's second law. As shown in Figure 4b, the roots of L_1 and L_3 exhibit negative stresses due to the above squeezing, in contrast, the roots of L_2 and L_4 display positive stresses caused by the above stretching, but both with an approximate equal absolute value of the stresses at the roots. In the ideal case, when applying a_y to the chip, V_{outy} increases with the increase of a_y, as shown in Figure 3c. Since L_1 and L_2 are squeezed and L_3 and L_4 are stretched under the action of a_x, the resulted deformations lead to the positive stresses exhibiting at the roots of L_1 and L_2, also causing the equal negative stress existing at the roots of L_3 and L_4, as shown in Figure 4a. In the ideal condition, the y-axis cross-sensitivity (S_{yx}) is ignorable under the action of a_x, as shown in Figure 3b. However, all of L_1, L_2, L_3, and L_4 are stretched when applying a_z to the chip, with approximately equal negative stresses at the roots of the four beams as shown in Figure 4c. From Figure 3d, the y-axis cross-sensitivity (S_{yz}) can be ignored under the action of a_z.

In order to reduce the effects of different beams thicknesses on the characteristics of the proposed sensor, a SOI wafer was used as the substrate of chip and MEMS technology was utilized to realize an accurate controlling of beams thicknesses. When applying acceleration along the z-axis, it is possible to form constant output voltages of the proposed sensor along x-axis and y-axis directions but no cross-sensitivity (S_{xz} and S_{yz}) due to the same deformations of the four L-shaped beams and the equal variations of the piezoresistors along x-axis and y-axis. Similarly, the cross-sensitivity (S_{yx}, S_{zx}, S_{zy}, and S_{xy}) equals zero, and the output voltages can be expressed as Equation (5).

$$
\begin{bmatrix} V_{outx} \\ V_{outy} \\ V_{outz} \end{bmatrix} = \begin{bmatrix} S_{xx} & 0 & 0 \\ 0 & S_{yy} & 0 \\ 0 & 0 & S_{zz} \end{bmatrix} \begin{bmatrix} a_x \\ a_y \\ a_z \end{bmatrix}
\tag{5}
$$

Though above design, the proposed sensor is possible to realize the measurement of acceleration along three-axis. The simulation result indicates that it is possible to improve the sensitive characteristics and reduce the sensitivity cross-interference.

3. Fabrication Technology

The chip was fabricated on a SOI wafer (the device layer is <100> orientation n-Si with resistivity of 0.1 Ω·cm, and the thickness of device layer, buried oxide, and handle substrate for SOI wafer are 100 μm, 0.5 μm, and 400 μm, respectively) by using MEMS technology. The main processing steps are shown in Figure 5.

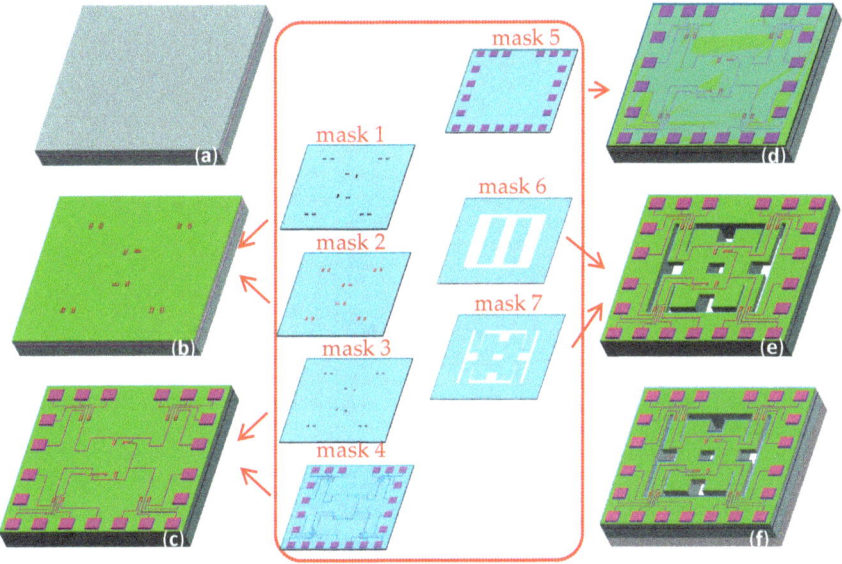

Figure 5. The main fabrication process of the proposed chip: (**a**) cleaning SOI wafer; (**b**) forming p- and p+ region; (**c**) forming contact hole and etching Al to form electrode; (**d**) forming pad; (**e**) etching the chip by inductively-coupled plasma (ICP) etching technology to form beam; and (**f**) bonding the back of the chip to the glass with hole.

(a) Cleaning the SOI wafer by using a standard cleaning method, and first oxidation to form a SiO_2 layer with a thickness of 50 nm by thermal oxidation method, taking as a buffer layer for the ion implantation.

(b) First photolithography to etch the SiO_2 layer by using wet etching technology and ion implantation to perform p- region; the ion injection dose and energy were 5×10^{13} cm^{-2} and 60 keV, respectively. Thereafter, second photolithography to etch windows of the piezoresistors and ion implantation to perform p+ region as piezoresistors, the ion injection dose and energy of 5×10^{15} cm^{-2} and 60 keV, respectively, and then annealing for 30 min at 1000 °C.

(c) Etching SiO_2 layer of 50 nm using wet etching technology and growing SiO_2 layer of 400–500 nm as insulating layer by using plasma enhanced chemical vapor deposition (PECVD) method, third photolithography to etch the top surface to perform a contact hole and fabricate metal Al of 500 nm on the top surface of substrate by a vacuum evaporation method, fourth photolithography to form the electrodes, and then metalizing at 420 °C for 30 min to achieve ohmic contact.

(d) Growing SiO_2 layer of 500–700 nm by using plasma enhanced chemical vapor deposition (PECVD) method to form passive layer and fifth photolithography to form pad.

(e) Etching the backside of the chip until the position of the buried oxide layer by inductively couple plasma (ICP) etching technology to form a silicon mass block. Thereafter, etching the front side of the chip until the same position of the buried oxide layer by ICP etching technology to release four L-shaped beams and the double beams.

(f) Bonding the glass sheet with the back of the chip using anodic bonding technology. Finally, the chip with an area of 4000 μm × 4000 μm was packaged on the printed circuit board (PCB) through internal lead bonding technology. As shown in Figure 6, the entire images of the chip were observed by using Olympus microscope.

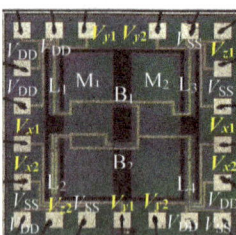

Figure 6. The photograph of the proposed three-axis acceleration sensor chip.

4. Results and Discussion

4.1. Testing System

As shown in Figure 7, a test system of a three-axis acceleration sensor was constructed, mainly consisting of a standard vibration table (Dongling ESS-050, Suzhou, China), a digital multimeter (Agilent 34410A, Agilent Technologies, Santa Clara, CA, USA), an oscilloscope (Agilent DSO-X 4154A, Agilent Technologies), and a programmable linear direct-current power (Rigol DP832A, RIGOL Technologies. Inc., Beijing, China) in a voltage range of 0 to 30 V. The testing system can supply an excitation frequency of 5–10000 Hz and an acceleration of 0–30 g, where an accelerometer (Dytran 3120B, Dytran Instrument, Inc., Chatsworth, CA, USA) is used as a reference. At room temperature, the chip is fixed on the surface of standard vibration table. On the basis of that, some relative properties are investigated, including the resonance frequency, the sensitivity, the cross-interference of sensitivity, and so on.

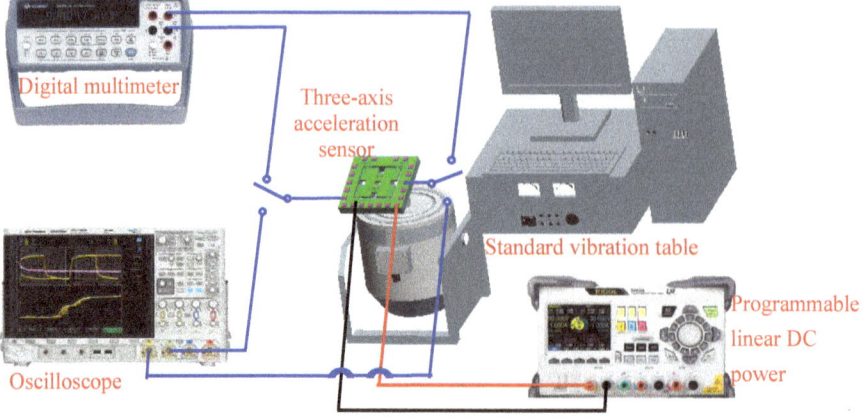

Figure 7. The test system of the three-axis acceleration sensor.

4.2. Resonance Characteristics

Under the conditions of vibration acceleration of 3.0 g, a supply voltage of 5.0 V, as well as an excitation frequency range from 100 Hz to 10000 Hz, the relationship curves of output voltage vs. excitation frequency of a_x, a_y, and a_z were obtained at room temperature, as shown in Figure 8. It can be seen that the output voltage changes with the increase of excitation frequency at a constant acceleration. As shown in Figure 8a, when increasing the excitation frequency along x-axis up to 8674 Hz, the output voltage begins to increase rapidly. Thereafter, when continually increasing the excitation frequency, the output voltage will sharply decrease. In view of the elastic theory analysis of beams, the four L-shaped beams exhibit a resonance at a frequency of 8674 Hz along the x-axis. By repeating the above testing process, similar resonance frequency characteristics at frequencies of 8707 Hz and 7840 Hz along the y-axis and z-axis can be observed in Figure 8b,c, respectively. It indicates that the elastic elements used to detect the acceleration along x-axis and y-axis can operate in the same way, thus achieving the approximate resonant frequency characteristics along x-axis and y-axis. Nevertheless, the different elastic structure along the z-axis leads to a resonant frequency of 7840 Hz lower than that of x-axis and y-axis.

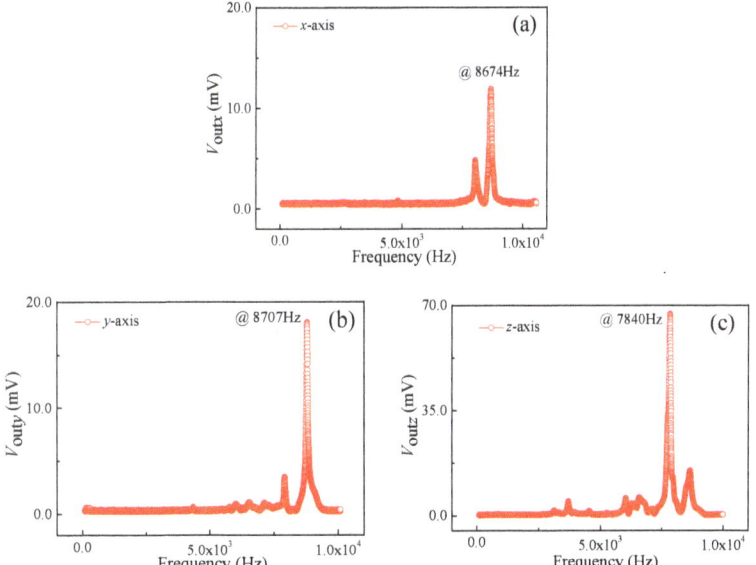

Figure 8. Resonance characteristics of the SOI three-axis acceleration sensor: (**a**) $a = a_x$; (**b**) $a = a_y$; and (**c**) $a = a_z$.

How the acceleration influences the output voltage at the three excitation frequencies were studied at a supply voltage of 5.0 V, including a middle resonance frequency, a low resonance frequency less than 100 Hz, and a high resonance frequency more than 100 Hz. During the testing process, the external acceleration was exerted from 0 g to 5.0 g, then from 5.0 g to 0 g with a step of 0.5 g as a cycle. The test was repeated for three cycles. The relationship curves between the output voltages of the sensor and the accelerations along x-axis, y-axis, and z-axis are shown in Figure 9a, Figure 10a, and Figure 11a, respectively. As shown in Figure 9a, the output voltage (V_{outx}) is approximately proportional to a_x at a constant V_{DD}, with a higher characteristic slope compared with the other two curves at the resonant frequency. In addition, the characteristic along x-axis is similar to that along y-axis and z-axis, as shown in Figures 10a and 11a. In view of that, the sensitivities of three-axis acceleration sensor along x-axis (S_{xx}), y-axis (S_{yy}), and z-axis (S_{zz}) can be calculated, that is, 4.18 mV/g, 5.88 mV/g, and 22.72 mV/g,

respectively. It indicates that the proposed sensor can realize the detection of a_x, a_y, and a_z, where S_{zz} is higher than S_{xx} and S_{yy}.

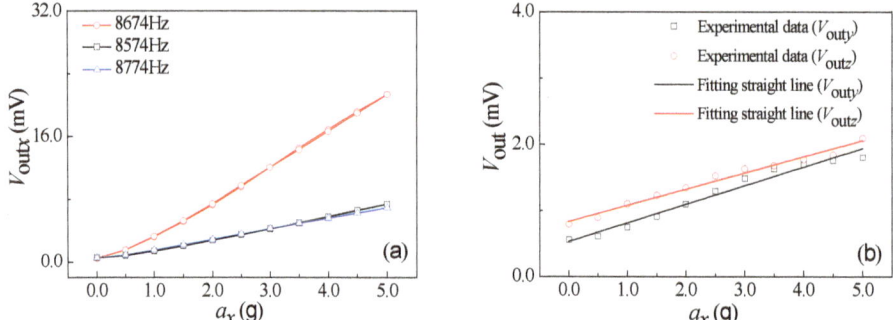

Figure 9. The relationship curves of output voltage and a_x. (**a**) The characteristic curves of differential frequency variation. (**b**) The relationship curves between V_{out} along nonsensitive axis and a_x.

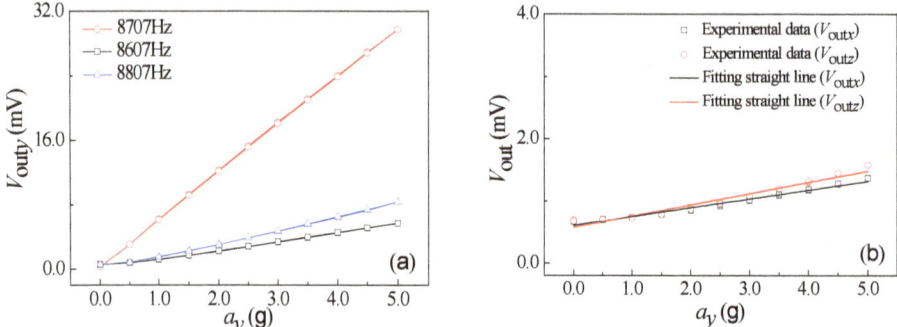

Figure 10. The relationship curves of output voltage and a_y. (**a**) The characteristic curves of differential frequency variation. (**b**) The relationship curves between V_{out} along nonsensitive axis and a_y.

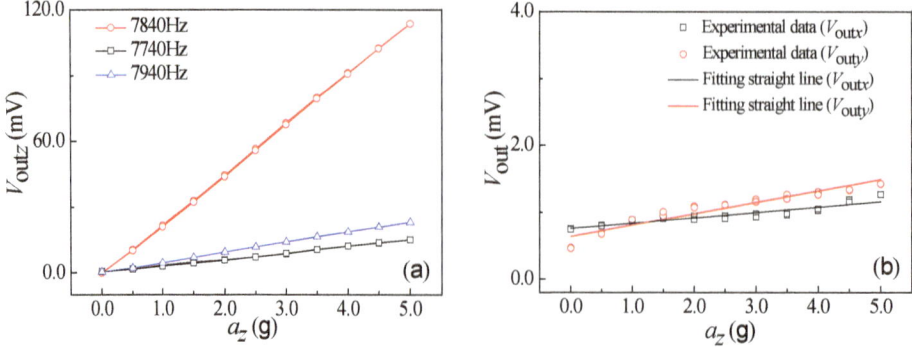

Figure 11. The relationship curves of output voltage and a_z. (**a**) The characteristic curves of differential frequency variation. (**b**) The relationship curves between V_{out} along nonsensitive axis and a_z.

In addition, how the acceleration influences the direction of nonsensitive axes were also investigated at a supply voltage of 5.0 V. During the testing process, the acceleration changed from 0 g to 5.0 g, with a step of 0.5 g at a resonance frequency. When applying acceleration along

nonsensitive axes, the output voltages of the sensor were detected. The similar cross sensitive characteristic curves along *x*-axis, *y*-axis, and *z*-axis are shown in Figures 9b, 10b and 11b, respectively, where no significant variations of V_{outy} and V_{outz} occurred with changing a_x. The experimental results show that the three-axis accelerometer realizes a low different cross-interference [18], i.e., the cross-interference between *z*-axis and *x*-axis as well as the *y*-axis are lower than that between *x*-axis and *y*-axis.

4.3. Sensitivity Characteristics

In addition, the sensitivity characteristics at the frequency of 160 Hz were investigated by varying the acceleration from 0 g to 5.0 g with a step of 0.5 g at a supply voltage of 5.0 V. Since the output signal of proposed sensor was small at a frequency of 160 Hz, the output signal was amplified by an instrumentation amplifier. In order to comparative analysis of sensitivity along sensitive axis and nonsensitive axis, we used the potentiometer to adjust the output voltages of proposed sensor without external acceleration in testing process. The relationship curves between the output voltage of proposed sensor and the external acceleration are shown in Figure 12.

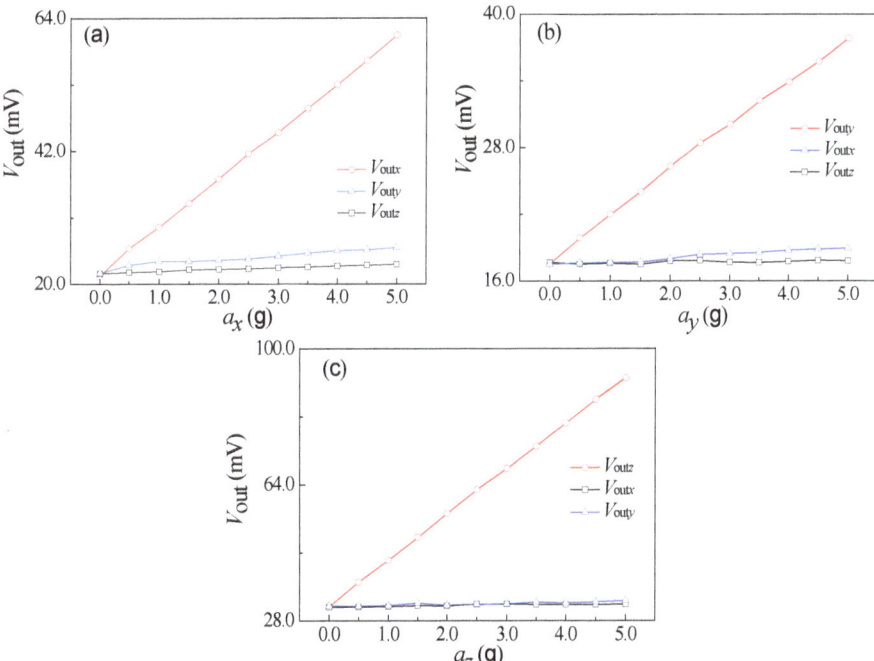

Figure 12. The relationship curves of output voltage and applied acceleration. (**a**) The relationship curves of output voltage and a_x. (**b**) The relationship curves of output voltage and a_y. (**c**) The relationship curves of output voltage and a_z.

Since the output voltage was small at a low frequency, it is necessary to amplify the output signals of the proposed sensor through an instrumentation amplifier, leading to an increase in zero drift. It can be seen that a_x is approximately proportional to V_{outx}, yet exhibits a very small effect on V_{outy} and V_{outz}, as shown in Figure 12a. When applying acceleration along *y*-axis, V_{outy} is approximately proportional to a_y, without significant changes of V_{outx} and V_{outz} with a_y, as shown in Figure 12b. From Figure 12c, V_{outz} is approximately proportional to a_z, with a constant V_{outx} and V_{outy} when changing a_z.

Based on the above data analysis, the sensitivity and the cross-sensitivity of the sensor can be calculated from Equation (4). Table 1 shows the characteristic parameters of the proposed sensor, clearly representing the sensitivities of 0.255 mV/g, 0.131 mV/g, and 0.404 mV/g prior to the amplification along *x*-axis, *y*-axis, and *z*-axis, and a minimum cross-interference of 2.2% (range of 0-5 g) along three directions.

Table 1. The characteristic parameters of three-axis acceleration sensor.

Characteristic Parameters / Acceleration Sensor	Resonant Frequency (Hz)	Bandwidth (Hz)	Sensitivity Along *x*-axis, *y*-axis and *z*-axis of Sensor at External Frequency of 160 Hz (mV/g)		
			a_x	a_y	a_z
Sensor along *x* axis	8674	100–7300	0.255	0.007	0.009
Sensor along *y* axis	8707	100–7500	0.027	0.131	0.018
Sensor along *z* axis	7840	100–6000	0.010	0.009	0.404

5. Conclusions

In conclusion, a SOI three-axis acceleration sensor was proposed in this work, consisting of two mass blocks, four L-shaped beams, and double beams in the middle. To detect the acceleration vector (a_x, a_y, and a_z) along three directions, three Wheatstone bridges were constructed by designing the piezoresistors on four L-shaped beams and two intermediate beams. In order to investigate the cross-interference of sensitivity for the proposed sensor, a sensitive element simulation model was built by using ANSYS finite element software. On the basis of that, the sensor chip was fabricated and packaged on a printed circuit board by using MEMS technology and SOI as wafer. At room temperature and V_{DD} = 5.0 V, the sensitivities of the sensor along *x*-axis, *y*-axis, and *z*-axis are 0.255 mV/g, 0.131 mV/g, and 0.404 mV/g, respectively at an excitation frequency of 160 Hz. The thickness of the beams can be accurately controlled by using an SOI wafer as a substrate and manufacturing by MEMS technology. The proposed sensor can realize the detection of acceleration along three-axis directions, with a minimum cross-interference of 2.2% in the *x*-axis, *y*-axis, and *z*-axis directions. The study on the SOI three-axis acceleration sensor supplies an innovative and promising solution to realize the detection of three-axis acceleration, and provide a feasible method to improve sensitivity and cross-interference for future work.

Author Contributions: X.Z. and D.W. conceived and designed the experiments; X.Z. and Y.W. performed the simulations and experiments; X.Z. and Y.W. analyzed the data; and X.Z. and Y.W. wrote the paper. The remaining part of the work has been done together.

Funding: This work is supported by the National Natural Science Foundation of China (grant numbers 61471159, 61006057), the Special Funds for Science and Technology Innovation Talents of Harbin in China under Grant 2016RAXXJ016, and the Modern Sensor Technology Innovation Team for College of Heilongjiang Province in China (grant number 2012TD007).

Conflicts of Interest: The authors declare no conflicts of interest.

References

1. Roy, A.L.; Bhattacharyya, T.K. Design, Fabrication and Characterization of High Performance SOI MEMS Piezoresistive Accelerometers. *J. Microsyst. Technol.* **2015**, *21*, 55–63. [CrossRef]

2. Roylance, L.M.; Angell, J.B. A batch-fabricated silicon accelerometer. *J. IEEE Trans. Electron Devices* **1979**, *26*, 1911–1917. [CrossRef]

3. Sun, C.M.; Tsai, M.H.; Liu, Y.C.; Fang, W. Implementation of a monolithic single proof-mass tri-axis accelerometer using CMOS-MEMS technique. *J. IEEE Trans. Electron Devices* **2010**, *57*, 1670–1679. [CrossRef]

4. Hu, X.; Mackowiak, P.; Bäuscher, M.; Ehrmann, O.; Lang, K.D.; Schneider-Ramelow, M.; Linke, S.; Ngo, H.D. Design and application of a high-G piezoresistive acceleration sensor for high-impact application. *J. Micromach.* **2018**, *9*, 266. [CrossRef] [PubMed]

5. Wang, H.; Yang, J.; Yan, S.; Jin, Z. Thin Layer SOI-FETs Used for Stress-Sensing and Its Application in Accelerometers. *J. IEEE Trans. Autom. Sci. Eng.* **2013**, *10*, 476–479. [CrossRef]

6. Kapti, A.O.; Muhurcu, G. Wearable acceleration sensor application in unilateral trans-tibial amputation prostheses. *J. Biocybern. Biomed. Eng.* **2014**, *34*, 53–62. [CrossRef]

7. Liu, H.; Pike, W.T. A micromachined angular-acceleration sensor for geophysical applications. *J. Appl. Phys. Lett.* **2016**, *109*, 173506. [CrossRef]

8. Hsieh, H.S.; Chang, H.C.; Hu, C.F.; Cheng, C.L.; Fang, W. A novel stress isolation guard-ring design for the improvement of a three-axis piezoresistive accelerometer. *J. Micromech. Microeng.* **2011**, *21*, 105006. [CrossRef]

9. Xu, Y.; Zhao, L.; Jiang, Z.; Ding, J.; Xu, T.; Zhao, Y. Analysis and design of a novel piezoresistive accelerometer with axially stressed self-supporting sensing beams. *J. Sens. Actuators A Phys.* **2016**, *247*, 1–11. [CrossRef]

10. Jung, H.I.; Kwon, D.S.; Kim, J. Fabrication and characterization of monolithic piezoresistive high-g three-axis accelerometer. *J. Micro Nano Syst. Lett.* **2017**, *7*, 1–5. [CrossRef]

11. Wang, P.; Zhao, Y.; Zhao, Y.; Zhang, Q.; Wang, Z. High performance piezoresistive accelerometer based on the slot etching in the EB(eight-beam) structure. In Proceedings of the 2017 IEEE 12th International Conference on Nano/Micro Engineered and Molecular Systems (NEMS), Los Angeles, CA, USA, 9–12 April 2017; pp. 406–409.

12. Marco, M.; James, N.; Chrysovalantis, P. Mechanical Structural Design of a MEMS-Based Piezoresistive Accelerometer for Head Injuries Monitoring: A Computational Analysis by Increments of the Sensor Mass Moment of Inertia. *J. Sens.* **2018**, *18*, 289.

13. Han, J.; Zhao, Z.; Niu, W.; Huang, R.; Dong, L. A low cross-axis sensitivity piezoresistive accelerometer fabricated by masked-maskless wet etching. *J. Sens. Actuators A Phys.* **2018**, *283*, 17–25. [CrossRef]

14. Geremias, M.; Moreira, R.C.; Rasia, L.A.; Moi, A. Mathematical modeling of piezoresistive elements. *J. Phy.* **2015**, *648*, 012012. [CrossRef]

15. Smith, C.S. Piezoresistance Effect in Germanium and Sillicon. *J. Phys. Rev.* **1954**, *94*, 42. [CrossRef]

16. Rui, W.U.; Wen, T.D. Silicon micro-acceleration sensor technologies. *J. Instrum. Tech. Sens.* **2007**, *36*, 8–10.

17. Yamada, K.; Nishihara, M.O.; Shimada, S.; Tanabe, M.A.; Shimazoe, M.; Matsuoka, Y. Nonlinearity of the piezoresistance effect of p-type silicon diffused layers. *J. IEEE Trans. Electron Devices* **2005**, *29*, 71–77. [CrossRef]

18. Liu, Y.; Zhao, Y.; Tian, B.; Sun, L.; Yu, Z.; Jiang, Z. Analysis and design for piezoresistive accelerometer geometry considering sensitivity, resonant frequency and cross-axis sensitivity. *J. Microsyst. Technol.* **2014**, *20*, 463–470. [CrossRef]

Article

Analysis of Kerr Noise in Angular-Rate Sensing Based on Mode Splitting in a Whispering-Gallery-Mode Microresonator

Zhaohua Yang *, Dan Li and Yuzhe Sun

School of Instrumentation Science and Opto-electronics Engineering, Beihang University, Beijing 100083, China; leedan05@foxmail.com (D.L.); sunyuzhe@buaa.edu.cn (Y.S.)
* Correspondence: yangzh@buaa.edu.cn; Tel.: +010-8233-8820

Received: 24 January 2019; Accepted: 20 February 2019; Published: 23 February 2019

Abstract: Whispering-gallery-mode (WGM) microresonators have shown their potential in high-precision gyroscopes because of their small volume and high-quality factors. However, Kerr noise can always be the limit of accuracy. Angular-rate sensing based on mode splitting treats backscattering as a measured signal, which can induce mode splitting, while it is considered as a main source of noise in conventional resonator optical gyroscopes. Meanwhile, mode splitting also provides superior noise suppression owing to its self-reference scheme. Kerr noise in this scheme has not been defined and solved yet. Here, the mechanism of the Kerr noise in the measurement is analyzed and the mathematical expressions are derived, indicating the relationship between the Kerr noise and the output of the system. The influence caused by Kerr noise on the output is simulated and discussed. Simulations show that the deviation of the splitting caused by Kerr noise is 1.913×10^{-5} Hz at an angular rate of 5×10^{6} °/s and the corresponding deviation of the angular rate is 9.26×10^{-9} °/s. It has been proven that angular-rate sensing based on mode splitting offers good suppression of Kerr noise.

Keywords: mode splitting; Kerr noise; angular-rate sensing; whispering-gallery-mode; optical microresonator

---

1. Introduction

In a whispering-gallery-mode (WGM) microresonator, the light wave is strongly confined in time and space by continuous total reflection, also described as a high Q factor and a small mode volume, respectively [1]. The light-matter interaction is enhanced, which makes the WGM microresonator extremely sensitive to weak signals, thereby making it useful for ultrasensitive detection [1,2]. It draws much attention to the field of sensors, including bio sensing, temperature sensing, pressure sensing, displacement sensing and gas sensing [3–12]. Modal coupling does not exist in the ideal WGM microresonator, where two WGMs (clockwise: CW and counterclockwise: CCW) propagate with a degenerate resonant frequency. While the degeneracy resonance will be split into two non-degenerate standing-wave modes (the symmetric mode (SM) and asymmetric mode (ASM)) if the symmetric resonator experiences any perturbation in the field distributions, such as structure defects, scattering around the cavity or rotation, namely, mode splitting [13–15]. Because of the successful demonstration in theory and experiments of nanoparticle detection by the above mode splitting, a new scheme for sensing has been developed [16–19].

Rotation of a resonator can also redistribute the internal modes in the frequency domain and give rise to mode splitting [18]. By detecting the frequency difference between the SM and ASM in the angular rate sensing based on mode splitting, good noise suppression is offered because of the same cavity environment experienced by the two modes. It was confirmed that in conventional

optical gyroscopes, as a kind of non-reciprocal noise, Kerr noise is a kind of main sources of system bias [20]. The optical Kerr effect is a third-order nonlinear effect, characterized by a change in the refractive index of the material in response to a light field, intuitively, with field intensity. As early as 1982, S. Ezekiel et al. observed a non-reciprocal bias in the interferometric fiber gyroscope, caused by the difference in power of two waves propagating in the opposite directions [21]. In conventional resonant optical gyroscopes, the output bias owing to Kerr noise is always proportional to the intensity mismatch [20,22]. The non-reciprocal noise of the optical path structure in the angular rate sensing based on mode splitting has been reported before [23]. However, Kerr noise has not been analyzed.

In this paper, the analysis of Kerr noise in angular rate sensing based on mode splitting in a WGM optical microresonator is developed, a new idea of introducing Kerr noise into angular-rate measurement is proposed and a mathematical model is built after the mechanism of the noise is analyzed theoretically. Simulations are also performed, showing that the deviation of splitting caused by Kerr noise is 1.913×10^{-5} Hz at an angular rate of 5×10^6 °/s; the corresponding deviation of angular rate is 9.26×10^{-9} °/s, which demonstrates that a good suppression of the Kerr noise is provided.

2. Principle and Theoretical Model

2.1. Theoretical Model of Angular-Rate Sensing Based on Mode Splitting

Backscattering is created, resulting in mode splitting owing to inevitable imperfections of the resonator itself, such as structural defects or surface contaminations, which are called intrinsic splitting and should be regarded as zero bias. However, the amount of intrinsic splitting is small, always less than $2/10^7$ of the laser frequency and difficult to measure accurately. Thus, a subwavelength particle is introduced into the resonator as Rayleigh scatter to induce splitting, which is considered as static splitting together with intrinsic splitting. The fiber-resonator coupled system is shown in Figure 1.

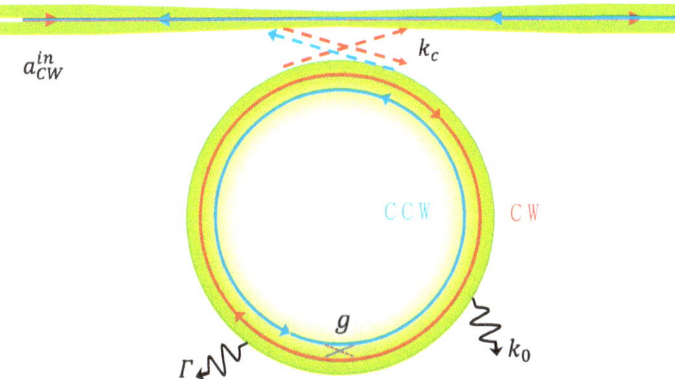

Figure 1. Sketch of the fiber-resonator coupled system, where k_0 is the intrinsic damping rate, k_c is the microresonator-taper coupling rate, Γ is the additional damping rate because of scattering loss and g is the coupling coefficient of the light scattered into the resonator. a_{CW}^{in} is the clockwise (CW) input field in the fiber taper.

Different splitting amounts can be obtained when the resonator rotates at disparate angular rates so that a correspondence can be established. Given a circular resonator with a radius R and a CCW rotation rate Ω, the frequency deviation of CW (CCW) per unit time is written as:

$$\Delta\omega_u = \frac{1}{2}\frac{\Delta\varphi}{\tau_r} = \frac{\omega_c R}{n_{\text{eff}} c}\Omega. \qquad (1)$$

Here, $\Delta\varphi = 8\pi^2 R^2 \Omega / \lambda c$ denotes the phase difference between CW and CCW per round owing to the rotation of the resonator based on the Sagnac effect. $\tau_r = 2\pi R n_{eff}/c$ is the time that light takes traveling one round in the resonator. c is the speed of light in vacuum, n_{eff} is the effective refracting. The rate equations of the coupled fiber-microresonator system in the presence of a particle with allowance for rotation can be expressed as [18]:

$$\frac{d}{dt}\begin{bmatrix} a_{CW} \\ a_{CCW} \end{bmatrix} = \begin{bmatrix} -\left(i(\omega_c + g + \Delta\omega_u) + \frac{k_{eff}}{2}\right) & -\left(ig + \frac{\Gamma}{2}\right) \\ -\left(ig + \frac{\Gamma}{2}\right) & -\left(i(\omega_c + g - \Delta\omega_u) + \frac{k_{eff}}{2}\right) \end{bmatrix}\begin{bmatrix} a_{CW} \\ a_{CCW} \end{bmatrix} - \begin{bmatrix} \sqrt{K_c} \\ 0 \end{bmatrix}a_{CW}^{in} \quad (2)$$

Here a_{cw} and a_{ccw} are the amplitudes of CW and CCW, k_{eff} is defined as the effective damping rate of the system. $k_{eff} = \Gamma + k_0 + k_c$, where k_0 is the intrinsic damping rate, k_c is the microresonator-taper coupling rate, Γ is the additional damping rate owing to scattering loss, g is the coupling coefficient of the light scattered in the resonator. a_{CW}^{in} is the CW input field in the fiber taper.

The imaginary part of the eigenvalue obtained by the state-space method represents the resonant frequency. By transforming the system matrix (2) into a diagonal matrix, the resonant frequencies of SM and ASM can be derived as:

$$\omega_{SM} = \omega_c + g + \frac{\Gamma g}{\sqrt{2\left[\frac{\Gamma^2}{4} - g^2 - \Delta\omega_u^2 + \sqrt{\left(\frac{\Gamma^2}{4} - g^2 - \Delta\omega_u^2\right)^2 + \Gamma^2 g^2}\right]}}, \quad (3)$$

$$\omega_{ASM} = \omega_c + g - \frac{\Gamma g}{\sqrt{2\left[\frac{\Gamma^2}{4} - g^2 - \Delta\omega_u^2 + \sqrt{\left(\frac{\Gamma^2}{4} - g^2 - \Delta\omega_u^2\right)^2 + \Gamma^2 g^2}\right]}}. \quad (4)$$

After simplification using $\delta_0 = 2g$, where δ_0 represents the value of static splitting, it is straightforward that the amount of splitting can be calculated from $\delta = |\omega_{SM} - \omega_{ASM}|$ using Equations (3) and (4) as

$$\delta = \sqrt{\delta_0^2 + \frac{\Omega^2\left(\frac{2\omega_c R}{n_{eff}c}\right)^2}{\left(1 + \frac{\Gamma^2}{\delta^2}\right)}}. \quad (5)$$

Additional damping owing to scattering loss can be ignored when the resonator is in the absorption-limited regime and the relationship between the splitting value [18] and the angular rate can be found as

$$\delta = 2\sqrt{g^2 + \Delta\omega_t^2} = \sqrt{\delta_0^2 + \Omega^2\left(\frac{2\omega_c R}{n_{eff}c}\right)^2}, \quad (6)$$

also written as $\Omega = \frac{n_{eff}c}{2\omega_c R}\sqrt{\delta^2 - \delta_0^2}$.

2.2. Kerr Effect-Induced Noise Model

It is generally accepted that, owing to the difference in the intensities of CW and CCW, the optical Kerr effect as a nonlinear optical effect can induce an offset in the output of a conventional gyroscope system, which can lead to a gross error in a high-precision navigation system. The refractive index of a silica optical microresonator will change under the action of an electric and magnetic field, as well as in the case of a strong light field. In the scheme of angular rate sensing based on mode splitting, the light field density at the edge of the cavity is high because of the strong confinement of the WGM microresonator, when CW and CCW travel concurrently because of scattering. However, the intensities are always different and will lead to a change in the refractive indices, which will subsequently affect

the propagation of resonant waves; thus, the Kerr effect-induced noise may also be one of the causes of non-reciprocity error in our scheme.

The Kerr effect in the path of CW (CCW) can be expressed as

$$n_{CW(CCW)} = n_1 + \gamma n_2 I_{CW(CCW)}/S \tag{7}$$

where $n_{cw(ccw)}$ is the refractive index of CW (CCW), n_1 and n_2 denote, respectively, the normal refractive index and nonlinear refractive index coefficient of the cavity, $I_{CW(CCW)}$ represents the field power of CW (CCW), γ is the polarization factor and $\gamma = 1$ is adopted here. S is the effective area of light concentration. In our scheme, the Kerr effect is introduced into the angular-rate sensing model and the time of CW and CCW light circling in the cavity is corrected as

$$\tau_{cw(ccw)} = \frac{2\pi R n_{cw(ccw)}}{c} = \frac{2\pi R \left(n_1 + \gamma n_2 I_{cw(ccw)}/S \right)}{c}. \tag{8}$$

It is clear that a time variation will appear owing to the imbalance of the field power. The frequency deviation of CW (CCW) per unit of time is given by

$$\Delta\omega_{tk} = \frac{1}{2}\left(\frac{\Delta\varphi}{2\tau_{cw}} + \frac{\Delta\varphi}{2\tau_{ccw}} \right) = \frac{\frac{8\pi^2 R^2}{\lambda c}\Omega}{\frac{4\times 2\pi R n_{cw}}{c}} + \frac{\frac{8\pi^2 R^2}{\lambda c}\Omega}{\frac{4\times 2\pi R n_{ccw}}{c}} = \frac{\omega_c R\Omega}{2c}\left(\frac{n_{cw}+n_{ccw}}{n_{cw}n_{ccw}} \right) = \\ \frac{\omega_c R\Omega}{2c}\left[\frac{2n_1+\gamma n_2(I_{cw}+I_{ccw})/S}{(n_1+\gamma n_2 I_{cw}/S)(n_1+\gamma n_2 I_{ccw}/S)} \right]. \tag{9}$$

Considering Kerr noise with respect to δ, the splitting amount δ_k becomes

$$\delta_k = \sqrt{\delta_0^2 + \Omega^2 \left\{ \frac{\omega_c R}{2c}\left[\frac{2n_1 + \gamma n_2(I_{cw} + I_{ccw})/S}{(n_1 + \gamma n_2 I_{cw}/S)(n_1 + \gamma n_2 I_{ccw}/S)} \right] \right\}^2}. \tag{10}$$

3. Simulation and Discussion

It is observed that the influence of Kerr noise on the system of angular-rate sensing based on mode splitting is owing to three factors: the angular rate of the microresonator, the light power distribution in the CW and CCW modes and the size of the cavity. The input laser power introduced for the evanescent field coupling to the left port of the fiber taper is usually fixed at one of the several commonly used values. It is not surprising that when we change the position and size of a particle relative to the cavity, the simulation result indicates that the power distribution of CW and CCW modes is greatly affected by scattering, as shown in Figure 2. Meanwhile, the size of the cavity also indirectly affects the effective cross-sectional area of the light concentration causing a slight change.

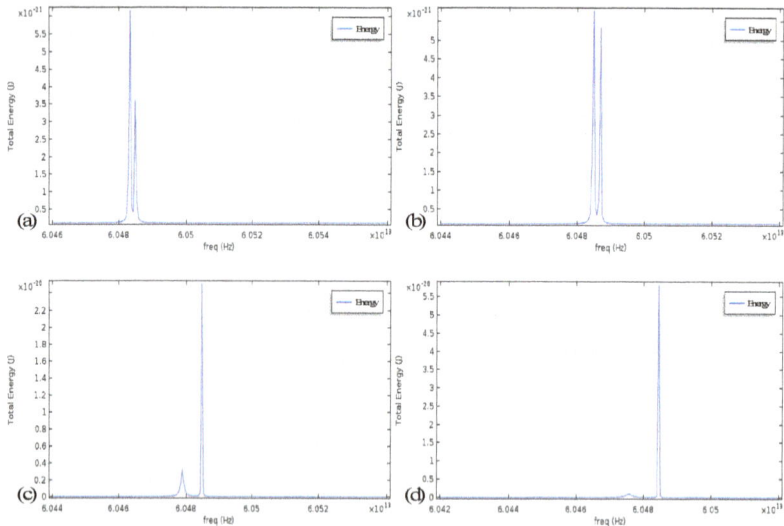

Figure 2. Light distribution of symmetric mode (SM) and asymmetric mode (ASM) in the cavity with mode splitting induced by different scattering: (**a**,**c**) Diameter 0.1 μm and 0.2 μm, tangent to the cavity; (**b**,**d**) Diameter 0.1 μm and 0.2 μm with a position of 0.05 μm further from the cavity than (**a**,**c**).

We compare the magnitude of the splitting of the original angular-rate measurement model with the case of taking into account Kerr noise, imparting different rotation rates to the cavity. The parameters are set as the conventional settings in the experiment, specific for $\lambda = 780$ nm, $R = 100$ μm, $n_0 = 1.445$, $p_{cw} = 2$ mw, $p_{ccw} = 8$ mw. As the values of the two splitting values are extremely close to each other, the two curves in the Figure 3 almost completely coincide, making it difficult to observe the differences, even if they are locally increased. It can be preliminarily concluded that Kerr noise there does not have a strong effect. To circumvent this problem, we extract the difference in the splitting values in two cases, which can be understood as a measurement error (system offset) caused by Kerr noise.

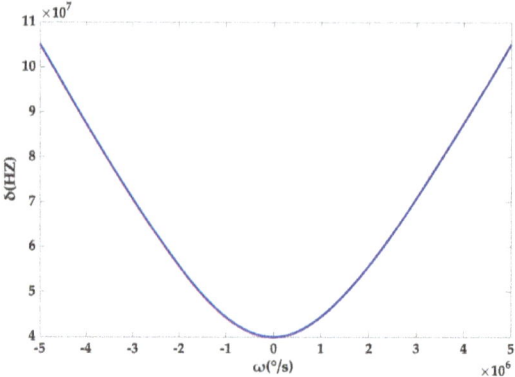

Figure 3. The contrast between the splitting value of the original model and the new model with Kerr noise introduced at different angular rates.

As can be seen from Figure 4, the offset of the system output, induced by Kerr noise, increases with the rotation rate of the cavity and the gradient increases and approximately approaches a constant.

However, it should be pointed out that the offset reaches only 1.913×10^{-5} Hz, even if the rotation rate is up to $5 \times 10^6 \, °/s$, which corresponds to the splitting amount of 1.05×10^8 Hz in Figure 3. This is also consistent with the previous result that the offset caused by Kerr noise is small.

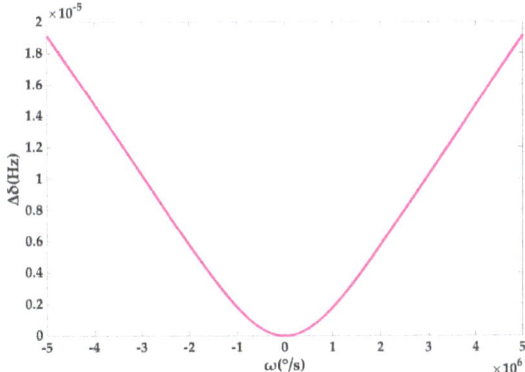

Figure 4. Splitting value deviation caused by Kerr noise at different angular rates.

The relationship between the Kerr noise and the rotation angular rate of the resonator at 780 nm, 1064 nm and 1550 nm is also given in Figure 5. The output offset caused by Kerr noise decreases with increasing wavelength, providing guidance for choosing a longer wavelength of 1550 nm for sensing. However, the deviation is also small.

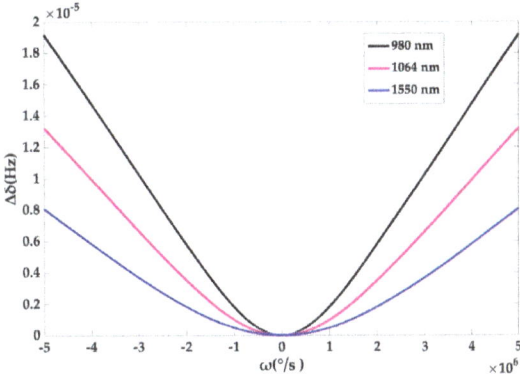

Figure 5. The relationship between the splitting value deviation and the angular rate at different wavelengths.

In conventional gyroscopes, the imbalance of the power in two directions is responsible for the strong nonlinear Kerr effect. Here we have taken all possible distribution ratios of the light power at a rotation rate of $5 \times 10^6 \, °/s$ to better demonstrate the Kerr noise error in Figure 6.

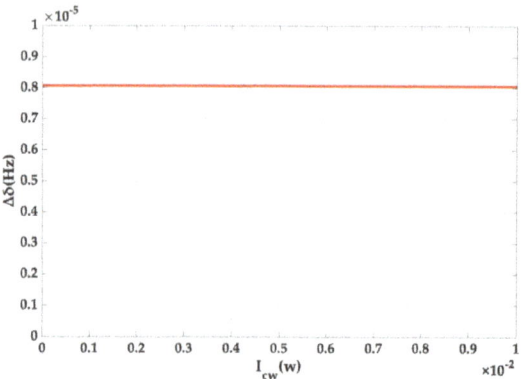

Figure 6. The relationship between the splitting value deviation and the angular rate for different distributions of light power.

The result is interesting but predictable. A small splitting value deviation caused by Kerr noise itself results in the insignificant change caused by the difference between the CW and CCW light power distributions and the image is shown as a horizontal line. It has been proven once again that the angular rate measurement system, based on mode splitting, well suppresses Kerr noise. All the simulation results presented above can reveal that the influence of Kerr noise on the output is weak.

4. Conclusions

In this paper, Kerr noise analysis in angular-rate sensing based on mode splitting in a WGM microresonator is performed. The mechanism of Kerr noise in the system was first analyzed. Subsequently, a theoretical model was constructed that considers Kerr noise while also providing a new idea on how to incorporate Kerr noise into a variety of angular-rate sensing models. Several simulations are carried out to visualize the influence on the output offset caused by Kerr noise. The deviation of splitting caused by Kerr noise is only 1.913×10^{-5} Hz at an angular rate of 5×10^6 °/s, the corresponding deviation of the angular rate is 9.26×10^{-9} °/s, which is a slight impact on the offset. Relevant parameters, such as wavelength and light power distribution, are also discussed.

Taken together with our previous analysis, we can conclude that the offset caused by Kerr noise is very small in our system of angular-rate sensing based on mode splitting in a WGM microresonator, indicating that the sensing scheme is more immune to Kerr noise. The results offer a great support to the good characteristic of the microcavity angular-rate sensing based on mode splitting and show a wide prospect of application using a WGM optical microresonator as the core component.

Author Contributions: Conceptualization, Z.Y.; methodology, Z.Y. and D.L.; software, D.L.; validation, Z.Y. and D.L.; writing, all the authors.

Funding: This research was funded by the Defense Industrial Technology Development Program (Grant number JCKY2016601C005) and the National Natural Science Foundation of China (Grant number 61473022).

Acknowledgments: We thank Lan Yang of Washington University Saint Louis, Lishuang Feng and Ming Ding of Beihang University and Yunfeng Xiao of Peking University for their theoretical and technical support.

Conflicts of Interest: The authors declare no conflict of interest.

References

1. Vahala, K.J. Optical Microcavities. *Nature* **2003**, *424*, 839–846. [CrossRef] [PubMed]
2. Chiavaioli, F.; Laneve, D.; Farnesi, D.; Falconi, M.C.; Conti, G.N.; Baldini, F.; Prudenzano, F. Long Period Grating-Based Fiber Coupling to WGM Microresonators. *Micromachines* **2018**, *9*, 366. [CrossRef] [PubMed]

3. Foreman, M.R.; Swaim, J.D.; Vollmer, F. Whispering gallery mode sensors. *Adv. Opt. Photonics* **2015**, *7*, 168–240. [CrossRef] [PubMed]
4. Vollmer, F.; Yang, L. Label-free detection with high-Q microcavities: A review of biosensing mechanisms for integrated devices. *Nanophotonics* **2012**, *1*, 267–291. [CrossRef] [PubMed]
5. Armani, A.M.; Kulkarni, R.P.; Fraser, S.E.; Flagan, R.C.; Vahala, K.J. Label-free, single-molecule detection with optical microcavities. *Science* **2007**, *317*, 783. [CrossRef] [PubMed]
6. Strekalov, D.V.; Thompson, R.J.; Baumgartel, L.M.; Grudinin, I.S.; Yu, N. Temperature measurement and stabilization in a birefringent whispering gallery mode resonator. *Opt. Express* **2011**, *19*, 14495–14501. [CrossRef] [PubMed]
7. Guan, G.; Arnold, S.; Otugen, M. Temperature Measurements Using a Micro-Optical Sensor Based on Whispering Gallery Modes. *AIAA J.* **2006**, *44*, 2385–2389. [CrossRef]
8. Manzo, M.; Ioppolo, T.; Ayaz, U.K.; Lapenna, V.; Ötügen, M.V. A photonic wall pressure sensor for fluid mechanics applications. *Rev. Sci. Instrum.* **2012**, *83*, 105003. [CrossRef] [PubMed]
9. Zamanian, A.H.; Ioppolo, T. Effect of wall pressure and shear stress on embedded cylindrical microlasers. *Appl. Opt.* **2015**, *54*, 7124–7130. [CrossRef] [PubMed]
10. Ma, Q.; Huang, L.; Guo, Z.; Rossmann, T. Spectral shift response of optical whispering-gallery modes due to water vapor adsorption and desorption. *Meas. Sci. Technol.* **2010**, *21*, 115206. [CrossRef]
11. Rubino, E.; Ioppolo, T. A Vibrometer Based on Magnetorheological Optical Resonators. *Vibration* **2018**, *1*, 239–249. [CrossRef]
12. Hallil, H.; Menini, P.; Aubert, H. Novel Microwave Gas Sensor using Dielectric Resonator with SnO$_2$ Sensitive Layer. *Procedia Chem.* **2009**, *1*, 935–938. [CrossRef]
13. Weiss, D.S.; Sandoghdar, V.; Hare, J.; Lefèvre-Seguin, V.; Raimond, J.-M.; Haroche, S. Splitting of high-Q Mie modes induced by light backscattering in silica microspheres. *Opt. Lett.* **1995**, *20*, 1835–1837. [CrossRef] [PubMed]
14. Gorodetsky, M.L.; Pryamikov, A.D.; Ilchenko, V.S. Rayleigh scattering in high-Q microspheres. *J. Opt. Soc. Am. B* **2000**, *17*, 1051–1057. [CrossRef]
15. Kippenberg, T.J.; Tchebotareva, A.L.; Kalkman, J.; Polman, A.; Vahala, K.J. Purcell-factor-enhanced scattering from Si nanocrystals in an optical microcavity. *Phys. Rev. Lett.* **2009**, *103*, 027406. [CrossRef] [PubMed]
16. Zhu, J.; Ozdemir, S.K.; Xiao, Y.F.; Li, L.; He, L.; Chen, D.-R.; Yang, L. On-chip single nanoparticle detection and sizing by mode splitting in an ultrahigh-Q microresonator. *Nat. Photonics* **2010**, *4*, 46–49. [CrossRef]
17. Chen, W.; Özdemir, Ş.K.; Zhao, G.; Wiersig, J.; Yang, L. Exceptional points enhance sensing in an optical microcavity. *Nature* **2017**, *548*, 192–196. [CrossRef] [PubMed]
18. Yang, Z.; Huo, J.; Han, X. Angular-rate sensing by mode splitting in a Whispering-gallery-mode optical microresonator. *Measurement* **2018**, *125*, 78–83. [CrossRef]
19. Li, Q.; Eftekhar, A.A.; Xia, Z.; Adibi, A. Unified approach to mode splitting and scattering loss in high-Q whispering-gallery-mode microresonators. *Phys. Rev. A* **2013**, *88*, 033816. [CrossRef]
20. Iwatsuki, K.; Hotate, K.; Higashiguchi, M. Kerr effect in an optical passive ring-resonator gyro. *J. Lightw. Technol.* **1986**, *LT-4*, 645–651. [CrossRef]
21. Ezekiel, S.; Davis, J.L.; Hellwarth, R.W. Intensity Dependent Nonreciprocal Phase Shift in a Fiberoptic Gyroscope. In *Fiber-Optic Rotation Sensors and Related Technologies*; Springer: Berlin/Heidelberg, Germany, 1982; pp. 332–336. [CrossRef]
22. Yu, H.; Zhang, C.; Feng, L.F.; Hong, L.H.; Wang, J. Research on Kerr-Effect-Induced Noise of Integrated Optical Gyroscope Based on Silicon on SiO$_2$ Waveguide Resonator. *Acta Opt. Sin.* **2011**, *31*, 1013003.
23. Yang, Z.; Xiao, Y.; Huo, J.; Shao, H. Analysis of nonreciprocal noise based on mode splitting in a high-Q optical microresonator. *Laser Phys.* **2018**, *28*, 015101. [CrossRef]

Article

A Novel Fault-Tolerant Navigation and Positioning Method with Stereo-Camera/Micro Electro Mechanical Systems Inertial Measurement Unit (MEMS-IMU) in Hostile Environment

Cheng Yuan [1,2], Jizhou Lai [1,2,*], Pin Lyu [1,2], Peng Shi [1,2], Wei Zhao [1,2] and Kai Huang [3]

[1] Navigation Research Center, College of Automation Engineering, Nanjing University of Aeronautics and Astronautics, Nanjing 211100, China; ycauto@nuaa.edu.cn (C.Y.); lvpin@nuaa.edu.cn (P.L.); ship@nuaa.edu.cn (P.S.); zhwac@nuaa.edu.cn (W.Z.)
[2] Key Laboratory of Internet of Things and Control Technology in Jiangsu province, Nanjing University of Aeronautics and Astronautics, Nanjing 211100, China
[3] Shanxi Baocheng Aviation Instrument Co., Ltd., AVIC, Baoji 721006, China; bj212hk@163.com
* Correspondence: laijz@nuaa.edu.cn; Tel.: +86-138-5147-5429

Received: 13 September 2018; Accepted: 22 November 2018; Published: 27 November 2018

Abstract: Visual odometry (VO) is a new navigation and positioning method that estimates the ego-motion of vehicles from images. However, VO with unsatisfactory performance can fail severely in hostile environment because of the less feature, fast angular motions, or illumination change. Thus, enhancing the robustness of VO in hostile environment has become a popular research topic. In this paper, a novel fault-tolerant visual-inertial odometry (VIO) navigation and positioning method framework is presented. The micro electro mechanical systems inertial measurement unit (MEMS-IMU) is used to aid the stereo-camera, for a robust pose estimation in hostile environment. In the algorithm, the MEMS-IMU pre-integration is deployed to improve the motion estimation accuracy and robustness in the cases of similar or few feature points. Besides, a dramatic change detector and an adaptive observation noise factor are introduced, tolerating and decreasing the estimation error that is caused by large angular motion or wrong matching. Experiments in hostile environment showing that the presented method can achieve better position estimation when compared with the traditional VO and VIO method.

Keywords: stereo visual-inertial odometry; fault tolerant; hostile environment; MEMS-IMU

1. Introduction

Visual navigation is an emerging technology that uses camera to capture images of the surrounding environment and processes these images to estimate ego-motion, recognize path, and make navigation decisions. The visual sensor is mature, low-cost and widely-used in robotics. Given that visual sensor is a passive sensor and does not rely on any external equipment except ambient light, one of the most important features of visual navigation is the autonomy. With the improvement of computational capabilities, visual navigation can be applied to many important applications in various fields, for instance, robot navigation [1], unmanned aerial vehicles [2], and virtual or augmented reality.

Visual odometry (VO) was first raised by Nister et al. [3] and it has become a widely-used pose estimation method. Typical VO detects and extracts feature points from a series of images that were captured by camera, then matches feature points and calculates relative pose to estimate the relative ego-motion of camera. VO can be classified based on the number of cameras into monocular VO, stereo (binocular) VO [4], and multi-camera VO [5]. The main difference is that stereo and multi-camera VO can get absolute scale information in application while monocular VO dose not, and therefore

requires a more complex initial process. Thus, the stereo VO is usually the preferable choice in practical navigation

Micro electro mechanical systems inertial measurement unit (MEMS-IMU) is also a common sensor in robots, unmanned aerial vehicles, and other moving carriers to estimate ego-motion [6,7]. It is mainly composed of accelerometers and gyroscopes, which are respectively used to obtain the acceleration and angular velocity of the carrier. Its high frequency provides precious motion information filling the interval gap of lower frequency associated vision sensors. Through using the two integrals of the acceleration and angular velocity, the attitude of the carrier can be measured. It also does not rely on any external information, can work in all conditions at any time, and has high data update rate, short-term accuracy and stability.

In recent years, visual and inertial information are usually combined to estimate the six degrees of freedom (6DOF) pose. When compared to VO, visual inertial odometry (VIO) [4,8–10] makes good use of the visual sensors and the inertial sensors, thereby acquiring more precise and robust 6DOF pose estimation. That also makes VIO play an essential role in autonomous navigation, especially in GPS-denied environment. Besides, more and more mobile robots are navigating through VIO, owing to the recent hardware improvements in mobile central processing units (CPUs) and graphics processing units (GPUs) (e.g., NVIDIA Jetson TX2 (NVIDIA corporation, Santa Clara, CA, USA)).

The mainstream of existing VIO approaches can be classified into loose coupling and tight coupling [2,5,9–11] by type of information fusion shown in Figure 1. When the system is loosely-coupled, both inertial and visual information are seen as independent measurements. The process of visual pose estimation, regarded as a black box, is only used to update a filter to restrain the inertial measurement unit (IMU) covariance propagation. By contrast, tight coupling considers the interaction of all measurements of sensors information before pose estimation, thereby achieving higher accuracy than loose coupling.

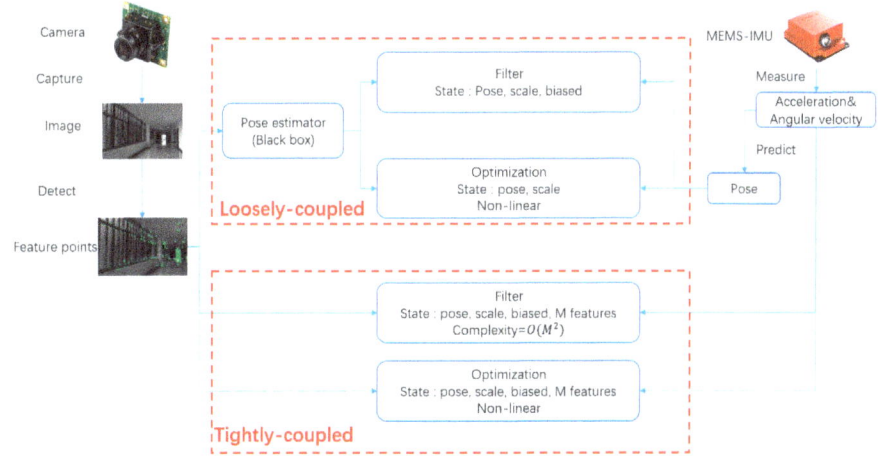

Figure 1. Loosely and tightly coupled visual inertial odometry (VIO).

Recently loosely-coupled stereo VIO systems are mostly based on Kalman filter and its derivatives. Tardif, et al. [12] proposed an EKF-based stereo VIO deployed on a moving vehicle. It used inertial information to predict the state and the stereo VO motion estimation as observations to get high frequency positioning information. Nevertheless, all of the states forecasted by inertial information, the covariance is sensitive to the IMU's bias and drift. Liu, et al. [13] presented a stereo VIO that carried out the orientation and position estimation with three filters. It fused the accelerometer and gyroscope to estimate a drift-free pitch and roll angle then fused VO and IMU to estimate motion. Nevertheless, its filtering architecture was complex and not in real-time. Schmid, et al. [14] proposed a real-time stereo VIO. It computed high quality depth images and estimated the ego-motion by key-frame based

VO and fused with the data of inertial information. However, it did not take the stereo VO's failure into account. All loosely-coupled stereo VIO systems share the disadvantage that the stereo VO's and IMU's covariance were independent and cannot reflect the entire error.

Recently tightly-coupled stereo VIO systems mainly use a filtering-based [15] or optimization-based [16] approach. Filtering-based methods propagated the mean and covariance in kalman-filtering framework, together with feature points and IMU's error. Sun, et al. [11] presented a filter-based stereo VIO system using the multi-state constraint kalman filter (MSCKF) [15] applied on an unmanned aerial vehicle. The system focused on lower computation costs. Ramezani, et al. [17] presented a stereo VIO system that was based on MSCKF and applied on vehicle, focusing on highly precise positioning. However, approaches above had high dimensional states vector and lack of robustness. The target of the optimization-based approach target was to minimize an energy function with a non-linear optimization by gauss-newton algorithm through frameworks, such as g2o [18] and ceres [19]. Usenko, et al. [4] presented a direct stereo VIO system estimated motion by minimizing a combined photometric and inertial energy function. It employed semi-dense depth maps instead of sparse feature points. Nevertheless, the inertial stability easily influenced by visual error and fault-tolerant method is simple consideration.

Subject to visual limitation, visual navigation is easily influenced when facing large scene changes that are caused by fast angular motion and low or dynamic light. To avoid positioning interruption, a fatal failure in robot navigation, current research mainly focuses on changing the feature descriptor to enhance the robustness of VO. Alismail, et al. [20] proposed new binary descriptors to achieve robust and efficient visual odometry with applications to poorly lit subterranean environments. However, the descriptors utilized information just from the images. When fast angular motion causes an image to be blurred or the environment is dark, the VO is doomed to fail. That will result in serious consequences.

To achieve satisfactory performance of VO withstanding all the limitations mentioned above, a fault-tolerant adaptive extended kalman filter (FTAEKF) framework integrated with a stereo-camera and a MEMS-IMU is proposed in this paper. The use of an EKF or one of its variants has been favored and extensively employed to fuse inertial and vision data, essentially to resolve pose estimation problem. When compared to traditional loose and tight VIO framework, both robustness and accuracy are under orders. Our main contributions are as follows:

- A stereo VIO with MEMS-IMU aided method is proposed in the framework. MEMS-IMU pre-integration constraint from prediction model is used to constrain a range of candidate feature points searching and matching. The constraint also set as to optimize the initial iterator pose to avoid local optimum instead of adding MEMS-IMU measurements error joint optimization.
- An adaptive method is introduced to adjust measurement covariance according to motion characteristic. Besides, a novel fault-tolerant mechanism is used to decide whether stereo VIO pose estimation is reliable by comparing it with MEMS-IMU measurements.

An improved stereo VIO method based on ORB-SLAM2 [21] (a visual-only stereo SLAM system demonstrated with its superior performance) is proposed in the framework. The framework can be easily integrated with any other stereo VO method. Because the computation process of MEMS-IMU pre-integration and initial iteration point prediction are mostly independent with the stereo VO.

The remainder of this paper is structured as follows: The definitions of coordinates and some symbols are presented in Section 2.1. The stereo VIO system aided by MEMS-IMU is introduced in Section 2.2. The FTAEKF is presented in Section 2.2.3. Experiment and evaluation of the proposed method are shown in Section 3, followed by discussion in Section 4.

2. Materials and Methods

2.1. Coordinates and Notations

The four coordinates that were used in our framework are shown in Figure 2, The world frame W is defined as ENU (east-north-up) by axes X_W, Y_W, and Z_W, with Z_W opposite to gravity, Y_W points

forward. The IMU frame, coincided with the body frame B also defined as ENU is attached to the center of MEMS-IMU with Z_B pointing upward and Y_B points forward. The camera frame C is set at the coordinate of left camera with Z_C forward and Y_C points downward. C is rigid relative pose with B. The relative pose is calibrated in advance.

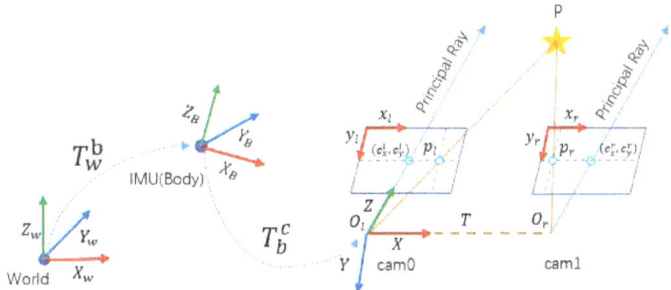

Figure 2. An illustration of coordinate system.

The rotation matrix of framework is modeled by ZYX Euler angles. To get from w to b, rotates about Z_W, Y_W, and X_W axes in turn, by the yaw angle ψ the pitch angle γ and the roll angle θ, respectively. The transformation matrix $\mathbf{T}$ is $\mathbf{T} = \begin{bmatrix} \mathbf{R} & \mathbf{t} \\ 0 & 1 \end{bmatrix}$, where $\mathbf{R} \in SO(3)$ denotes the rotation matrix, and the rotation matrix $\mathbf{R}_w^c$ represents from w to c. $\mathbf{t} = \left(p_x, p_y, p_z\right)^T$ denotes the translation vector. Vectors in the camera, body and world frames are defined as $(\cdot)^c$, $(\cdot)^b$ and $(\cdot)^w$, respectively. The transformation matrix from w to b is $\mathbf{T}_w^b$, b to c is $\mathbf{T}_b^c$.

2.2. Framework of Fault-Tolerant with Stereo-Camera and MEMS-IMU

The pipeline of the proposed framework is illustrated in Figure 3. The aim of the proposed framework is to get robust and precise motion estimation in a hostile environment. The loop closing and full bundle adjustment in ORB-SLAM2 are not involved in this paper. Our contributions are mainly on the dark red block and the red arrow.

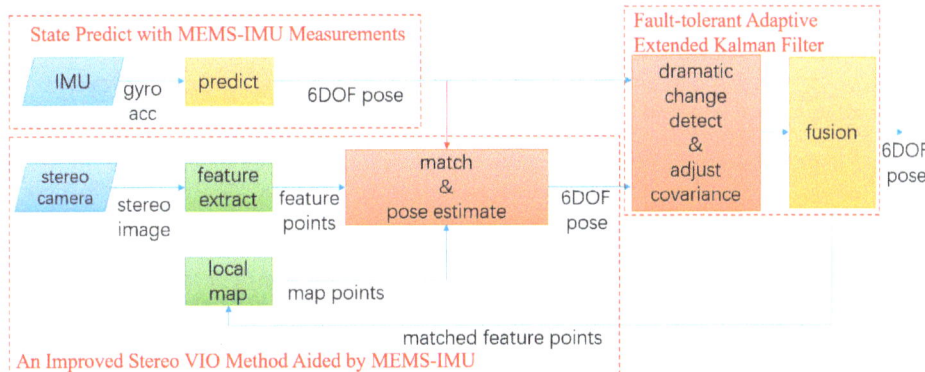

Figure 3. Framework of the proposed method. (the dark red blocks and red line are the difference between traditional VIO framework. the blue blocks represent the source from micro electro mechanical systems inertial measurement unit (MEMS-IMU) and stereo-camera. the green blocks represent traditional VO and dark yellow blocks represent MEMS-IMU measurements aided).

The stereo-camera and MEMS-IMU are tightly-coupled based on FTAEKF. The pre-integration of MEMS-IMU measurement confines the range of searching and matching feature points, and fault tolerance. Different from the traditional VIO method, the pre-integration of MEMS-IMU measurements is used to optimize the initial iterate point of pose estimation. It is also used to decide whether the result of pose estimation is credible to detect fault. Besides, to reflect the accumulated drift error, the observation covariance is adaptive according to motion characteristics. It combines the good properties of both loosely-coupled and tightly-coupled approaches. In this framework, the independence of stereo VO maximized. The framework has a good level of fault tolerance. It can function properly, even under stereo VIO failure, and then recover the whole system. This is because the framework allows a limited amount of independence and stereo VIO system avoids scale ambiguity in the monocular VO system. The details are described below.

2.2.1. State Predict with MEMS-IMU Measurements

The framework of FTAEKF is based on an iterated EKF where the state prediction is driven by IMU measurements. The system states $x \in \mathbb{R}^{16 \times 1}$ of VIO consists of number of states:

$$\mathbf{x} = \left(\mathbf{q}^w, \mathbf{p}^w, \mathbf{v}^w, \boldsymbol{\beta}_g^b, \boldsymbol{\beta}_a^b \right)^T \tag{1}$$

Namely, $\mathbf{q}^w = (q_0, q_1, q_2, q_3)^T$ is the attitude in quaternions, reflecting the world frame (W) to the body frame (B). $\mathbf{p}^w = (px^w, py^w, pz^w)^T$ is the position and $\mathbf{v}^w = \left(v_x^w, v_y^w, v_z^w \right)$ is the velocity expressed in the world frame, $\boldsymbol{\beta}_g^b$ and $\boldsymbol{\beta}_a^b$ are the biases of three-axis gyroscopes and three-axis accelerometers, respectively. The measurements from gyroscope and accelerometer are denoted as η_{wb}^b and $\mathbf{a}_{wb}^b$, respectively.

The prediction model vector $\dot{x} = (\dot{\mathbf{q}}^w, \dot{\mathbf{p}}^w, \dot{\mathbf{v}}^w, \dot{\boldsymbol{\beta}}_g^b, \dot{\boldsymbol{\beta}}_a^b)^T$ is defined as:

$$\begin{aligned}
\dot{\mathbf{q}}^w &= \tfrac{1}{2}\Omega(\hat{\eta}_{wb}^b)\mathbf{q}^w \\
\dot{\mathbf{p}}^w &= \mathbf{v}^w \\
\dot{\mathbf{v}}^w &= \mathbf{C}_b^w \left(\mathbf{a}_{wb}^b - \boldsymbol{\beta}_a^b \right) + \mathbf{g}^w \\
\dot{\boldsymbol{\beta}}_g^b &= 0 \\
\dot{\boldsymbol{\beta}}_a^b &= 0
\end{aligned} \tag{2}$$

with $\mathbf{C}_b^w$ representing the rotation matrix from B to W, the instantaneous angular velocity of B relative to W expressed in coordinate frame B $\hat{\eta}_{wb}^b$ and the quaternion update matrix $\Omega(\hat{\eta}_{wb}^b)$ are defined as: $\hat{\eta}_{wb}^b = \eta_{wb}^b - \boldsymbol{\beta}_g^b$,

$$\Omega\left(\hat{\eta}_{wb}^b \right) = \begin{bmatrix} 0 & -\hat{\eta}_{wbx}^b & -\hat{\eta}_{wby}^b & -\hat{\eta}_{wbz}^b \\ \hat{\eta}_{wbx}^b & 0 & -\hat{\eta}_{wbz}^b & \hat{\eta}_{wby}^b \\ \hat{\eta}_{wby}^b & \hat{\eta}_{wbz}^b & 0 & -\hat{\eta}_{wbx}^b \\ \hat{\eta}_{wbz}^b & -\hat{\eta}_{wby}^b & \hat{\eta}_{wbx}^b & 0 \end{bmatrix} \tag{3}$$

In proposed framework, the pre-integration of MEMS-IMU measurements is obtained through the prediction model.

2.2.2. An Improved Stereo VIO Method Aided by MEMS-IMU

In this part, the pre-integration of MEMS-IMU measurements is used to aid the stereo VO system. The stereo VIO system that was employed in this paper is based on ORB-SLAM2 with good performance.

Both original feature based VO and VIO use brute-force or bag of words (BOW) matchers to match extracted feature points within reference frame and current frame These matchers take the descriptor of one feature in current frame and are matched to all other features in reference frame using hamming distance calculation. The closest one is returned. As a result, the pose estimation produced error when false matching occurred frequently in a hostile environment due to the close hamming distance of similar descriptor. In our approach, the MEMS-IMU measurements are pre-integrated to aid stereo VIO through constraining matching and predicting initial iteration pose. The process of this part shown in Figure 4.

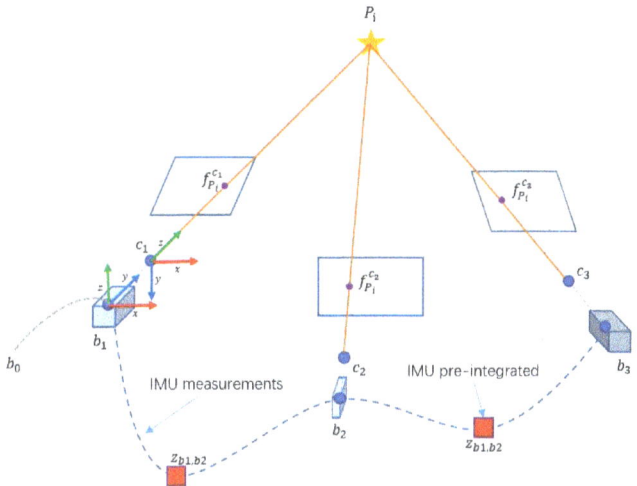

Figure 4. The process of improved stereo VIO method aided by MEMS-IMU. The inertial measurement unit (IMU) measurements are pre-integrated to predict position of feature points.

Traditionally, the initial frame pose of stereo VO is configured as world frame. However, it hardly reflects physical truth. As shown in Figure 4, VIO initialized coordinate with MEMS-IMU forward as initial heading and aligns geographic coordinate system through gravity. The stereo VIO pose is compensated by $\mathbf{T}_w^{b_1}$ from the MEMES-IMU measurement.

$$\mathbf{T}_w^{b_1} = \left[\begin{array}{cc} \mathbf{R}_w^{b_1} & \mathbf{t}_w^{b_1} \\ 0 & 1 \end{array} \right], \mathbf{R} \in SO(3), \mathbf{t} \in \mathbb{R}^{3 \times 1} \tag{4}$$

where $\mathbf{R}_w^{b_1}$ is the rotation matrix and $\mathbf{t}_w^{b_1}$ are the translation matrix from w to b_1 when VIO obtains the first image. The time interval between the image and closest MEMS-IMU measurement can be ignored due to high frequency of MEMS-IMU and low dynamic condition in beginning.

When the first stereo image is retrieved from camera, ORB feature points are extracted and matched with left and right image to estimate the depth through epipolar and disparity constraints. Then initial three-dimensional (3D) feature points in C are generated and projected based on initial pose. When a new frame was obtained from the stereo-camera, the 3D feature points are reconstructed then matched to the reference frame 3D feature points with ORB descriptors. In order to avoid the false matching caused by similar descriptors in a hostile environment. We introduce MEMS-IMU pre-integration constraint, which confined the searching and matching region to get more correct matching.

As shown in Figure 5, a point P_i is observed by two consequent frames that obtain two feature points $f_{P_i}^{c_1}$, $f_{P_i}^{c_2}$. The feature point in current frame can be project to last frame with MEMS-IMU

pre-integration. The coordinates in the pixel coordinates of both feature points $f_{P_i}^{c_1}$ and $f_{P_i}^{c_2c_1}$ are close after reprojection. We can match within bounds to decrease the workload and possibility of error.

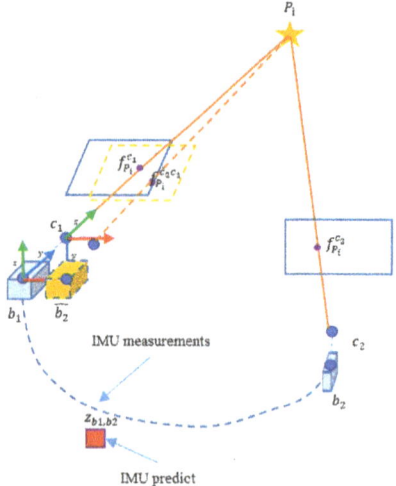

Figure 5. An illustration of predicting searching region with pre-integrating measurements of MEMS-IMU.

In our approach, the MEMS-IMU pre-integration is obtained with the prediction model. MEMS-IMU measurements between two consequent frames at discrete time $k - m$, k predict MEMS-IMU pre-integration $\Delta\xi_{k-m,k}^{imu} = \left(\Delta\mathbf{q}_{k-m,k}^{w}, \Delta\mathbf{p}_{k-m,k}^{w}\right)^{T}$:

$$\Delta\xi_{k-m,k}^{imu} = \sum_{i=k-m}^{k} \left[\; \tfrac{1}{2}\Omega\left(\hat{\eta}_{wb}^{b}\right)q_i^{w} \qquad \tfrac{1}{2}(v_{i-1}^{w} + v_i^{w}) \;\right]^{T} \Delta t \tag{5}$$

where $\mathbf{v}_i^{w}$ denotes the velocity in w at time i, $\hat{\eta}_{wb}^{b}$ denotes the instantaneous angular velocity of B and $\mathbf{q}_i^{w}$ denotes the quaternions from w to b at time i.

To reflect the motion of the camera, the pre-integration $\Delta\xi_{k-m,k}^{imu}$ needs to align with C:

$$\mathbf{T}(\Delta\xi_{k-m,k}^{cam}) = \mathbf{T}_{b}^{c}\mathbf{T}(\Delta\xi_{k-m,k}^{imu})\mathbf{T}_{b}^{c\,-1}$$
$$\mathbf{T}(\Delta\xi_{k-m,k}^{imu}) = \begin{bmatrix} \mathbf{R}_{k-m,k}^{b} & \mathbf{t}_{k-m,k}^{b} \\ 0 & 1 \end{bmatrix}, \mathbf{T}(\Delta\xi_{k-m,k}^{cam}) = \begin{bmatrix} \mathbf{R}_{k-m,k}^{c} & \mathbf{t}_{k-m,k}^{c} \\ 0 & 1 \end{bmatrix} \tag{6}$$

where $\mathbf{T}(\Delta\xi_{k-m,k}^{cam})$ denotes the transformation matrix from time $k - m$ to k in c, $\mathbf{T}_{b}^{c}$ is the transformation matrix from b to c. $\mathbf{R}_{k-m,k}^{b}$ is the quaternions $\Delta\mathbf{q}_{k-m,k}^{w}$ expressed in rotation matrix, $\mathbf{t}_{k-m,k}^{b} = C_{w}^{b}\Delta\mathbf{p}_{k-m,k}^{w}$ is the translation vector in B, where C_{w}^{b} is the rotation matrix from w to b.

After getting the coarse pose estimation of camera $\hat{\mathbf{T}}(\Delta\xi_{k-m,k}^{cam})$, we can predict the camera pose by equation:

$$\hat{\mathbf{T}}(\xi_{k}^{cam}) = \mathbf{T}(\Delta\xi_{k-m,k}^{cam})\mathbf{T}(\xi_{k-m}^{cam}) = \begin{bmatrix} \hat{\mathbf{R}}(\xi_{k}^{cam}) & \hat{\mathbf{t}}(\xi_{k}^{cam}) \\ 0 & 1 \end{bmatrix} \tag{7}$$

For each 3D feature point of current frame, the matched feature points should near it. After predicting the coarse pose estimation, we project each feature point of current frame into the initial camera frame. The search for candidates only in a small range of each 3D feature points in local map. The range depends on the bias and noise of the MEMS-IMU. We do BOW matching between each

feature point and its candidates to get matched feature point. Due to the confinement of the region, the error and the time consuming in searching and matching will reduce.

After getting the matched result, bundle adjustment optimization is performed to optimize the camera pose by minimizing the reprojection error between the matched 3D feature points $\mathbf{F}^i \in \mathbb{R}^3$ in map and feature points $\mathbf{f}^i \in \mathbb{R}^3$ in current frame. The $i \in \chi$ is a set of matched points:

$$\{\mathbf{R}, \mathbf{t}\} = \underset{\mathbf{R}, \mathbf{t}}{\operatorname{argmin}} \sum_{i \in \chi} \rho(\left\| f_{(.)}^i - \pi_{(.)}(RF^i + t) \right\|_{\Sigma}^2) \tag{8}$$

where the ρ is the robust Huber cost function and $\sum$ is the covariance matrix associated to the scale of feature points, which is one when with stereo-camera. $\pi_{(.)}$ is the projection functions monocular π_m, rectified stereo π_s are defined, as follows:

$$\pi_m\left(\begin{bmatrix} X \\ Y \\ Z \end{bmatrix}\right) = \begin{pmatrix} f_x\dfrac{X}{Z} + c_x \\ f_y\dfrac{X}{Z} + c_y \end{pmatrix}, \pi_s\left(\begin{bmatrix} X \\ Y \\ Z \end{bmatrix}\right) = \begin{pmatrix} f_x\dfrac{X}{Z} + c_x \\ f_y\dfrac{X}{Z} + c_y \\ f_x\dfrac{X-b}{Z} + c_x \end{pmatrix} \tag{9}$$

where (f_x, f_y) is focal length, (c_x, c_y) is the principal point and b is the baseline, all is known in advanced.

However, the bundle adjustment to minimize the reprojection error is nonlinear. It cannot always get a global optimal point. As shown in Figure 6, VO falls into local optimum easily because the initial iteration point is last frame pose.

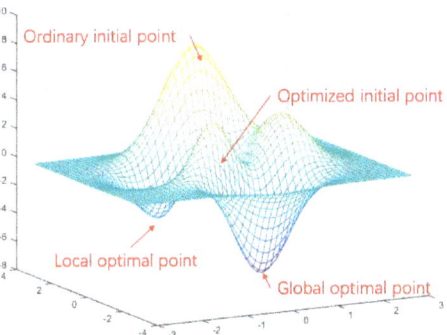

Figure 6. An illustration of association between initial point and result of optimization

In our approach, the initial iteration pose is set as prediction of MEMS-IMU pre-integration $\mathbf{R} = \hat{\mathbf{R}}(\xi_k^{cam})$ and $\mathbf{t} = \hat{\mathbf{t}}(\xi_k^{cam})$ to get close to global optimal point. Then, stereo VIO 6DOF pose estimation is optimized in order to avoid local optimum.

2.2.3. Fault-Tolerant Adaptive Extended Kalman Filtering

In this part, the FTAEKF is introduced to tolerant wrong stereo VIO pose estimation limited by the visual principle in a hostile environment.

1. Fault-tolerance with dramatic change detection

In some extreme cases, with fast motion in hostile environment, a large error of VIO pose estimation occurs because of the limited number in matched feature points or similar descriptor. The matcher matches feature points simply depending on the hamming distance.

Therefore, a fault-tolerant method with MEMS-IMU measurements is introduced through dramatic change detection.

One way to detect the sudden step change, by comparing the number of matched points with threshold after eliminating exterior point in bundle adjustment, has been proposed before. However, this is an indirect technique. In some scenario, the number of matched points is large enough, but they mostly matched with wrong feature points and significant estimation error still occurs in this direction. Sudden step change detecting in VIO mostly consider setting a transformation threshold between two consequent frames. They all only detected faults without isolation lead to failure of the system.

In this paper, a new approach using the detection function to detect and isolate dramatic change was proposed. As an accurate pose can be estimated from MEMS-IMU during a short period, the framework considered the MEMS-IMU pre-integration $\hat{\mathbf{T}}(\Delta \xi_{k-m,k}^{cam})$ as a reference. It compares to final relative VIO pose estimation $\mathbf{T}\left(\Delta \xi_{k-m,k}^{cam}\right) = \mathbf{T}(\xi_k^{cam})\mathbf{T}\left(\Delta \xi_{k-m}^{cam}\right)^{-1}$ between time k and $k-1$ to detect dramatic change. If the value of detection function $f_d \geq 1$, then the dramatic change detection is deemed to occur. The detection function f_d is defined as:

$$\Delta \mathbf{T}\left(\Delta \xi_{k-m,k}^{cam}\right) = \mathbf{T}\left(\Delta \xi_{k-m,k}^{cam}\right)\hat{\mathbf{T}}(\Delta \xi_{k-m,k}^{cam})^{-1} = \begin{bmatrix} \Delta \mathbf{R}_{k-m,k}^{cam} & \Delta \mathbf{t}_{k-m,k}^{cam} \\ 0 & 1 \end{bmatrix}$$

$$f_d = \sqrt{\frac{\left(\Delta \mathbf{t}_{k-m,k}^{cam} - \mathbf{t}_{k-m,k}^{imu}\right)^T \left(\Delta \mathbf{t}_{k-m,k}^{cam} - \mathbf{t}_{k-m,k}^{imu}\right)}{E_{\epsilon t}^2} \cdot \frac{\epsilon \psi_{k-m,k}^2 + \epsilon \theta_{k-m,k}^2 + \epsilon \gamma_{k-m,k}^2}{E_{\epsilon \psi}^2 + E_{\epsilon \theta}^2 + E_{\epsilon \gamma}^2}} \tag{10}$$

where the $\Delta \mathbf{T}\left(\Delta \xi_{k-m,k}^{cam}\right)$ is the transformation difference estimation between pre-integration of MEMS-IMU measurements and VIO. $\epsilon \psi_{k-m,k}$, $\epsilon \theta_{k-m,k}$, and $\epsilon \gamma_{k-m,k}$ are defined as: $\epsilon \gamma_{k-m,k} = \Delta \gamma_{k-m,k}^{imu} - \Delta \gamma_{k-m,k}^{cam}$, $\epsilon \theta_{k-m,k} = \Delta \theta_{k-m,k}^{imu} - \Delta \theta_{k-m,k}^{cam}$, and $\epsilon \psi_{k-m,k} = \Delta \psi_{k-m,k}^{imu} - \Delta \psi_{k-m,k}^{cam}$. Where $\Delta \gamma_{k-m,k}^{imu}$, $\Delta \theta_{k-m,k}^{imu}$, and $\Delta \psi_{k-m,k}^{imu}$ are the incremental relative attitude change estimated by MEMS-IMU measurements, $\Delta \gamma_{k-m,k}^{cam}$, $\Delta \theta_{k-m,k}^{cam}$, and $\Delta \psi_{k-m,k}^{cam}$ are the incremental relative attitude change estimated by VIO.

The threshold $E_{\epsilon t}$, $E_{\epsilon \psi}$, $E_{\epsilon \theta}$, and $E_{\epsilon \gamma}$ are set up according to the drift of motion estimation by prediction using MEMS-IMU during one period of slam procedure, which is from discrete time $k - m$ to k. As a more reliable pose can be estimated from MEMS-IMU during a short period of time, the transformation difference estimation between MEMS-IMU prediction and stereo VIO system estimation should be within this range.

In consideration of the drift of estimation by MEMS-IMU, the threshold $E_{\epsilon t}$, $E_{\epsilon \psi}$, $E_{\epsilon \theta}$, and $E_{\epsilon \gamma}$ change adaptively. As continuous change detected in hostile environment increases, $E_{\epsilon t}$, $E_{\epsilon \psi}$, $E_{\epsilon \theta}$, and $E_{\epsilon \gamma}$ are growing. $E_{\epsilon t}$, $E_{\epsilon \psi}$, $E_{\epsilon \theta}$, and $E_{\epsilon \gamma}$ are to be reinitialized with the original value if no environmental transition is detected.

2. Covariance adaptive filtering

Due to the change and accumulation of error in each process of pose estimation from VIO, the observation covariance from VIO is set to dynamic dependent upon the distance and motion characteristics to achieve better positioning accuracy. The observation covariance is adjusted to better represent practical situations.

VIO is a dead-reckon algorithm in which the error of stereo VIO pose estimation is accumulated by distance. A factor λ_d, related to the distance of stereo VIO d^{cam} reflect the error accumulating is introduced:

$$d^{cam} = \sum_{i=1}^{k-1} \sqrt{t\left(\Delta \xi_{i,i+1}^{cam}\right)^T t\left(\Delta \xi_{i,i+1}^{cam}\right)}$$

$$\lambda_d = \sigma d^{cam} \tag{11}$$

where $t\left(\xi_{i,i+1}^{cam}\right)$ is the camera translation vector between time k and $k + 1$ in C, σ is dependent on characteristics of the stereo VIO system.

Besides, the precision of stereo VIO pose estimation is also influenced obviously by motion characteristics. The field of view changes fast and the same feature points are reduced speedily when great angular change is made in a short time. MEMS-IMU measurements are more suitable and precise for the estimation and VIO is no longer reliable. Thus, a factor λ_a is introduced to adapt the specialties of MEMS-IMU and stereo VIO.

$$\lambda_a = \sum_{i=k-n}^{k} \sqrt{\hat{\eta}_{wb,i}^{b}{}^{T} \hat{\eta}_{wb,i}^{b}} \tag{12}$$

where $\hat{\eta}_{wb,i}^{b}$ is $\hat{\eta}_{wb}^{b}$ at time i, n is the size of the slide window.

When filtering, the error state vector used to correct the predicted state in filter is defined as follows:

$$\delta\mathbf{X} = \left(\delta\mathbf{q}^w, \delta\mathbf{p}^w, \delta\mathbf{v}^w, \delta\boldsymbol{\beta}_g^b, \delta\boldsymbol{\beta}_a^b \right)^T \tag{13}$$

where, $\delta\mathbf{X}$ is the state vector composed by quaternions, position, velocity, and bias error.

With no dramatic change detecting in perceived environment, the predicted states are corrected by measurements information obtained from stereo VIO pose estimation. As no drift pitch or roll angle can be obtained through gravity correction, the observation model in proposed FTAEKF is as follows:

$$
\begin{aligned}
\mathbf{Z}_k &= \mathbf{H}_k \delta\mathbf{X}_k + \boldsymbol{\mu}_k \\
\boldsymbol{\mu}_k &= \left[\begin{array}{cccc} \lambda_d \varepsilon_{p_x}^r & \lambda_d \varepsilon_{p_y}^r & \lambda_d \varepsilon_{p_z}^r & \lambda_a \varepsilon_\psi^r \end{array} \right]^T \\
\mathbf{Z}_k &= \left(\widetilde{x}_k^w - \overline{x}_k^w, \widetilde{y}_k^w - \overline{y}_k^w, \widetilde{z}_k^w - \overline{z}_k^w, \widetilde{\psi}_k^w - \overline{\psi}_k^w \right)^T \\
\overline{\psi}_k^w &= \tan^{-1}\left(\frac{2\left(q_{1,k}^w * q_{2,k}^w + q_{0,k}^w * q_{3,k}^w \right)}{1 - 2\left(q_{2,k}^w * q_{2,k}^w + q_{3,k}^w * q_{3,k}^w \right)} \right) \\
\mathbf{H}_k &= \left[\begin{array}{cccccc} \mathbf{0}_{3\times1} & \mathbf{0}_{3\times1} & \mathbf{0}_{3\times1} & \mathbf{0}_{3\times1} & \mathbf{I}_{3\times3} & \mathbf{0}_{3\times9} \\ \dfrac{\partial\overline{\psi}_k^w}{\partial q_{o,k}^w} & \dfrac{\partial\overline{\psi}_k^w}{\partial q_{1,k}^w} & \dfrac{\partial\overline{\psi}_k^w}{\partial q_{2,k}^w} & \dfrac{\partial\overline{\psi}_k^w}{\partial q_{3,k}^w} & \mathbf{0}_{1\times3} & \mathbf{0}_{1\times9} \end{array} \right]
\end{aligned}
\tag{14}
$$

where $\mathbf{Z}_k$ is the observation, $\widetilde{x}_k^w$, $\widetilde{y}_k^w$, $\widetilde{z}_k^w$, and $\widetilde{\psi}_k^w$ are the observation position and yaw in the world frame from the stereo VIO pose estimation, respectively, $\overline{x}_k^w$, $\overline{y}_k^w$, $\overline{z}_k^w$, and $\overline{\psi}_k^w$ are the predicted position and yaw in the world frame from IMEMS-MU mechanization, respectively, $\mathbf{H}_k$ is the observation matrix and $\boldsymbol{\mu}_k$ is the observation noise, which is adaptive.

When dramatic change occurred, MEMS-IMU measurements pre-integration will be used as pose estimation to isolate and tolerate fault. Since the pose estimated with MEMS-IMU during a short period of time is with sufficient accuracy, the stereo VIO system is reinitialized based on the MEMS-IMU pose in W at the closest time. The λ_a and λ_d is also reinitialized. That makes the framework with the ability to navigate even when stereo VIO system failed.

After filtering, the new matched feature points are projected to initial c to update the local map. The position of the same feature is represented using the average of position value.

When the dramatic change is detected, the local map points are cleared and the initial pose is set to MEMS-IMU pose in w with the closest time.

3. Results

3.1. Experiment Setup

3.1.1. Equipment

The equipment that we employed was based on commercial off the shelf shown in Figure 7. It consists of a ZED stereo camera, a Xsens MTI-G-710 MEMS-IMU, and a NVIDIA Jetson TX2. The ZED stereo camera resolution is set to 1280×720, baseline is 12cm and the frame rate at 15 HZ.

The Xsens MTI-G-710 can measure the acceleration and angular velocity in body frame running at 200 HZ. The MEMS-IMU was mounted on left camera of ZED that was calibrated in advanced. The processing platform is NVIDIA Jetson TX2 with dual-core NVIDIA Denver2 and quad-core ARM Cortex-A57 running on Ubuntu 16.04. The Novatel OEM6 GPS receiver worked with GPS-RTK running at 1HZ as outdoor reference. All of the sensors were connected with TX2 through USB cable and the implementation is based on C++ with Robot Operating System (ROS) Kinetic. The sensors are mounted on a tripod with three rollers.

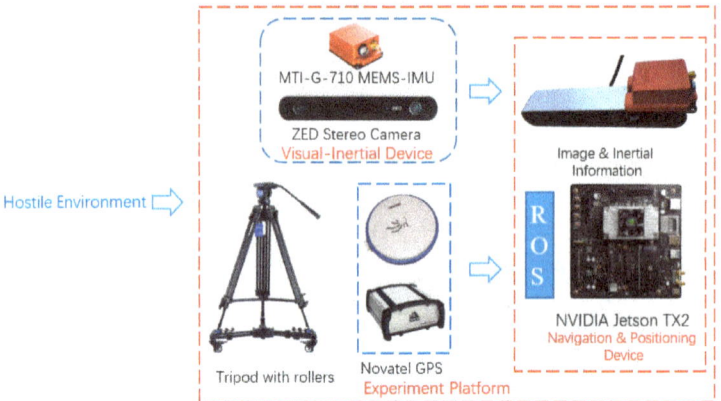

Figure 7. An illustration of platform. It consisted of ZED camera, MTI MEMS-IMU, Novatel GPS and Jetson TX2.

3.1.2. Experiment Environment Description

In order to evaluate the performance of the proposed method under a hostile environment, the experiments were carried out in the corridor outside the laboratory and a tennis court in campus, as shown in Figures 8 and 9. For the corridor, the wall of the corridor was sparse-feature. The make part of descriptors were similar. Ambient lighting in the corridor is unsatisfactory in some places, as it is bright near the window but is considerably darker elsewhere. The corridor plan is known in advance with the floor that consisted of fixed size tiles. Each tile is a square with sides of 60 cm. We pushed the tripod along the tile edge and obtained the ideal trajectory reference through a corridor plan. Some artificial mark points located at door and corner have been set in advance to evaluate the performance more comprehensively. It is regarded as the ideal path to evaluate the performance of the proposed framework. The yaw angle of MTI that was fused with magnetic is regarded as yaw angle reference. For the tennis court, the color of the ground was also simple and surrounded by similar meshes. The outdoor distance of feature was far beyond indoor environment. The reference of pose was obtained through GPS-RTK. Both environments can be considered as the hostile environment.

3.2. Experiments Results

We carried out a semi-physical simulation experiment to verify the performance of our proposed framework. The data was collected with the equipment and processed in platform. The proposed framework is compared against ORB-SLAM2, MSF-EKF [22], and VINS-Mono [23] in the experiments. The MSF-EKF based on the modular-sensor fusion framework by the University of Zurich is widely used to loosely couple inertial information and visual information. Moreover, the tightly-coupled VINS-Mono is high-performance and robust by the Hong Kong University of Science and Technology. Because the methods was multi-threaded and contained some random processing, the data took the 3σ bounds of results to eradicate any discrepancies.

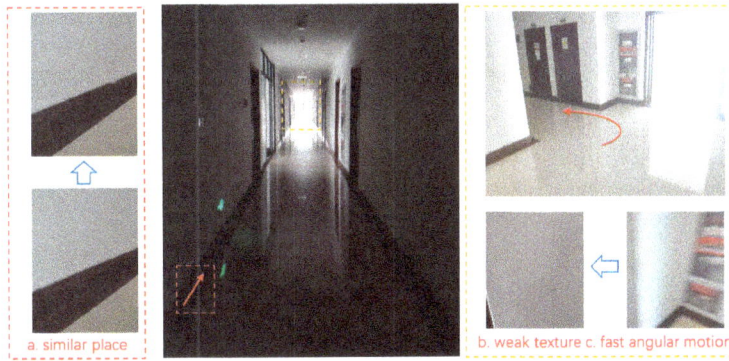

Figure 8. An illustration of the corridor where experiment carried on.

Figure 9. An illustration of the tennis court where experiment carried on.

3.2.1. Experiment I: In Corridor

In experiment I, we pushed the tripod along the tile edge in the corridor. The experiment intended to assess the comprehensive performance of the proposed framework in an indoor hostile environment.

The red line is the ideal trajectory, as shown in Figure 10. The time at passing the mark points was recorded. The estimation of motion and yaw angle from different methods shown in Figure 11a,b. The position is projected onto X-Y plane. It was clear to see our proposed method achieved more accurate pose estimation. In addition, the value of fault illustrated seven dramatic changes that were detected by FTAKF in the experiment I in Figure 12a and the adaptive observation covariance is shown in Figure 12b. Moreover, the value of mean error and root mean square error (RMSE) of yaw angle and motion estimation from different methods, as shown in Figures 13 and 14.

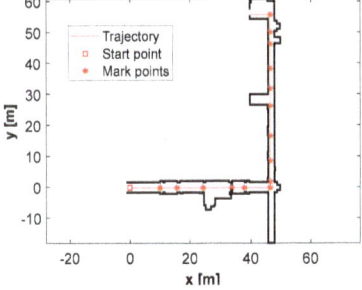

Figure 10. An illustration on the corridor plan, the ideal trajectory, and markers.

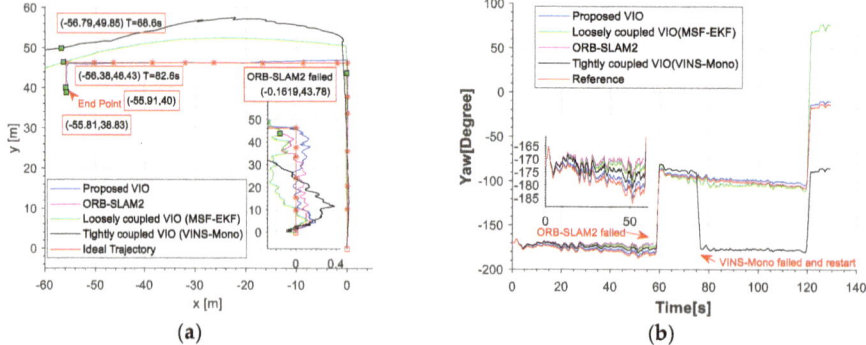

Figure 11. (**a**) An illustration of motion estimation results from different methods. (**b**) An illustration of Yaw angle estimated by different methods. ORB-SLAM2 failed due to few feature points and the noise of MEMS-IMU propagated speedily without measurements. The MEMS-IMU was meeting a corner causing fast angular velocity at 120 s. The noise of the gyroscopes propagated more speedily that causing sudden change in yaw angle difference with MSF-EKF.

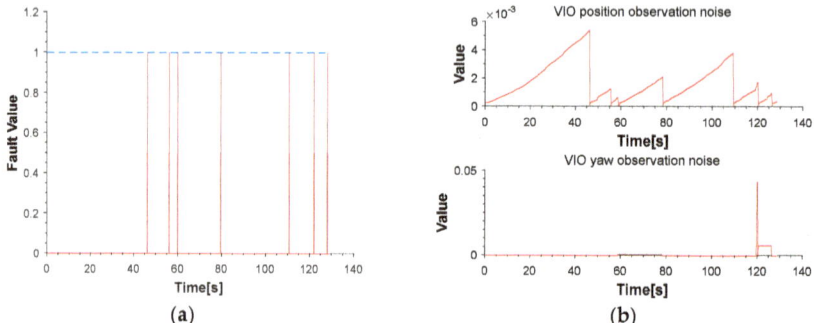

Figure 12. (**a**) The value of fault detect function demonstrates the dramatic change. (**b**) An illustration of the value of position and yaw observation noise.

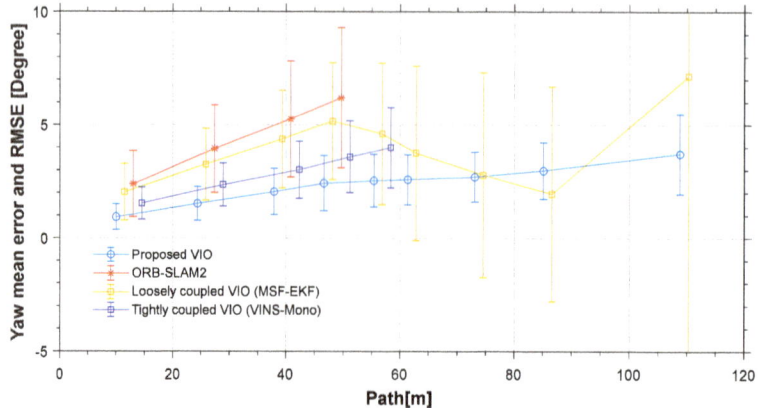

Figure 13. An illustration of value of yaw angle mean and RMSE from different methods. (VINS-Mono without output before initialization and value after system failing are ignored).

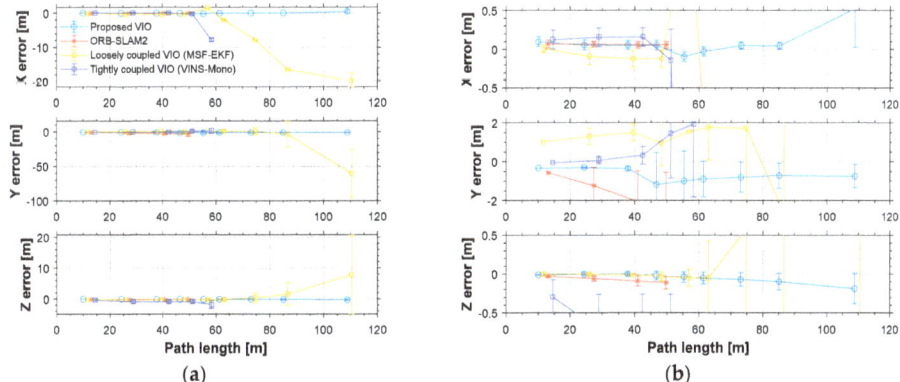

Figure 14. (**a**) The value of mean error and root mean square error (RMSE) of motion estimation from different methods. (**b**) An illustration of partial enlargement.

3.2.2. Experiment II: In tennis court

In experiment II, we pushed the tripod along the edge of the tennis court. The experiment intended to evaluate the performance of the proposed framework in an outdoor hostile environment under the RTK position and heading reference.

The red line is RTK trajectory as shown in Figure 15 with time synchronized through ROS. The estimation of motion and yaw angle from different methods shown in Figure 15a,b. Our proposed method achieved more accurate pose estimation. The value of fault illustrated six dramatic changes was detected by FTAKF in the experiment II in Figure 16a and the adaptive observation covariance is shown in Figure 16b. The value of mean error and RMSE of yaw angle and motion estimation from different methods shown in Figures 17 and 18.

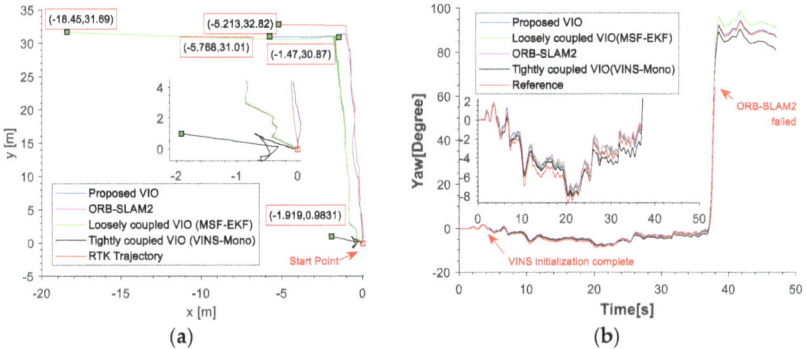

Figure 15. (**a**) An illustration of motion estimation results from different methods (**b**) An illustration of yaw angle estimated by different methods. (VINS-Mono without output before initialization).

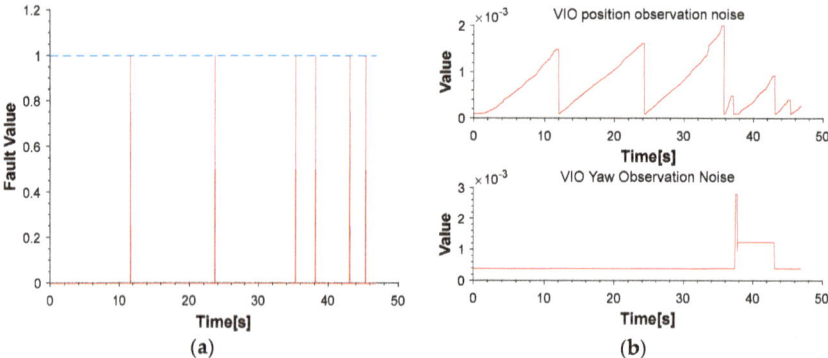

Figure 16. (**a**) The value of fault detect function demonstrates the dramatic change. (**b**) An illustration of the value of position and yaw observation noise.

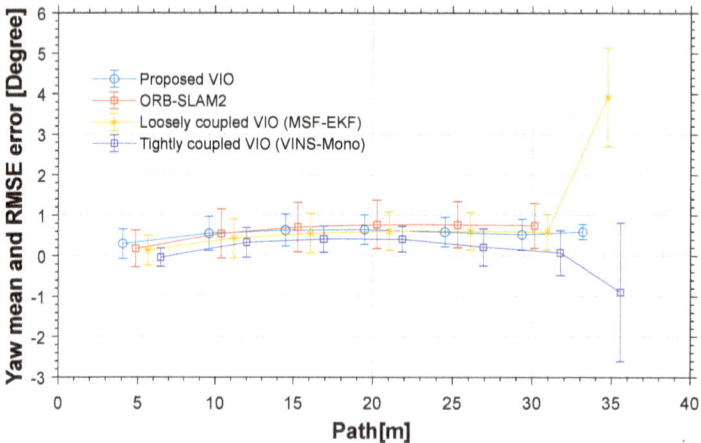

Figure 17. An illustration of value of yaw angle mean and RMSE from different methods. VINS-Mono without output before initialization.

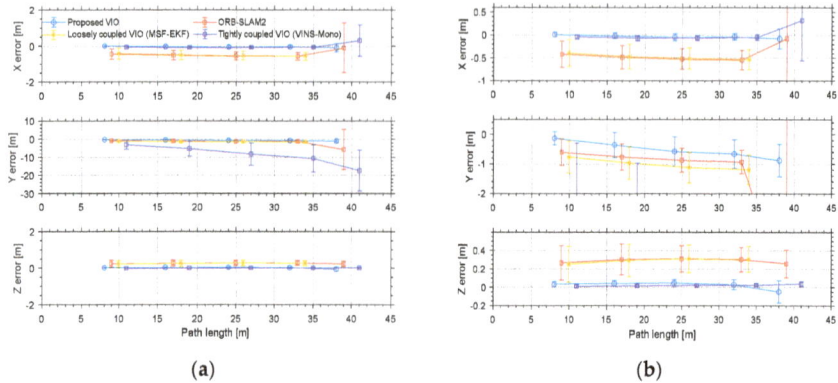

Figure 18. (**a**) The value of mean error and RMSE error of motion estimation from different methods. (**b**) An illustration of partial enlargement.

3.3. Experimental Analysis

3.3.1. Accuracy Analysis

In the experiments, the accuracy of the proposed algorithm in the reconstructed trajectory is calculated as the RMSE with mark points and RTK references in Tables 1 and 2. Moreover, the Euclidean distance between the last position of the estimated camera trajectory and the expected end point were calculated in Tables 3 and 4. Value marked with an asterisk (*) was obtained before failure.

Table 1. RMSE (m) of motion estimation in different methods. (Value marked with an asterisk (*) was obtained before VO failure.)

Length (m)	Proposed Error	ORB-SLAM2 Error	MSF-EKF Error	VINS-Mono Error
Experiment I: 108.8	0.43(0.58 *)	0.94 *	16.57 (0.90 *)	1.80 *
Experiment II: 38	0.6(0.53 *)	0.75 *	3.94 (0.6 *)	0.88 (0.08 *)

Table 2. RMSE (°) of yaw angle estimation in different methods. (Value marked with an asterisk (*) was obtained before VO failure.)

Yaw Angle Change (°)	Proposed Error	ORB-SLAM2 Error	MSF-EKF Error	VINS-Mono Error
Experiment I: 180	4.52 (2.9 *)	3.13 *	21.84 (3.10 *)	3.0 *
Experiment II: 90	0.19 (0.38 *)	0.55 *	1.21 (0.44 *)	1.72 (0.56 *)

Table 3. Length accuracy (m).

Length (m)	Proposed Error	ORB-SLAM2 Error	MSF-EKF Error	VINS-Mono Error
Experiment I: 108.8	0.92, 0.8%	194.3, 179.9%	55.88, 51.4%	67.36, 67.4%
Experiment II: 38	1.89, 4.98%	4.22, 11.1%	13.3, 35.0%	32.0, 84.2%

Table 4. Yaw angle accuracy (°).

Yaw Angle Change (°)	Proposed Error	ORB-SLAM2 Error	MSF-EKF Error	VINS-Mono Error
Experiment I: 180	1.8, 1%	176.3, 97.9%	68.5, 38.1%	62.3, 34.6%
Experiment II: 90	0.37, 0.4%	22.17, 25%	3.98, 4.04%	5.95, 6.61%

The accuracy for the experiments was depicted in above tables. The true length of different trajectories is, respectively, 108.8 m and 38 m, and the changes of reference yaw angle are 180° and 90°. As shown in Figures 11 and 15, the stereo-camera and MEMS-IMU experienced different motions with smooth motion, fast rotational, and translational motion of indoor and outdoor. As both mean error and root mean square error of ORB-SLAM2, MSF-EKF, and VINS-Mono were larger than the proposed method in hostile environment. It is clearly seen that the estimated results from the proposed method in Experiment I and II were more accurate and robust than those from ORB-SLAM2, MSF-EKF, and VINS-Mono in Figures 13, 14, 17 and 18. Pose estimation of both VO and VIO without fault tolerance were failed or divergent, which may cause fatal problems in robot navigation.

3.3.2. Inertial Aided Matching and Fault Tolerance Analysis

Figures 11 and 15 shows the pose estimation of two experiments from four different methods. ORB-SLAM2, MSF-EKF, and VINS-Mono produced large error in both position and yaw angle estimation under hostile environments. During experiments, systems including ORB-SLAM2 and VINS-Mono were in poor performance due to few feature or similar feature in hostile environment.

Moreover, ORB-SLAM2 failed because the number of feature points at corner lower than threshold. The failure of ORB-SLAM2 also caused divergence of MSF-EKF without VO output as measurement.

With the number of feature points decreasing, the part of cost function occupied by each feature points was increasing. In addition, the influence of mismatch was increased, resulting in the divergence of a system. VINS-Mono failed by detecting much large translation between two frames in experiment I. For experiment II, the feature points in starting position of tennis court were too similar and far to produce enough disparity between two consequent frames. This situation caused the error in direction of x axis with ORB-SLAM2 and false initialization with VINS-Mono which tracking feature points through optical flow method.

The pre-integration of measurements of MEMS-IMU could constrain the region of matching to reduce incorrect candidate points that achieve better match result, as shown in Figure 19. Besides, the dramatic changes was detected shown in Figures 12 and 16, were isolated in the proposed framework that able to navigate properly in hostile environment. In addition, the adaptive noise of measurements shown in Figures 12 and 16 make the proposed framework obtained more accurate pose estimation than traditional loosely-coupled VIO, such as MSF-EKF.

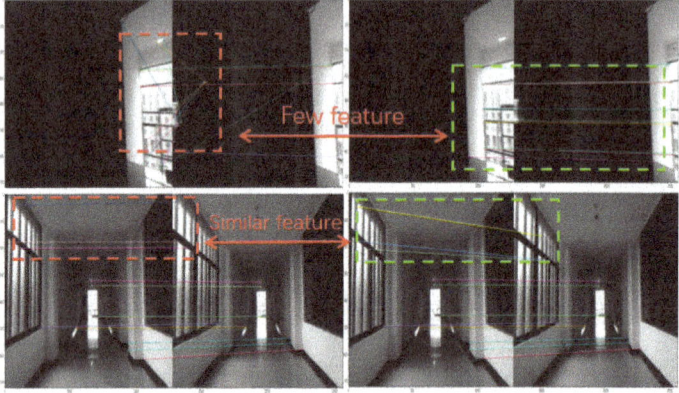

Figure 19. An illustration of wrong matching in hostile situation. Left image represented matching with all feature points in references frame and right confined matching by pre-integration.

4. Conclusions

In this work, a novel fault-tolerant framework with stereo-camera and MEMS-IMU was proposed to obtain robust and precise positioning information in a hostile environment. MEMS-IMU measurements predict the camera motion and adaptive observation covariance noise are taken in the framework. It makes stereo VO motion estimation more precise when meeting hostile environment. A fault-tolerant mechanism is also introduced to detect and isolate the dramatic change in order to achieve more robust positioning information.

When comparing to traditionally loosely-coupled VIO systems that are not considered to detect the wrong measurements, our proposed method introduced an adaptive noise according to motion characteristics that obtain more precise positional information. For the tightly-coupled VIO systems, which introduced inertial error to obtain more robust and accurate positioning results, the relation between inertial error and visual error is not considered, which leads to the influence of inertial error estimation after the error of visual matching, resulting in the instability of the whole system. Our proposed framework isolated visual error, which was detected by comparing with more reliable inertial error, made the whole system more reliable and stable. The framework also maintains a certain degree of independence between framework and stereo VO system that can be easily integrated with other stereo VO system. By evaluating the results of experiments, the proposed VIO system has achieved a satisfactory performance in state estimation in a hostile environment.

In our future work, we hope to apply the inertial information to graph-pose optimization in order to realize the function of loop detection and optimization in hostile environment. We also hope to employ the method in more challenging environments.

Author Contributions: C.Y. and P.S. proposed the original idea and wrote this paper; W.Z. and P.L. performed the experiments, analyzed the data; J.L. and K.H. participated in design of the experimental demonstration, revised the paper and gave some valuable suggestions.

Funding: This research was funded by [Jiangsu provincial SixTalent Peaks] grant number [2015-XXRJ-005], [Jiangsu Province Qing Lan Project], [National Natural Science Foundation of China] grant number [61703207], [Jiangsu Provincial Natural Science Foundation of China] grant number [BK20170801], and [Aeronautical Science Foundation of China] grant number [2017ZC52017].

Conflicts of Interest: The authors declare no conflicts of interest.

References

1. Liu, Z.; El-Sheimy, N.; Yu, C.; Qin, Y.; Liu, Z.; El-Sheimy, N.; Yu, C.; Qin, Y. Motion Constraints and Vanishing Point Aided Land Vehicle Navigation. *Micromachines* **2018**, *9*, 249. [CrossRef] [PubMed]
2. Weiss, S.; Achtelik, M.W.; Lynen, S.; Chli, M.; Siegwart, R. Real-time onboard visual-inertial state estimation and self-calibration of MAVs in unknown environments. In Proceedings of the IEEE International Conference on Robotics and Automation, Saint Paul, MN, USA, 14–18 May 2012; pp. 957–964.
3. Nister, D.; Naroditsky, O.; Bergen, J. Visual odometry. In Proceedings of the IEEE Computer Society Conference on Computer Vision and Pattern Recognition (CVPR), Washington, DC, USA, 27 June–2 July 2004; Volume 1, pp. I-652–I-659.
4. Usenko, V.; Engel, J.; Stückler, J.; Cremers, D. Direct visual-inertial odometry with stereo cameras. In Proceedings of the IEEE International Conference on Robotics and Automation (ICRA), Stockholm, Sweden, 16–21 May 2016; pp. 1885–1892.
5. Vidal, A.R.; Rebecq, H.; Horstschaefer, T.; Scaramuzza, D. Ultimate SLAM? Combining Events, Images, and IMU for Robust Visual SLAM in HDR and High-Speed Scenarios. *IEEE Robot. Autom. Lett.* **2018**, *3*, 994–1001. [CrossRef]
6. Corrêa, D.; Santos, D.; Contini, L.; Balbinot, A. MEMS Accelerometers Sensors: An Application in Virtual Reality. *Sens. Transducers Tor.* **2010**, *120*, 13–26.
7. Wang, J.; Zeng, Q.; Liu, J.; Meng, Q.; Chen, R.; Zeng, S.; Huang, H. Realization of Pedestrian Seamless Positioning Based on the Multi-Sensor of the Smartphone. *Navig. Position. Timing* **2018**, *1*, 28–34. [CrossRef]
8. Tian, Y.; Chen, Z.; Lu, S.; Tan, J. Adaptive Absolute Ego-Motion Estimation Using Wearable Visual-Inertial Sensors for Indoor Positioning. *Micromachines* **2018**, *9*, 113. [CrossRef] [PubMed]
9. Mur-Artal, R.; Tardos, J.D. Visual-Inertial Monocular SLAM with Map Reuse. *IEEE Robot. Autom. Lett.* **2017**, *2*, 796–803. [CrossRef]
10. He, Y.; Zhao, J.; Guo, Y.; He, W.; Yuan, K.; He, Y.; Zhao, J.; Guo, Y.; He, W.; Yuan, K. PL-VIO: Tightly-Coupled Monocular Visual–Inertial Odometry Using Point and Line Features. *Sensors* **2018**, *18*, 1159. [CrossRef] [PubMed]
11. Sun, K.; Mohta, K.; Pfrommer, B.; Watterson, M.; Liu, S.; Mulgaonkar, Y.; Taylor, C.J.; Kumar, V. Robust Stereo Visual Inertial Odometry for Fast Autonomous Flight. *IEEE Robot. Autom. Lett.* **2018**, *3*, 965–972. [CrossRef]
12. Tardif, J.P.; George, M.; Laverne, M.; Kelly, A.; Stentz, A. A new approach to vision-aided inertial navigation. In Proceedings of the IEEE/RSJ International Conference on Intelligent Robots and Systems, Taipei, Taiwan, 18–22 October 2010; pp. 4161–4168.
13. Liu, Y.; Xiong, R.; Wang, Y.; Huang, H.; Xie, X.; Liu, X.; Zhang, G. Stereo Visual-Inertial Odometry With Multiple Kalman Filters Ensemble. *IEEE Trans. Ind. Electron.* **2016**, *63*, 6205–6216. [CrossRef]
14. Schmid, K.; Lutz, P.; Tomić, T.; Mair, E.; Hirschmüller, H. Autonomous Vision-based Micro Air Vehicle for Indoor and Outdoor Navigation. *J. Field Robot.* **2014**, *31*, 537–570. [CrossRef]
15. Mourikis, A.I.; Roumeliotis, S.I. A Multi-State Constraint Kalman Filter for Vision-aided Inertial Navigatio. Proceedings IEEE International Conference on Robotics and Automation, Roma, Italy, 10–14 April 2007; pp. 3565–3572.
16. Forster, C.; Carlone, L.; Dellaert, F.; Scaramuzza, D. On-Manifold Preintegration for Real-Time Visual–Inertial Odometry. *IEEE Trans. Robot.* **2017**, *33*, 1–21. [CrossRef]

17. Ramezani, M.; Khoshelham, K. Vehicle Positioning in GNSS-Deprived Urban Areas by Stereo Visual-Inertial Odometry. *IEEE Trans. Intell. Veh.* **2018**, *3*, 208–217. [CrossRef]

18. Kümmerle, R.; Grisetti, G.; Strasdat, H.; Konolige, K.; Burgard, W. G2o: A general framework for graph optimization. In Proceedings of the IEEE International Conference on Robotics and Automation, Shanghai, China, 9–13 May 2011; pp. 3607–3613.

19. Wielicki, B.A.; Barkstrom, B.R.; Harrison, E.F.; Lee, R.B.; Smith, G.L.; Cooper, J.E. Clouds and the Earth's Radiant Energy System (CERES): An Earth Observing System Experiment. *Bull. Am. Meteor. Soc.* **1996**, *77*, 853–868. [CrossRef]

20. Alismail, H.; Kaess, M.; Browning, B.; Lucey, S. Direct Visual Odometry in Low Light Using Binary Descriptors. *IEEE Robot. Autom. Lett.* **2017**, *2*, 444–451. [CrossRef]

21. Mur-Artal, R.; Tardós, J.D. ORB-SLAM2: An Open-Source SLAM System for Monocular, Stereo, and RGB-D Cameras. *IEEE Trans. Robot.* **2017**, *33*, 1255–1262. [CrossRef]

22. Lynen, S.; Achtelik, M.W.; Weiss, S.; Chli, M.; Siegwart, R. A robust and modular multi-sensor fusion approach applied to MAV navigation. In Proceedings of the IEEE/RSJ International Conference on Intelligent Robots and Systems, Tokyo, Japan, 3–7 November 2013; pp. 3923–3929.

23. Qin, T.; Li, P.; Shen, S. VINS-Mono: A Robust and Versatile Monocular Visual-Inertial State Estimator. *arXiv* **2017**, arXiv:1708.03852. [CrossRef]

Review

Monolithic Multi Degree of Freedom (MDoF) Capacitive MEMS Accelerometers

Zakriya Mohammed [1,*], Ibrahim (Abe) M. Elfadel [2] and Mahmoud Rasras [3]

[1] Department of Electrical and Computer Engineering, New York University-Tandon School of Engineering, Brooklyn, NY 11201, USA
[2] Department of Electrical and Computer Engineering, Khalifa University, Abu Dhabi 54224, UAE; ibrahim.elfadel@ku.ac.ae
[3] Engineering Department, New York University, Abu Dhabi 129118, UAE; mrasras@nyu.edu
* Correspondence: zm775@nyu.edu

Received: 9 October 2018; Accepted: 12 November 2018; Published: 16 November 2018

Abstract: With the continuous advancements in microelectromechanical systems (MEMS) fabrication technology, inertial sensors like accelerometers and gyroscopes can be designed and manufactured with smaller footprint and lower power consumption. In the literature, there are several reported accelerometer designs based on MEMS technology and utilizing various transductions like capacitive, piezoelectric, optical, thermal, among several others. In particular, capacitive accelerometers are the most popular and highly researched due to several advantages like high sensitivity, low noise, low temperature sensitivity, linearity, and small footprint. Accelerometers can be designed to sense acceleration in all the three directions (X, Y, and Z-axis). Single-axis accelerometers are the most common and are often integrated orthogonally and combined as multiple-degree-of-freedom (MDoF) packages for sensing acceleration in the three directions. This type of MDoF increases the overall device footprint and cost. It also causes calibration errors and may require expensive compensations. Another type of MDoF accelerometers is based on monolithic integration and is proving to be effective in solving the footprint and calibration problems. There are mainly two classes of such monolithic MDoF accelerometers, depending on the number of proof masses used. The first class uses multiple proof masses with the main advantage being zero calibration issues. The second class uses a single proof mass, which results in compact device with a reduced noise floor. The latter class, however, suffers from high cross-axis sensitivity. It also requires very innovative layout designs, owing to the complicated mechanical structures and electrical contact placement. The performance complications due to nonlinearity, post fabrication process, and readout electronics affects both classes of accelerometers. In order to effectively compare them, we have used metrics such as sensitivity per unit area and noise-area product. This paper is devoted to an in-depth review of monolithic multi-axis capacitive MEMS accelerometers, including a detailed analysis of recent advancements aimed at solving their problems such as size, noise floor, cross-axis sensitivity, and process aware modeling.

Keywords: accelerometer; multi-axis sensing; capacitive transduction; inertial sensors; three-axis accelerometer; micromachining; miniaturization

---

1. Introduction

An accelerometer is a mechanical sensor which measures various modes of accelerations whether they are constant (gravity), time varying (vibrations), or quasi static (tilt). The miniaturization of these sensors was triggered with the advent of microelectromechanical systems (MEMS) technology in the late 1960s and early 1970s [1]. MEMS have had a great positive impact on growing the applications of accelerometers to domains ranging from automotive to biomedical [2,3]. Apart from accelerometers,

many sensors have used MEMS technology for miniaturization, including pressure sensors, gyroscopes, micromirrors, and microphones [4].

To accurately determine the motion and position of an object in space, a microsystem must have 10 degrees of freedom (DOF) [5]. This condition can be fulfilled by using a combined system of three-axis accelerometer (3 DOF), three-axis gyroscope (3 DOF), three-axis magnetometer (3 DOF), and a barometer (1 DOF). Therefore, for a high precision inertial navigation system, an accelerometer with three-axis sensing is desired. Most of the reported accelerometers are either single-axis or two-axis and for sensing motion in three directions, assembling two or three of these accelerometers is typically undertaken. The simplest assembly approach is orthogonal mounting and packaging of three single-axis accelerometers. However, there are many drawbacks in such assembly, including larger device footprint, higher packaging cost, and increased chances of misalignment errors [6,7]. These misalignment errors require corrections and compensations which further increases the cost.

Monolithic three-axis accelerometers seem to solve many of the issues related to expensive packaging, misalignment errors, and size. There are several approaches for the monolithic implementation of a three-axis accelerometer, including

1. Single chip integration of three proof masses, each sensing a particular axis.
2. Monolithic fabrication of two proof masses, one for in-plane sensing (X and Y) and the other for out-of-plane sensing (Z-axis).
3. Single proof mass designed to sense all the three directions.

Monolithic three-axis accelerometers that are composed of multiple proof masses have been reported since 1990s [8]. These devices have very low cross-axis sensitivity but suffer from high Brownian noise and have relatively large form factor [8]. On the other hand, it has been found that with the use of a single proof mass for three-axis sensing, a 50% reduction in the chip size can be achieved [9,10]. Even though their Brownian noise is low, the single proof-mass accelerometers suffer from very high cross-axis sensitivity. Moreover, complex innovative designs are needed for sensing all the three directions with only a single proof mass. The main objective of the present paper is to survey the reported monolithic multi-axis accelerometers and analyze in detail their structures and key MEMS design decisions that have enabled them to overcome the reported sensing challenges.

2. Applications of Multi-Axis Accelerometers

Miniaturized multi-axis accelerometers are mainly used in inertial measurement units (IMUs), along with gyroscopes and magnetometers for position and motion sensing. However, there use is increasing in many consumer electronic applications. Accelerometers are incorporated in electronics such as digital cameras, smart phones, notebooks, and video games. Apple iPhone 3G [11], Google Nexus One [12], Nokia N97 [13], and Nintendo Wii [14], all had three-axis accelerometers. In the current generation of these smart devices, the board size to mount the sensors is decreasing while the number of mounted sensors is increasing. Therefore, it is necessary to constantly pursue research for reducing the accelerometer footprint while enhancing its performance. In these smart gadgets, the accelerometer performs various functions, including flipping the display according to changes in gadget orientations, image stabilization while taking photos, better user experience while playing video games, and detecting whether the gadget is at rest of under free fall. This feature of detecting free fall is used to protect data in a notebook by quickly turning off the hard drive during accidental drops.

In industry, three-axis accelerometers are used for motion control, robot positioning, finding incline angles of bulky structure, and vibration monitoring. In healthcare, they are used to monitor fibrillation and arrhythmias during heart surgery. Heavy vehicles such as pickup trucks and sports utility vehicles have a very high center of gravity making them more susceptible to rollover accidents. Therefore, a three-axis accelerometer is used to detect rollovers and deploy side airbags. Some of the advanced applications of multi-axis accelerometers include electronic stability control, automotive headlight leveling and vehicle alarm.

3. Accelerometer Operating Principle

An accelerometer can be modeled as a second order spring-mass-damper system (Figure 1). When an acceleration (a) is applied to proof mass (m) suspended by springs with a spring constant (k), and having a damping (b), then the force ($F_{applied}$) acting on the proof mass is given by:

$$F_{applied} = ma_{applied} \tag{1}$$

The force exerted by springs and damping in the system can be defined as:

$$F_{spring} = kx \tag{2}$$

$$F_{damping} = b\dot{x} \tag{3}$$

Applying Newton's second law which states that the algebraic sum of all the forces equals the inertial force of the proof mass, we get:

$$F_{applied} - F_{spring} - F_{damping} = m\ddot{x} \tag{4}$$

$$m\ddot{x} + b\dot{x} + kx = F_{applied} = ma_{applied} \tag{5}$$

The transfer function $H(s)$ of the system is given by:

$$ms^2x(s) + bsx(s) + kx(s) = F(s) = ma(s) \tag{6}$$

$$s^2x(s) + \frac{b}{m}sx(s) + \frac{k}{m}x(s) = \frac{F(s)}{m} = a(s) \tag{7}$$

$$H(s) = \frac{x(s)}{a(s)} = \frac{1}{s^2 + \frac{b}{m}s + \frac{k}{m}} = \frac{1}{s^2 + \frac{\omega_0}{Q}s + \omega_0^2} \tag{8}$$

In Equation (8), ω_0 is the resonance frequency and Q is the quality factor given by:

$$\omega_0 = \sqrt{k/m} \tag{9}$$

$$Q = \frac{m\omega_0}{b} \tag{10}$$

Accelerometers work in the low frequency domain ($\omega \ll \omega_0$) with their mechanical sensitivity calculated by setting $s = 0$ in the transfer function $H(s)$ to get

$$\frac{x}{a} \sim \frac{m}{k} = \frac{1}{\omega_0^2} \tag{11}$$

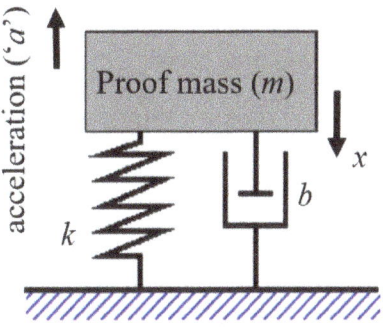

Figure 1. Model of accelerometer.

In order to have a large sensing bandwidth, we need a high resonant frequency which can be achieved by reducing the size of the proof mass and increasing the stiffness of the springs. However, this reduces the sensitivity of the device. Therefore, there is a tradeoff between the sensitivity and bandwidth.

4. Specifications of Accelerometers

MEMS accelerometers are used for various kinds of applications and therefore their specifications are application dependent. For example, in seismic measurements, accelerometers with an operation range greater than ±0.1 g, frequency range of 0–1 Hz, and resolution less than 1 µg are required. On the other hand, in shock or impact sensing, they require a range of 10,000 g, a resolution less than 1 g, and a bandwidth of 50 kHz. In this section, we give a brief overview of the specifications of an accelerometer and the design parameters on which they depend. Accelerometers are typically characterized by their Brownian noise, sensitivity, frequency response, resolution, nonlinearity, range, cross-axis sensitivity, and shock resistance.

4.1. Brownian Noise

One of the most important factor to be considered during the design is the Brownian noise. It limits the minimum achievable resolution of an accelerometer. Brownian noise is given by:

$$\sqrt{\frac{a_n^2}{\Delta f}} = \frac{\sqrt{4K_B T b}}{m} = \sqrt{\frac{4K_B T \omega}{mQ}} \tag{12}$$

where

a_n = Brownian equivalent acceleration noise
Δf = Bandwidth
K_B = Boltzmann constant
T = Absolute temperature in Kelvin

From Equation (12), it is clear that lower noise can be achieved with larger proof mass and higher quality factor. In a single proof-mass three-axis accelerometer, a relatively large proof mass is used to sense acceleration in all the three directions. Therefore, it will have lesser noise compared to three-axis accelerometer formed by the integration of three smaller one-axis accelerometers. The noise floor in the later can also be reduced by increasing the individual size of each proof mass but this will drastically increase the overall footprint.

4.2. Sensitivity

The sensitivity of an accelerometer is defined as the output voltage signal generated per unit input acceleration in 'g'. It is sometimes referred to as scale factor and denoted by 'S'. The general units are mV/g. For a triaxial accelerometer, the axial sensitivities are independent along the X, Y and Z axes are denoted by X_S, Y_S and Z_S.

$$X_S = \frac{Output\ Volage\ generated\ (\text{mV})}{input\ acceleration\ along\ X - axis\ (\text{g})} \tag{13}$$

$$Y_S = \frac{Output\ Volage\ generated\ (\text{mV})}{input\ acceleration\ along\ Y - axis\ (\text{g})} \tag{14}$$

$$Z_S = \frac{Output\ Volage\ generated\ (\text{mV})}{input\ acceleration\ along\ Z - axis\ (\text{g})} \tag{15}$$

4.3. Cross-Axis Sensitivity

Cross-axis sensitivity is the output voltage generated due to an acceleration orthogonal to a sensitive axis. Cross-axis sensitivity is generally expressed in percentage of the sensitivity i.e., ratio of the measured voltage in the cross-axis direction to the measured voltage in the sensing axis. For a tri-axial accelerometer, each axis has two cross-axis sensitivities. For example, in the case of X-direction sensing axis, there is cross-axis sensitivity due to Y-axis acceleration $(X_s)_{AY}$ and Z-axis acceleration $(X_s)_{AZ}$.

$$(X_S)_{AY} = \frac{Output\ Volage\ generated\ (mV)}{input\ acceleration\ along\ Y-axis\ (g)} \tag{16}$$

$$(X_S)_{AZ} = \frac{Output\ Volage\ generated\ (mV)}{input\ acceleration\ along\ Z-axis\ (g)} \tag{17}$$

A three-axis single proof-mass accelerometer can move freely in the three directions and the proof-mass displacement is directly proportional to the output voltage. It is therefore prone to high cross-axis sensitivity. On the other hand, a single-axis accelerometer has high stiffness in the cross direction and thus has very low cross-axis sensitivity. Therefore, the monolithic integration of multiple proof masses has a similar advantage.

4.4. Dynamic Range and Nonlinearity

The dynamic range of the accelerometer is the maximum dynamic acceleration that can be measured accurately. It is given in '$\pm$g'.

The output response of an ideal accelerometer is linear with the input acceleration. The nonlinearity of the accelerometer, therefore, measures the deviation in the output signal with respect to the ideal linear sensitivity behavior. It is expressed in terms of full-scale range as

$$\%\ Non\ linearity = \frac{Maximum\ deviation\ (g)}{Full\ scale\ range\ (g)} \times 100 \tag{18}$$

4.5. Frequency Response and the Bandwidth

The frequency response gives the dependence of accelerometer sensitivity on frequency. It also gives the amplitude and phase responses of the accelerometer. The sensitivity of an accelerometer remains constant below the resonant frequency. The range of frequencies in which the sensitivity remains constant within a tolerance band of $\pm$3 dB is the 3 dB bandwidth of the accelerometer

5. Types of Accelerometers

Depending on the transduction mechanism employed to convert the proof-mass displacement due to acceleration into a measurable signal, accelerometers can be classified as Piezoresistive [15], Piezoelectric [16], capacitive, resonant [17], optical [18], thermal [19], and tunneling [20]. The advantages and disadvantages of these transductions are explained in Figure 2.

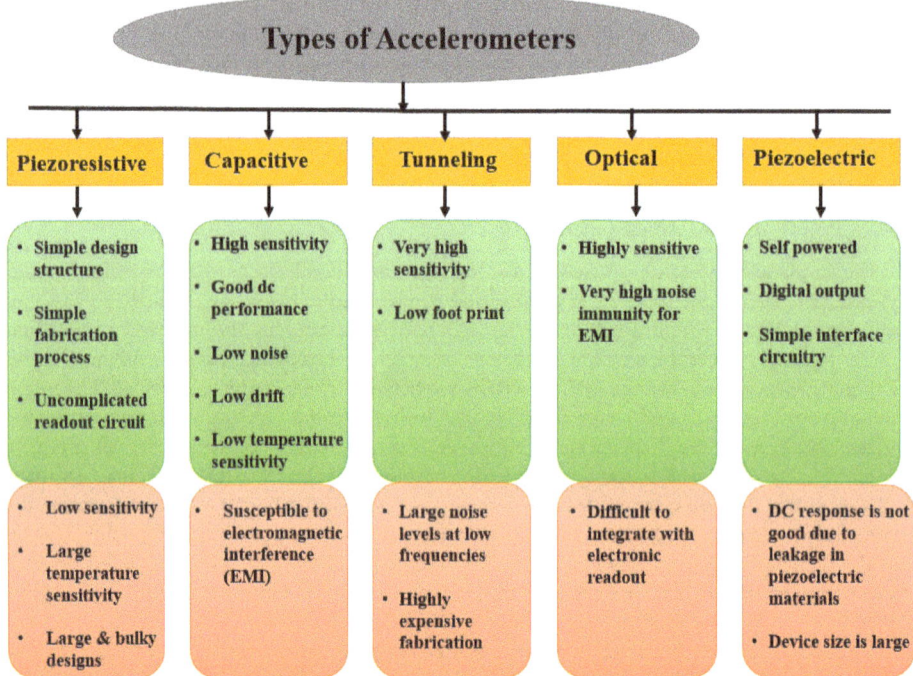

Figure 2. Advantages and disadvantages of various transduction schemes.

5.1. Capacitive Accelerometers

In capacitive accelerometers, the displacement in the proof mass due to acceleration is converted to a proportional capacitance change, which is later converted and amplified into a voltage signal. There are rotor electrode plates attached to the proof mass and stator electrode plates attached to the substrate. The design of a capacitive accelerometer is accomplished so as to have a simultaneous capacitance increase and decrease with the same acceleration with differential sensing traditionally used for quantifying the acceleration. Differential sensing increases the sensitivity by a factor of 2.

In order to fabricate capacitive accelerometers, there are two basic processes: surface and bulk micromachining. In surface micromachining, the accelerometer structure is fabricated on top of the substrate [21,22]. This is done by using various film deposition techniques similar to that of complementary metal-oxide-semiconductor (CMOS) fabrication. Therefore, the main advantage of this method lies in excellent CMOS compatibility. In this process, the first step is to deposit and pattern a sacrificial layer on the substrate, followed by the deposition and patterning of a structural layer on top. The sacrificial layer is subsequently etched, releasing a suspended mechanical structure. The devices fabricated using surface-micromachining suffers from high noise due to the thin structural layer thickness, and high internal stresses.

In contrast, bulk-micromachining uses etching of the bulk silicon substrate to create a suspended structure within the wafer [23]. The etching can be done using either wet (isotropic/anisotropic) or dry etching techniques. In isotropic etching, the etch rate is the same in all directions while in anisotropic etching the rate differs according to crystal orientation. For high aspect ratio structures, reactive ion etching (RIE) or deep reactive ion etching (DRIE) techniques are used. The structures realized using the bulk micromachining process have the advantages of low noise due to thick structures and good stability but have the drawbacks of higher cost and complex fabrication.

Apart from these two basic fabrication processes, there are new fabrication techniques that are meant to overcome the various drawbacks. Some of them utilize advantages of both surface and bulk micromachining while others use nonstandard methods for surface micromachining such as electroplating through resist molds [24]. Electroplating mitigates the disadvantage of surface micromachining by increasing the thickness of the structure. There are also reported developments for monolithically fabricating a three-axis accelerometer, a three-axis gyroscope, and a three-axis magnetometer in a single chip [25,26].

5.1.1. In-Plane Capacitive Accelerometers

The in-plane acceleration (X-axis, Y-axis or both) is generally sensed using multiple comb electrodes attached to the proof mass (rotors) and combs fixed to the substrate (stators). Since the capacitance is directly proportional to overlap area between the combs and inversely proportional to finger gap, the accelerometer can be designed to generate capacitance change with acceleration in either of two ways i.e., change in the overlap area between combs (Area Change) [27] or change in the gap between rotor and stator combs (Gap Closing). In the first approach, the capacitance changes linearly with the displacement. However, it results in a very small fractional change in capacitance. This approach is not used very often because of its low sensitivity. Figure 3a shows a block diagram of the area changeable accelerometer. With the application of an external acceleration, the proof mass is displaced, causing capacitance change in the right and left-hand side combs. On one side, the capacitance 'C_1' increases while on the other, capacitance 'C_2' decreases under the same acceleration. The stator fingers are excited with a differential voltage (opposite polarity) to produce an output signal which is proportional to the capacitance difference (C_1–C_2).

The second approach, which is based on the change in the gap spacing between the two plates [28,29], creates a relatively larger change in capacitance. Hence it is easier to sense but its response is nonlinear. In this case, the capacitance change is inversely proportional to the square of the finger gap, thus causing a large capacitance change with acceleration. The nonlinearity can be reduced if the displacement is made very small compared with the gap spacing. Linear output also simplifies the implementation of the readout circuit. Figure 3b shows the implementation of a gap change accelerometer. The configuration is differential. Figure 3c shows the equivalent circuit diagram, which is common for gap and area changeable accelerometers.

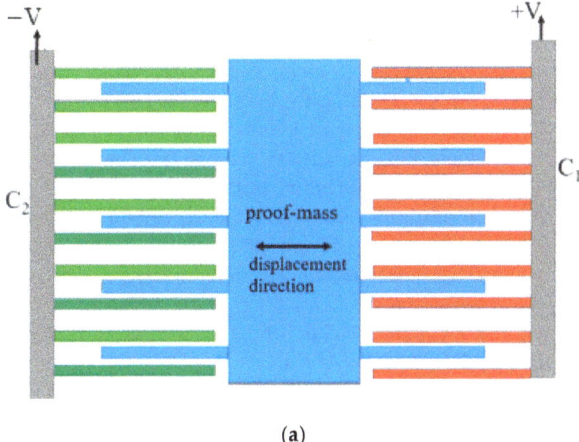

(a)

Figure 3. *Cont.*

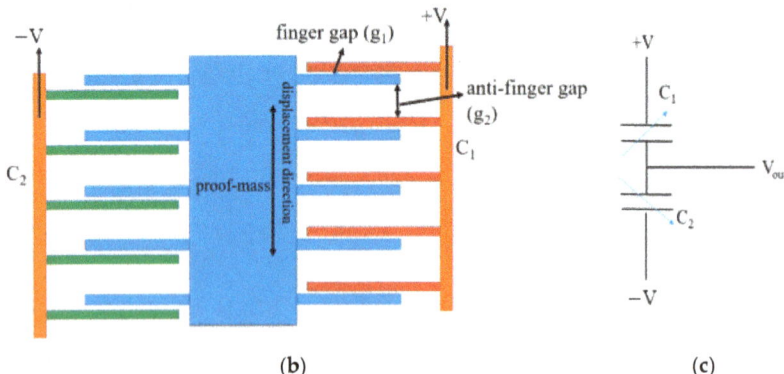

Figure 3. Sensing scheme of (**a**) area change accelerometer (**b**) gap change accelerometer (**c**) equivalent circuit.

There is also room for further improvement in gap-change accelerometers using interdigitated fingers [30,31]. Figure 4a illustrates the block diagram of a gap-change accelerometer using inter-digitated fingers. The fabrication of interdigitated fingers is complex because it requires isolation between the top and bottom stator electrodes and wiring resources to connect electrode contacts. All the top stator fingers are wired together (red fingers) to form a single capacitance plate. Similarly, all the bottom fingers are wired together (green fingers) to form another single capacitance plate. This is done so as to form a fully differential capacitive bridge (Figure 4b) in order to improve sensitivity and for canceling the offsets.

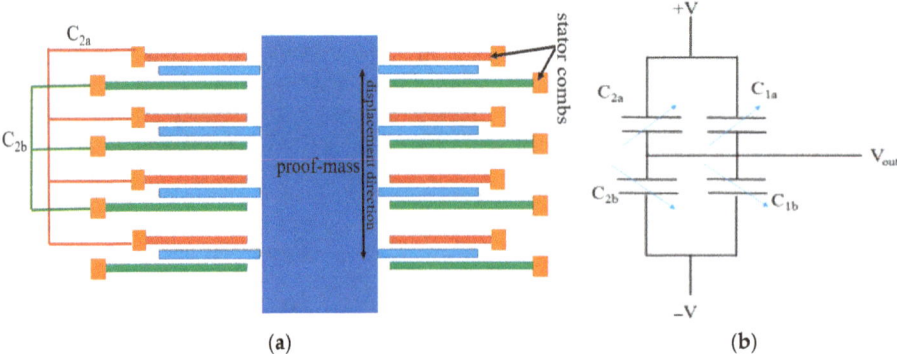

Figure 4. Sensing scheme of (**a**) gap changeable fully differential accelerometer (**b**) equivalent circuit.

The working of two-axis, in-plane accelerometer with a single proof mass is similar to one-axis accelerometer. However, the suspension system is designed to facilitate the displacement in the X- as well as Y-direction while having dedicated combs to sense both in-plane accelerations. Due to this, there is high cross-axis sensitivity along the X- and Y-axis. However, new methodologies are currently being developed to overcome this challenge [32]. As with the one-axis accelerometers, the 2-axis accelerometers can be area-change, gap-change, partially differential or fully differential. Figure 5 shows a sample of 2-axis accelerometer [33]. It is a gap change differential accelerometer. The Y-axis acceleration creates a gap change and therefore a capacitance changes in the Y-axis combs while the X-axis acceleration creates capacitance change in the X-axis combs. Voltages in the X- and Y-axis comb electrodes are modulated with different frequencies and demodulated at the output to measure the in-plane acceleration.

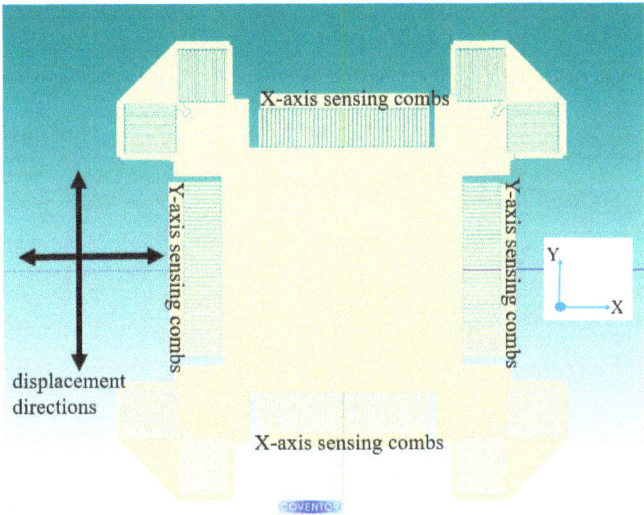

Figure 5. Dual axis gap change differential accelerometer.

5.1.2. Z-Axis Capacitive Accelerometers

For sensing out-of-plane acceleration, the arrangement of a large single-stator electrode separated from the proof mass by narrow air gaps is generally used. Here, the entire proof mass acts as a rotor electrode. This stator and rotor pair acts as a parallel plate capacitor creating a capacitance change with out-of-plane displacement under applied acceleration. However, this arrangement is non-differential. In order to create a differential capacitance, two electrode plates are used [34–36]. Figure 6a shows such an arrangement with the proof-mass suspended between the top and bottom stator plates. When acceleration is applied, one stator gap decreases, thus increasing capacitance C+ while the other stator gap increases, thus decreasing capacitance C−. It is also possible to create differential capacitance changes with a single electrode plate. Figure 6b demonstrates a torsional Z-axis accelerometer with a single electrode plate. Here torsional springs are used to displace a non-uniform proof mass, that is, a proof mass with a nonuniform mass distribution creating a heavy side and a light side. Such non-uniform mass distribution is typically achieved by a non-uniform perforation of the proof mass using etching techniques (Figure 7). The Z-axis acceleration creates a torsional see-saw motion that results in a differential capacitance change [37,38].

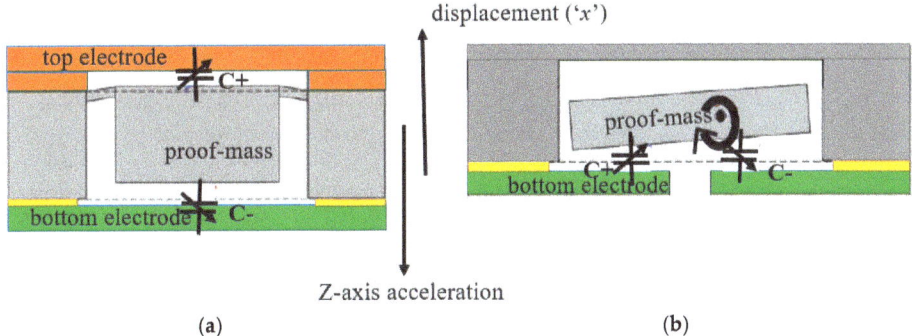

Figure 6. Sensing scheme of (**a**) vertical Z-axis accelerometer (**b**) torsional Z-axis accelerometer.

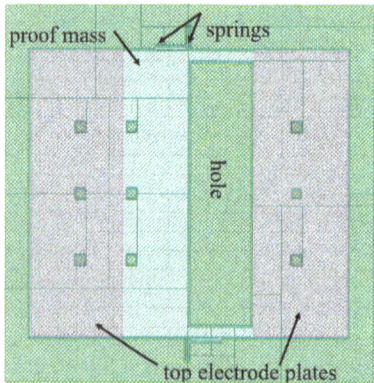

Figure 7. Layout of a torsional Z-axis accelerometer [31].

6. Development of Monolithic Multi-Axis Capacitive Accelerometers

6.1. Multiple Proof-Mass Monolithic Integrated Accelerometers

Table 1 summarizes the performance of tri-axial multiple proof-mass accelerometers.

One of the simplest methods to design a three-axis accelerometer is to fabricate three individual accelerometers monolithically on a single chip. One of the earlier designs using this method was reported by M. Lemkin and B.E. Boser in 1999 [8]. The device consists of three individual proof masses, each measuring acceleration in a particular direction, fabricated using surface micromachining. The configuration used is differential (half-bridge). The X-direction (and Y-direction) sensing proof-mass uses comb fingers while for Z-axis sensing, a reference structure is attached to the substrate. The overall die size is 4 mm × 4 mm, including the readout electronics. The three individual proof masses are designed to be very small which resulted in high Brownian noise. One main advantage of Lemkin-and-Boser design is the use of a sigma-delta (ΣΔ) Modulated force-feedback loop to provide the output in digital form. In order to stabilize the proof mass after acceleration, a control signal in a negative feedback loop is used. Thus, through the control and stabilization of deflections, measurement nonlinearities are minimized. This is because feedback control extends the bandwidth of the sensor beyond its natural frequency. However, this design is not suitable for high 'g' applications. Since in high 'g', the force generated is not sufficient to bring the proof mass back into equilibrium.

In the same year, Y. Matsumoto and his collaborators demonstrated an accelerometer using an SOI fabrication process [39]. The double challenge they addressed is that of after-rinse stiction during the fabrication process and in-use stiction during operation whether it is due to high 'g' shock accelerations or high bias electrostatic forces due to applied voltages at the stators. Stiction is caused when rotor plates come in contact with the stator plates, resulting in output saturation and possibly permanent failure. In [39], the authors added a photoresist-buried plug and a side stopper, which removes 'after-rinse stiction' resulting in a more than 90% manufacturing yield. A fluorocarbon film with plasma polymerization has been used to prevent in-use stiction [40].

S. Butefisch et al. reported a three-axis bulk micromachined accelerometer (four prototypes) with four proof masses oriented orthogonal to each other [41]. Among these designs, one has the proof mass suspended by a single beam while in other a modified proof mass (triangular shape) is suspended with stiffer suspension (double beam). The latter design is of higher quality due to low cross-axis effects. The remaining two designs are improvements of the single-beam and double-beam designs. Three proof masses, each rotated by 90° in the wafer plane, were sufficient to detect acceleration in all directions, making the forth one redundant. Most probably the fourth proof mass is used to make the design more symmetric. Each proof mass measures (length × width × height) 1000 μm × 1000 μm × 300 μm, making the overall die size without the readout circuits 9000 μm × 9000 μm × 1300 μm.

Some compensation schemes are used to reduce the size but still the device footprint is quite large. As the size of an individual proof mass is large, the device sensitivity is high at about 210 mV/g and 990 mV/g for a single beam and a double beam, respectively. Other exemplary designs using four proof masses are reported in [42,43] where a more compact accelerometer (2.5 mm × 2.0 mm × 6 µm) with highly symmetric sensitivity is proposed. This was at the expense of a maximum sensitivity of 1.51 fF/g, which is very low.

A low-noise three-axis accelerometer integrating three individual proof masses was reported in [44]. The fabrication was based on both bulk and surface micromachining. By using this process, the authors of [44] were able to fabricate a thick device (475 µm) with narrow sense gap (<1.5 µm). Due to the thick structural layer and large device size, the total measured noise floor was 1.6 µg/$\sqrt{\text{Hz}}$ for the X- and Y-direction and 1.08 µg/$\sqrt{\text{Hz}}$ for Z-direction. The work of [44] is an integration of the authors' previously reported in-plane accelerometer [45] and out-of-plane accelerometer [46].

In 2013, Y.C. Liu et al. have demonstrated monolithic three-axis accelerometer with multiplexed read-out circuit [47]. Due to the tight integration, the authors were able to achieve a smaller chip size. Prior to their three-axis accelerometer chip, the three-axis readout circuitry consisted of three different circuits, each connected to a proof-mass for one-axis sensing. This triplication of readout circuits results in an increase of not only the overall footprint but also of the power consumption. The in-plane acceleration is detected by comb fingers and the out-of-plane acceleration is detected using top and bottom electrode plates. The wiring is done to implement fully differential configuration which enhances the signal-to-noise ratio (SNR).

A sandwich three-axis bulk-micromachined accelerometer with three individual proof masses is proposed by S. Tez and T. Akin [48,49]. The design consists of comb fingers proof masses for in-plane sensing and electrode plates (top and bottom) for Z-axis sensing. The overall die size is 12 mm × 7 mm × 1 mm and the structural thickness of the device is 35 µm. The main focus of [48,49] is to reduce the cross-axis sensitivity and achieve low noise in a reasonable measurement range. A Double-Glass, Modified Silicon-on-Glass (DGM-SOG) process is used for fabrication. Due to the multiple stacking of glass-silicon-glass, individual in-plane and out-of-plane proof–masses are implemented. The top glass layer also acts as top electrode for the Z-axis proof mass. The same authors proposed a similar sandwiched, three-axis accelerometer [50] where the Z-axis proof mass (2 mm × 2 mm) and its electrode area are perforated to reduce damping. Again, three individual proof masses are used to sense acceleration in three directions. The lateral accelerometer has combs attached to its proof mass for in-plane sensing. Two such proof-masses (2.7 mm × 4.2 mm) are used that are oriented orthogonally to each other.

Table 1. Performance summary of tri-axial multiple proof-mass accelerometers.

Ref	Year	Author	Device Size (mm × mm)	Range ('±g')	Sensitivity X, Y, and Z	Noise (µg/√Hz) X, Y, and Z	Nonlinearity X, Y, and Z	Cross-Axis Sensitivity X, Y, and Z
[8]	1999	Lemkin, M.	4 × 4 (including read out)	1.9	Digital Output (0.4 fF/bit)	110, 160, 990	-	-
[39]	1999	Matsumoto, Y.	5 × 5	-	25 fF/g, 25 fF/g, 100 fF/g			<10%
[41]	2000	Butefisch, S	9 × 9	-	210 mV/g, 990 mV/g	-	$R^2 = 0.997$ $R^2 = 0.99$	-
[43]	2005	Rodjegard, H.	2.5 × 2	-	1.27 fF/g, 1.27 fF/g, 0.82 fF/g	-	-	0.12 fF/g
[44]	2005	Chae, J.	7 × 9	1	6.8 pF/g, 6.8 pF/g, 2.9 pF/g	1.6, 1.6, 1.08	-	-
[47]	2013	Liu, Y.C.	1.57 × 1.73	0.01–2	105 mV/g, 127 mV/g, 58 mV/g	400, 210, 940	1%, 0.5%, 2.4%	3%, 2.3%, 8.8%
[49]	2015	Tez, S.	12 × 7	10 (X, Y) +12, −7 (Z)	-	5.4, 5.5, 12.6	0.34%, 0.28%, 0.41%	<1%
[50]	2016	Aydemir, A.	11.8 × 4.8	4 / 71 (X, Y) 231 (Z) estimated	70.2 mV/g, 70.4 mV/g, 21.6 mV/g	13.9, 13.2, 17.8	0.26%, 0.28%, 0.3%	<1%

6.2. Single-Proof-Mass 3-Axial Accelerometers

Table 2 summarizes the design characteristics of all the single proof-mass accelerometers surveyed in this section.

One of the earliest demonstrations of using a single proof mass for three-axis sensing was performed by T. Mineta et al. [51]. The structure utilizes a bulk-micromachined, Glass-Silicon-Glass process with a sensor made of a Pyrex glass plate. The sensor has no comb fingers and only electrode plates are used for sensing both in-plane and out-of-plane acceleration. The X and Y direction acceleration cause tilting in the proof mass and the capacitance is changed with respect to the fixed electrode plates. In turn, the Z-axis acceleration causes a parallel shift. The overall chip size was 10 mm × 10 mm × 1 mm with the best achieved sensitivity being 40 mV/g. This is likely due to due to the low performance of the readout circuit. Furthermore, the cross-axis sensitivity is as high as 10%.

A surface-micromachined, single, proof-mass three-axis accelerometer with integrated electronics is reported by Lemkin and Boser [10]. The thickness of the device is 2.3 μm. It uses comb fingers for in-plane acceleration detection and one bottom electrode plate for Z-axis acceleration sensing. Comb fingers are laid in the common centroid geometry, which causes off-axis acceleration to be rejected as a common-mode, first-order signal. Furthermore, quad symmetry of the proof mass around the Z-axis is adopted to minimize cross-axis sensitivity. The design has equal compliance in the three directions with almost equal resonant frequencies. The proof mass size is 500 μm × 500 μm. The chip consists of three separate readout circuits for X-, Y- and Z-axis, yielding an overall size of 4 mm × 4 mm. In order to have high performance, the operation is closed loop, i.e., there are three individuals ΣΔ feedback loops with three readout circuits designed for the three proof masses. The results indicate a sensitivity of 0.24 fF/g for in-plane motion and 0.82 fF/g for out-of the plane motion with the maximum noise floor being 0.76 mg/$\sqrt{Hz}$. The maximum cross-axis sensitivity is only 1.58% (as inferred from [10]). Due to the use of very small proof masses the device suffers from poor sensitivity and high noise.

A theoretically zero-cross-axis sensitive, single proof-mass, bulk-micromachined three-axis accelerometer is reported by Li et al. [52]. This is accomplished by using a highly symmetrical quad beam structure. There are no comb fingers in the design, and all the three-axis sensing is implemented by placing electrode plates on the top of the proof mass. The acceleration in the Z-direction causes the proof mass to move in the Z-axis while X-directional acceleration causes a rotation around the Y-axis and translation along the X-axis. The capacitances are changed with respect to fixed electrode plates placed on the top. The proof mass measures 1.8 mm × 1.8 mm in size with a structural thickness of 0.4 mm. The theoretical sensitivity is around 6–8 fF/g, which corresponds to 30–37 mV/g of measurements. The maximum cross-axis sensitivity was found to be less than 5%.

In 2003, H. Xie et al. have proposed a very compact, monolithically integrated three-axis accelerometer with a readout circuit [53]. The design consists of a large outer proof-mass in which the Z-axis proof -mass is embedded. The comb fingers are placed all around the proof mass for in-plane sensing but inside the proof mass for Z-axis sensing. The sensor uses the side-wall capacitance of the comb fingers to detect three-axis acceleration. In order to create a side-wall capacitance, three metal lines are used and are interconnected to form fully differential bridge configuration. The design uses a single crystalline silicon (SCS) CMOS-MEMS process for to achieve high resolution with a small size. The overall die measures only 1 mm × 1 mm and is able to calculate a very low-noise floor of only 50 μg/$\sqrt{Hz}$. However, the design uses Al/SiO$_2$ thin film spring beams for suspending the Z-axis proof-mass, thus making it more sensitive to temperature. Furthermore, the etching steps to create electrical isolation introduce undercuts on the sensing combs. In a MEMS accelerometer, the undercut problem is quite common and may be due to a variety of reasons [54]. This causes the capacitive gaps to increase between the rotor and stator combs.

H. Qu et al. have addressed the undercut and thermal sensitivity problems and reported their findings in [55]. They have used the same SCS-based CMOS-MEMS process as above. In order to avoid undercuts they have sacrificed one interconnect layer. The design consists of crab-leg suspended outer proof-mass for lateral sensing with an unbalanced proof-mass embedded inside it and suspended

using a torsional spring. Again, capacitive combs wired to form a side-wall capacitance are used in a fully differential configuration. The design has nonetheless suffered undesirable undercuts due to overheating. The same authors have further addressed these issues and reported on their results in [56] where they have demonstrated a much more robust three-axis accelerometer with a readout circuit. In particular, they have used the same SCS CMOS-MEMS process and mitigated the above problems by improving the DRIE post-processing. The design was also identical with some changes in the parameters (spring length, proof-mass area...).

In 2009, Analog Devices reported a very low-cost three-axis accelerometer for consumer electronics [57]. The proposed method to reduce the cost was to use a single proof-mass accelerometer for the three-axis sensing. Also, a two-chip solution was chosen in which instead of monolithically integrating the accelerometer with readout circuit, separate chips for MEMS and electronics were used. The Analog Devices accelerometer was known as ADXL335. For fabricating this MEMS sensor, a new process was developed based on surface micromachining. From comprehending the chip micrograph, it can be concluded that comb fingers are used for in-plane sensing. For Z-axis sensing, the proof mass acts as one electrode plate, which changes the capacitance with respect to a reference electrode plate, making Z-axis sensing not fully differential. In an overall chip size of 4 mm × 4 mm × 1.45 mm, the device achieved a scale factor of 300 mV/g with a noise-limited resolution of 150 $\mu g/\sqrt{Hz}$.

An implementation of a novel single proof-mass three-axis accelerometer was reported by C.M. Sun et al. [58]. In their approach, one proof mass (Z-axis) and two supporting frames (X- and Y-axis) are used. There are an inner proof mass for Z-axis sensing, intermediate frame for Y-axis sensing and outer frame for X-axis sensing. The inner proof mass is connected to the Y-axis frame using V-shaped springs. The Y-axis frame is connected to the X-axis frame using serpentine springs, and the X-axis frame is connected to the substrate using the same type of serpentine springs. These two sets of springs are flexible only in one direction to reduce cross-axis sensitivity. Similarly, the V-shaped Z-axis springs contribute to the reduction of cross-axis sensitivity. There are three sets of comb fingers that are micromachined on the X-, Y-, and Z-proof masses. The intermediate proof mass acts as an outer frame for the inner proof mass, and the X-axis proof mass acts as an outer frame for intermediate proof mass. The entire sensing is through gap-change comb fingers with no electrode plates being used. The Z-proof mass is designed to move in an out-of-plane direction with Z-axis acceleration causing a capacitance change in the comb fingers. Theoretically, any other acceleration (X- or Y-axis) causes no capacitance change in the Z-electrode combs and therefore no cross-axis sensitivity. The same applies for the in-plane motion of the inner and outer frames. The overall chip size along with the readout circuit is 1.78 × 1.38 mm². In the acceleration range of 0.8–6 g, the results indicate a sensitivity of 0.53 mV/g, 0.28 mV/g and 0.2 mV/g for the X-, Y- and Z-axis respectively. The cross-axis sensitivity ranges from 1–8.3% and the nonlinearity is between 2.5% and 3.5% for all the three axes. The Z-axis proof mass is the smallest causing a high noise floor of 357 $mg/\sqrt{Hz}$, followed by the Y-axis (271 $mg/\sqrt{Hz}$) and the X-axis (120 $mg/\sqrt{Hz}$).

A compact three-axis accelerometer with very low cross-axis sensitivity was reported by Y.W. Hsu et al. in [59]. Three spring-mass systems were integrated into one structure using linkage springs with an overall foot-print of 1.3 × 1.28 mm². Silicon-On-Glass (SOG) bulk micromachining was used to fabricate the sensor. An inner proof mass is used for the Y-direction, an intermediate proof mass for the X-direction, and outer proof mass for the Z-direction. The in-plane sensing is done using comb fingers while for the out-of-plane Z-axis sensing, two electrode plates are used. With the Z-axis acceleration, the out-of-balance proof mass undergoes a torsional movement that generates a capacitive difference with respect to reference plates. The device is symmetric and has a sensitivity of 1.4442 V/g, 1.241 V/g, and 1.434 V/g in X-, Y-. and Z-direction, respectively. The noise floor and cross-axis sensitivity for the in-plane X- and Y-direction are 138 $\mu g/\sqrt{Hz}$ (0.28%) and 159 $\mu g/\sqrt{Hz}$ (0.7%), respectively, while for the Z-direction, the noise floor is 49 $\mu g/\sqrt{Hz}$ (0.54%). The accelerometer is packaged with readout circuits and measures 4 mm × 4 mm × 1.2 mm. This design has achieved excellent performance figures that are attributed to the DRIE process with high aspect ratio and a highly symmetric design.

Table 2. Performance summary of three-axis single proof-mass accelerometers.

Ref	Year	Author	Device Size (mm × mm)	Range ('±g')	Sensitivity X, Y, and Z	Noise (μg/√Hz) X, Y, and Z	Nonlinearity X, Y, and Z	Cross-Axis Sensitivity X, Y, and Z
[51]	1996	Mineta, T.	10 × 10	-	-	-	-	10%
[10]	1997	Lemkin, M.A.	4 × 4 (including read out)	11-X-axis, 11-Y-axis, 5.5-Z-axis	0.24 fF/g, 0.24 fF/g, 0.82 fF/g	730, 730, 760	-	1.58% (calculated)
[52]	2001	Li, G.	1.8 × 1.8 (only proof mass)	-	30 mV/g, 30 mV/g, 37 mV/g	-	-	<5%
[53]	2003	Xie, H.	1 × 1 (including readout)	-	-	50 (estimated)	-	-
[56]	2008	Qu, H.	4 × 4 (including readout)	1	520 mV/g, 460 mV/g, 320 mV/g	12, 14, 110	-	2.38%, 2.26%, 4.73% Maximum values
[57]	2009	Hollocher, D.	4 × 4 (including read out)	3	300 mV/g	150, 150, 300	0.3%	1%
[58]	2010	Sun, C.M.	1.78 × 1.78 (including read out)	0.8–6	0.53 mV/g, 0.28 mV/g, 0.2 mV/g	120,000, 271,000, 357,000	2.64%, 3.15%, 3.36%	<7.4%, <8.05%, <8.3% Max values
[59]	2010	Hsu, Y.W.	1.3 × 1.28	1	1.44 mV/g, 1.24 mV/g, 1.4 mV/g	138, 159, 49	0.52%, 0.56%, 0.24%	0.28%, 0.7%, 0.54% Max values
[60]	2012	Tsai, M.H.	0.4 × 0.4 (only proof mass)	0–1	14.7 mV/g, 15.4 mV/g, 14.6 mV/g	2100, 2000, 2100	3.2%, 1.4%, 2.8%	6.6%, 5.4%, 5.3% Maximum values
[61]	2013	Lo, S.C.	1.7 × 1.7 (only proof mass)	0.1–3	4.31 mV/g, 4.3 mV/g, 3.48 mV/g	-	2.72%, 2.57%, 2.91%	6.8%, 6.8%, 9.0%
[62]	2014	Serrano, D.E.	0.45 × 0.45 (only proof mass)	6	6 mV/g, 5 mV/g, 11 mV/g	13, 13, 30	0.5%, 0.5%, 1%	3% (maximum)

A very compact three-axis accelerometer (400 μm × 400 μm) was reported by M.H. Tsai et al. [60]. In this design, gap-change comb fingers are used to sense acceleration is each of the three directions. The comb fingers are distributed not only along the length and width of the proof mass but also along its thickness. This arrangement offers a larger number of fingers in a small space with the vertical fingers drastically improving sensitivity in the Z-direction. Further, the process is rich in interconnect resources that are used to connect the fingers in an interdigitated fashion in order to create fully differential configurations for all the three directions. The sensitivity is close to 15 mV/g, which corresponds to a capacitance change of approximately 2.6 fF/g. The noise floor is 2.1 mg/$\sqrt{\text{Hz}}$ and the maximum cross-axis sensitivity is less than 6.6%.

A novel three-axis polysilicon rib proof-mass accelerometer was proposed by S-C Lo et al. [61]. The design is implemented using two poly-Si trench refill processes, which provides comb electrodes with high aspect ratio, thus increasing the sensing capacitance by 30 folds. For sensing in-plane acceleration, gap-change comb fingers are used, while for out-of-plane sensing, gap-change plate electrodes are used. The sensing is differential in all three directions. For Z-axis sensing, a novel method is implemented which uses movable and fixed lower and upper electrodes.

A three-axis accelerometer specifically designed for an inertial measurement unit (IMU) was reported by D.E. Serrano et al. [62]. In their design, a three-axis pendulum accelerometer is proposed to operate in vacuum. The rationale of this proposal is based on the fact that the IMU gyroscope must be packaged under vacuum, and so the integrated accelerometer itself can be packaged under the same condition. The sensor is designed for the quasi-static domain which requires high damping. This is achieved by increasing the squeeze-film damping through the reduction of the comb finger gaps. The design consists of a pendulum-like structure composed of a 450 μm × 450 μm × 40 μm single-crystal silicon proof mass anchored to the substrate by a cross-shaped polysilicon spring. This type of structure is said to be effective in reducing the footprint. The tethers that compose the spring are attached to the mass using a self-aligned process that prevents offsets in the center-of-mass. Such offsets are strong contributors to cross-axis sensitivity. Four pick-off electrodes placed on the top of the moving structure are multiplexed to read out changes in capacitance generated by the X-, Y- and Z-axis components. In the presence of acceleration along the X-axis, the tethers act as torsional springs, allowing the mass to tilt. This causes a differential change in capacitance.

6.3. Comparison of Single-Proof-Mass and Multiple-Proof-Mass Accelerometers

The performance of accelerometers generally scales with device size. The larger devices have higher sensitivity and lower noise floor. Therefore, in order to effectively compare accelerometers, we use normalized metrics such as sensitivity per unit area and noise-area product. For top performing sensors, the value of sensitivity per unit area should be high while the value of the noise-area product should be low. Some accelerometers have different sensitivities and noise figures along different axes. Since we are comparing three-axis accelerometers we will consider the lowest sensitivity and highest noise floor reported. The sensitivity is either expressed in fF/g or mV/g. For most of the devices, the datasheet sensitivity is given in mV/g. The mV/g gain of the off-the-shelf readout circuit MS3110 IC is used to convert the capacitive sensitivity (fF/g) into output voltage sensitivity (mV/g) for the accelerometers where voltage sensitivity (mV/g) is not reported. The output voltage of MS3110 IC is given by:

$$V_{out} = Gain * V2PS * 1.14 * \frac{\Delta C}{C_F} + V_{ref}$$

where

Gain = 2 or 4 (we will take 2 for our calculation)
V2PS = 2.25
ΔC = Capacitive Sensitivity in fF/g
C_F = 1.5 pF
V_{ref} = 0.5 or 2.25 (this is an offset that we will ignore in our comparisons)

Table 3 gives a comparison overview of various single proof-mass and multiple-proof-mass accelerometers. For a single-axis accelerometer, Chae et al. [44] has the highest sensitivity per unit area and lowest noise-area product. This was made possible by fabricating devices using a process that uses both surface and bulk micromachining techniques. The next best performance is demonstrated by the device reported in [47]. It has used a multiplexed readout circuit for reducing the overall footprint.

In the case of single proof-mass accelerometers, Tsai et al. [60] has maximum sensitivity per unit area while [62] has a minimum noise area product. The accelerometer of [61] uses a novel method to fabricate comb fingers in the thickness direction, which improves Z-axis sensitivity. The vacuum packaging in the case of [62] has helped the device to achieve a minimum noise-area product. The main problem which is faced by most single-proof mass accelerometers is with respect to Z-axis sensing. The overall performance is slightly lower than expected due to low sensitivity and high noise for Z-direction sensing.

Table 3. Comparison of accelerometers.

Multiple Proof-Mass Accelerometers				Single Proof-Mass Accelerometers			
Ref	Year	Sensitivity/Area (mV/mm^2)	Noise $\times$ Area $(\mu g \; mm^2/\sqrt{Hz})$	Ref	Year	Sensitivity/Area (mV/mm^2)	Noise $\times$ Area $(\mu g \; mm^2/\sqrt{Hz})$
[8]	1999	-	15,840	[10]	1997	0.0513	12,160
[39]	1999	3.42	-	[52]	2001	9.259	-
[41]	2000	2.59	-	[53]	2003	-	50
[43]	2005	0.56	-	[56]	2008	20	1760
[44]	2005	157	100	[57]	2009	18	4800
[47]	2013	21.35	2553	[58]	2010	0.06	1,131,118
[49]	2015	-	1058	[59]	2010	0.745	264
[50]	2016	0.381	1008	[60]	2012	91.25	336
-	-	-	-	[61]	2013	1.204	-
-	-	-	-	[62]	2014	24	6

7. Conclusions

In this paper, we have given an overview of monolithic, multi-axis accelerometers. We have discussed various challenges associated with multi-axis sensor design and fabrication and have provided an overview of accelerometer principles, with focus on the design options of the proof mass, sensing comb elements, fabrication process, and read out circuitry. Research on monolithic three-axis accelerometer has been on-going since 1996, and one can observe significant progress has been achieved. From MEMS accelerometers with large footprint, nonlinear, high-noise, and high cross-axis sensitivity devices, the technology has evolved into devices that are compact, highly linear, with high sensitivity, low cross-axis sensitivity, and low µg resolution.

Our literature survey has shown that the majority of accelerometers which use a single proof mass for sensing three-axis acceleration are of very small footprint and are low cost. Unfortunately, with small size come undesirable effects such as undercut of comb fingers during electrical routing. However, such effects along with nonlinearity, high cross-axis sensitivity, and noise may be solved with various innovative techniques already proven to be effective. Maintaining device symmetry was one of the targets for most of the reported devices to reduce off-axis sensitivity. In addition, to this, the concept of embedding the Z-axis proof mass in an XY-sensing frame was found to be widely used. This type of interconnected structures is structurally simpler in terms of suspension design and reduces the overall complexity. However, the use of small Z-axis proof mass increases the Brownian noise in the out-of-plane direction, thus lowering overall performance metrics.

On the other end of the design spectrum, the majority of multiple proof-mass, monolithic accelerometers with large device footprints have superior sensitivity, linearity, and noise floor. The main problem with multiple proof-mass designs is the large size required to obtain good performance. Conversely, most devices with small footprint suffer from poor sensitivity and high noise floor.

Funding: This work was funded by the Mubadala Development Company, Abu Dhabi, UAE, the Economic Development Board, Singapore, and GLOBALFOUNDRIES, Singapore, under the framework of the Twinlab

Micromachines **2018**, *9*, 602

project with participation of A*STAR Institute of Microelectronics (IME), Singapore, Khalifa University, Abu Dhabi, UAE, and GLOBALFOUNDRIES, Singapore.

Conflicts of Interest: The authors declare no conflict of interest.

References

1. Volant Technologies Web Site—Accelerometer and Pressure Sensor MEMS History. Available online: http://terahz.org/_html/22SensorChronology.html (accessed on 1 August 2018).
2. Luczak, S.; Oleksiuk, W.; Bodnicki, M. Sensing tilt with MEMS accelerometers. *IEEE Sens. J.* **2006**, *6*, 1669–1675. [CrossRef]
3. Perez, R.; Costa, Ú.; Torrent, M.; Solana, J.; Opisso, E.; Caceres, C.; Tormos, J.M.; Medina, J.; Gómez, E.J. Upper Limb Portable Motion Analysis System Based on Inertial Technology for Neurorehabilitation Purposes. *Sensors* **2010**, *10*, 10733–10751. [CrossRef] [PubMed]
4. Qu, H. CMOS MEMS Fabrication Technologies and Devices. *Micromachines* **2016**, *7*, 14. [CrossRef] [PubMed]
5. Ayazi, F. Multi-DOF Inertial MEMS: From Gaming to Dead Reckoning. In Proceedings of the 16th International Solid-State Sensors, Actuators and Microsystems Conference (Transducers), Beijing, China, 5–9 June 2011.
6. Honeywell. *ASA7000, Micromachined Accelerometer, Data Sheet*; Honeywell: Morris Plains, NJ, USA, 2001.
7. I. O. Inc. *Si-FlexTM SF3000L Low-Noise Tri-Axial Accelerometer*; I. O. Inc.: Palm Bay, FL, USA, 2004.
8. Lemkin, M.; Boser, B.E. A three-axis micromachined accelerometer with a CMOS position-sense interface and digital offset-trim electronics. *IEEE J. Solid-State Circuits* **1999**, *34*, 456–468. [CrossRef]
9. Lemkin, M.A.; Ortiz, M.A.; Wongkomet, N.; Boser, B.E.; Smith, J.H. A 3-Axis Surface Micromachined ΣΔ Accelerometer. In Proceedings of the 1997 IEEE International Solids-State Circuits Conference, San Francisco, CA, USA, 8 February 1997.
10. Lemkin, M.A.; Boser, B.E.; Auslander, D.; Smith, J.H. A 3-axis force balanced accelerometer using a single proof-mass. In Proceedings of the International Conference on Solid State Sensors and Actuators Conference, Chicago, IL, USA, 19 June 1997.
11. "One-Third of Mobile Phones to Use Accelerometers by 2010, Spurred by iPhone and Palm Pre," News, iSupply Corp., El Segundo, CA, USA. Available online: http://www.isuppli.com/News/Pages/One-Third-of-Mobile-Phonesto-Use-Accelerometers-by-2010-Spurred-by-iPhone-and-Palm-Pre.aspx (accessed on 20 June 2010).
12. "Nexus One the Google Phone Is Coming," Article, Examiner.com, Denver, CO, USA. Available online: http://www.examiner.com/x-33316-Boulder-Technology-Examinery2009m12d15-Nexus-OneThe-Google-Phone-is-coming (accessed on 20 June 2010).
13. "Nokia beats Apple to Compass-in-Phone," Article, MEMS Industry Group, Pittsburgh, PA, USA. Available online: http://memsblog.wordpress.com/2009/12/03/nokia-beats-apple-to-compassin-phone/ (accessed on 20 June 2010).
14. "Analog Devices and Nintendo Collaboration Drives Video Game Innovation with iMEMS Motion Signal Processing Technology," Press Release, Analog Devices, Inc., Nordwood, MA, USA. Available online: http://www.analog.com/en/pressrelease/May092006ADINintendoCollaboration/press.html (accessed on 20 June 2010).
15. Yazdi, N.; Ayazi, F.; Najafi, K. Micromachined Inertial Sensors. *Proc. IEEE* **1998**, *86*, 1640–1659. [CrossRef]
16. Yazıcıoğlu, R.F. Surface Micromachined Capacitive Accelerometers Using MEMS Technology. Master's Thesis, Middle East Technical University, Ankara, Turkey, 2003.
17. Seshia, A.A.; Palaniapan, M.; Roessing, T.A.; Howe, R.T.; Gooch, R.W.; Schimert, T.R.; Montague, S. A Vacuum Packaged Surface Micromachined Resonant Accelerometer. *J. Microelectromech. Syst.* **2002**, *11*, 784–793. [CrossRef]
18. Baldwin, C.; Niemczuk, J.; Kiddy, J.; Slater, T. Review of fiber optic accelerometers. In Proceedings of the IMAC XXIII Conference & Exposition on Structural Dynamics, Society for Experimental Mechanics, Orlando, FL, USA, 31 January–3 February 2005.
19. Milanovi, V.; Bowen, E.; Tea, N.; Suehle, J.; Payne, B.; Zaghloul, M.; Gaitan, M. Convection based Accelerometer and Tilt Sensor Implemented in Standard CMOS. In Proceedings of the International Mechanical Engineering Congress and Exposition, San Francisco, CA, USA, 15–20 November 1998.

20. Lui, C.H.; Kenny, T.H. A High-Precision Wide-Bandwidth Micromachined Tunneling Accelerometer. *J. Microelectromech. Syst.* **2001**, *10*, 425–433.

21. Xie, H.; Fedder, G.K. A CMOS Z-Axis Capacitive Accelerometer with Comb-Finger Sensing. In Proceedings of the IEEE Micro Electro Mechanical Systems Conference, Miyazaki, Japan, 23–27 January 2000.

22. Jiang, X.; Wang, F.; Kraft, M.; Boser, B.E. An Integrated Surface Micromachined Capacitive Lateral Accelerometer with 2 μG/rt-Hz Resolution. In Proceedings of the Solid State Sensor and Actuator Workshop, Hilton Head Island, SC, USA, 2–6 June 2002.

23. Puers, R.; Reyntjens, S. Design and Processing Experiments of a new Miniaturized Capacitive Triaxial Accelerometers. *Sens. Actuators A* **1998**, *68*, 324–328. [CrossRef]

24. Qu, W.; Wenzel, C.; Jahn, A. One-mask Procedure for the Fabrication of Movable High-Aspect-Ratio 3d Microstructures. *J. Microelectromech. Syst.* **1998**, *8*, 279–283. [CrossRef]

25. Robin, L.; Mounier, E. Inertial sensor market moves to combo sensors and sensor hubs. *MEMS' Trends Mag.* **2013**, *16*, 16–18.

26. Ocak, I.E.; Cheam, D.D.; Fernando, S.N.; Lin, A.T.; Singh, P.; Sharma, J.; Kwong, D.L. A Monolithic 9 Degree of Freedom (DOF) Capacitive Inertial MEMS Platform. In Proceedings of the IEEE International Electron Devices Meeting, San Francisco, CA, USA, 15–17 December 2014.

27. Weigold, J.W.; Najafi, K.; Pang, S.W. Design and Fabrication of Submicrometer, Single Crystal Si Accelerometer. *J. Microelectromech. Syst.* **2001**, *10*, 518–524. [CrossRef]

28. Chau, K.H.L.; Lewis, S.R.; Zhao, Y.; Howe, R.T.; Bart, S.F.; Marcheselli, R.G. An integrated force-balanced capacitive accelerometer for low-g applications. *Sens. Actuators A* **1996**, *54*, 472–476. [CrossRef]

29. Chae, J.; Kulah, H.; Najafi, K. A hybrid Silicon-On-Glass (SOG) lateral micro-accelerometer with CMOS readout circuitry. In Proceedings of the IEEE International Conference on MEMS, Las Vegas, NV, USA, 24 January 2002.

30. Benmessaoud, M.; Nasreddine, M.M. Optimization of MEMS capacitive accelerometer. *Microsyst. Technol.* **2013**, *19*, 713–720. [CrossRef]

31. Aydin, O.; Akin, T. A bulk-micromachined fully-differential MEMS accelerometer with interdigitated fingers. In Proceedings of the IEEE Sensors Conference, Taipei, Taiwan, 28–32 October 2012.

32. Mohammed, Z.; Gill, W.A.; Rasras, M. Double-Comb-Finger Design to Eliminate Cross-Axis Sensitivity in a Dual-Axis Accelerometer. *IEEE Sens. Lett.* **2017**, *1*, 1–4. [CrossRef]

33. Mohammed, Z.; Dushaq, G.; Chatterjee, A.; Rasras, M. Bi-axial highly sensitive ±5g polysilicon based differential capacitive accelerometer. In Proceedings of the 17th International Conference on Thermal, Mechanical and Multi-Physics Simulation and Experiments in Microelectronics and Microsystems (EuroSimE), Montpellier, France, 18–20 April 2016.

34. Seidel, H.; Riedel, H.; Kolbeck, R.; Mück, G.; Kupke, W.; Königer, M. Capacitive silicon accelerometer with highly symmetrical design. *Sens. Actuators A Phys.* **1990**, *21*, 312–315. [CrossRef]

35. Matsumoto, Y.; Iwakiri, M.; Tanaka, H.; Ishida, M.; Nakamura, T. A capacitive accelerometer using SDB-SOI structure. *Sens. Actuators A* **1996**, *53*, 267–272. [CrossRef]

36. Chen, W.; Huo, M.; Lin, Y.; Liu, X.; Zhang, R. A novel Zaxis capacitive accelerometer using SOG structure. In Proceedings of the 6th International Conference on Electronics Packaging Technology, Shenzhen, China, 30 August–2 September 2005.

37. Lee, I.; Yoon, G.H.; Park, J.; Seok, S.; Chun, K.; Lee, K. Development and analysis of the vertical capacitive accelerometer. *Sens. Actuators A* **1996**, *119*, 8–18. [CrossRef]

38. Mohammed, Z.; Elfadel, I.M.; Rasras, M. High dynamic range Z-axis hybrid spring MEMS capacitive accelerometer. In Proceedings of the IEEE Symposium on Design, Test, Integration & Packaging of MEMS and MOEMS (DTIP), Rome, Italy, 22–25 May 2018.

39. Matsumoto, Y.; Nishimura, M.; Matsuura, M.; Ishida, M. Three-axis SOI capacitive accelerometer with PLL C–V converter. *Sens. Actuators A Phys.* **1999**, *75*, 77–85. [CrossRef]

40. Matsumoto, Y.; Yoshida, K.; Ishida, M. Fluorocarbon film for protection from alkaline etchant and elimination of in-use stiction. In Proceedings of the International Solid State Sensors and Actuators, Chicago, IL, USA, 19 June 1997.

41. Butefisch, S.; Schoft, A.; Buttgenbach, S. Three-axes monolithic silicon low-g accelerometer. *J. Microelectromech. Syst.* **2000**, *9*, 551–556. [CrossRef]

42. Rodjeg, H.; Andersson, G. Design optimization of three-axis accelerometers based on four seismic masses. In Proceedings of the IEEE Sensors, Orlando, FL, USA, 12–14 June 2002.

43. Rödjegård, H.; Johansson, C.; Enoksson, P.; Andersson, G. A monolithic three-axis SOI-accelerometer with uniform sensitivity. *Sens. Actuators A Phys.* **2005**, *123–124*, 50–53. [CrossRef]

44. Chae, J.; Kulah, H.; Najafi, K. A Monolithic Three-Axis Micro-g Micromachined Silicon Capacitive Accelerometer. *J. Microelectromech. Syst.* **2005**, *14*, 235–242. [CrossRef]

45. Chae, J.; Kulah, H.; Najafi, K. An In-Plane High-Sensitivity, Low-Noise Micro-g Silicon Accelerometer with CMOS Readout Circuitry. *J. Microelectromech. Syst.* **2004**, *13*, 628–635. [CrossRef]

46. Yazdi, N.; Ayazi, F.; Najafi, K. An All-Silicon Single-Wafer Micro-g Accelerometer with a Combined Surface and Bulk Micromachining Process. *J. Microelectromech. Syst.* **2000**, *9*, 1–8. [CrossRef]

47. Liu, Y.C.; Tsai, M.H.; Li, S.S.; Fang, W. A Fully-Differential, Multiplex-Sensing Interface Circuit Monolithically Integrated with Tri-Axis Pure Oxide Capacitive CMOS-MEMS Accelerometers. In Proceedings of the 17th International Conference on Solid-State Sensors, Actuators and Microsystems, Barcelona, Spain, 16–20 June 2013.

48. Tez, S.; Akin, T. Fabrication of a sandwich type three axis capacitive MEMS accelerometer. In Proceedings of the IEEE Sensors, Baltomire, MD, USA, 3–6 November 2013.

49. Tez, S.; Aykutlu, U.; Torunbalci, M.M.; Akin, T. A Bulk-Micromachined Three-Axis Capacitive MEMS Accelerometer on a Single Die. *J. Microelectromech. Syst.* **2015**, *24*, 1264–1274. [CrossRef]

50. Aydemir, A.; Terzioglu, Y.; Torunbalci, M.M.; Akin, T. A new design and a fabrication approach to realize a high performance three axes capacitive MEMS accelerometer. *Sens. Actuators A Phys.* **2016**, *244*, 324–333. [CrossRef]

51. Mineta, T.; Kobayashi, S.; Watanabe, Y.; Kanauchi, S.; Nakagawa, I.; Suganuma, E.; Esashi, M. Three-axis capacitive accelerometer with uniform axial sensitivities. *J. Micromech. Microeng.* **1996**, *6*, 431–435. [CrossRef]

52. Li, G.; Li, Z.; Wang, C.; Hao, Y.; Li, T.; Zhang, D.; Wu, G. Design and fabrication of a highly symmetrical capacitive triaxial accelerometer. *J. Micromech. Microeng.* **2001**, *11*, 48–54. [CrossRef]

53. Xie, H.; Pan, Z.; Frey, W.; Fedder, G. Design and fabrication of an integrated CMOS-MEMS 3-axis accelerometer. In Proceedings of the 2003 Nanotechnology Conference, San Francisco, CA, USA, 23–27 February 2003.

54. Mohammed, Z.; Dushaq, G.; Chatterjee, A.; Rasras, M. An optimization technique for performance improvement of gap-changeable MEMS accelerometers. *Mechatronics* **2018**, *54*, 203–216. [CrossRef]

55. Qu, H.; Fang, D.; Xie, H. A Single-Crystal Silicon 3-axis CMOS-MEMS Accelerometer. In Proceedings of the IEEE Sensors Conference, Vienna, Austria, 24–27 October 2004.

56. Qu, H.; Fang, D.; Xie, H. A Monolithic CMOS-MEMS 3-Axis Accelerometer with a Low-Noise, Low-Power Dual-Chopper Amplifier. *IEEE Sens. J.* **2008**, *8*, 1511–1518.

57. Hollocher, D.; Zhang, X.; Sparks, A.; Bart, S.; Sawyer, W.; Narayanasamy, P.; Mhatre, R. A Very Low Cost, 3-axis, MEMS Accelerometer for Consumer Applications. In Proceedings of the IEEE Sensors, Christchurch, New Zealand, 25–28 October 2009.

58. Sun, C.M.; Tsai, M.H.; Liu, Y.C.; Fang, W. Implementation of a monolithic single proof-mass tri-axis accelerometer using CMOS-MEMS technique. *IEEE Trans. Electron. Dev.* **2010**, *57*, 1670–1679. [CrossRef]

59. Hsu, Y.W.; Chen, J.Y.; Chien, H.T.; Chen, S.; Lin, S.T.; Liao, L.P. New capacitive low-g triaxial accelerometer with low cross-axis sensitivity. *J. Micromech. Microeng.* **2010**, *20*, 055019. [CrossRef]

60. Tsai, M.H.; Liu, Y.C.; Fang, W. A Three-Axis CMOS-MEMS Accelerometer Structure with Vertically Integrated Fully Differential Sensing Electrodes. *J. Microelectromech. Syst.* **2012**, *21*, 1329–1337. [CrossRef]

61. Lo, S.C.; Chan, C.K.; Lai, W.C.; Wu, M.; Lin, Y.C.; Fang, W. Design and Implementation of A Novel Poly-Si Single Proof-Mass Differential Capacitive-Sensing 3-Axis Accelerometer. In Proceedings of the 17th International Conference on Solid-State Sensors, Actuators and Microsystems, Barcelona, Spain, 16–20 June 2013.

62. Serrano, D.E.; Jeong, Y.; Keesara, V.; Sung, W.K.; Ayazi, F. Single Proof-Mass Tri-Axial Pendulum Accelerometers Operating in Vacuum. In Proceedings of the IEEE 27th International Conference on Micro Electro Mechanical Systems (MEMS), San Francisco, CA, USA, 26–30 January 2014.

Article

Design and Analysis of the Elastic-Beam Delaying Mechanism in a Micro-Electro-Mechanical Systems Device

Fufu Wang [1], Lu Zhang [1,*], Long Li [2], Zhihong Qiao [1] and Qian Cao [1]

[1] Key Laboratory of Space Utilization, Technology and Engineering Center for space Utilization, Chinese Academy of Sciences, Beijing 100094, China; wangfufu2004@sina.com (F.W.); qiaozhihong@csu.ac.cn (Z.Q.); caoqian@csu.ac.cn (Q.C.)

[2] Qian Xuesen Laboratory of Space Technology, NO. 104 Youyi Road, Haidian District, Beijing 100094, China; lilong@qxslab.cn

* Correspondence: zhanglu@csu.ac.cn; Tel.: +86-135-8187-5465

Received: 17 October 2018; Accepted: 31 October 2018; Published: 2 November 2018

Abstract: The delaying mechanism is an important part of micro-electro-mechanical systems (MEMS) devices. However, very few mechanical delaying mechanisms are available. In this paper, an elastic-beam delaying mechanism has been proposed innovatively through establishing a three-dimensional model of an elastic-beam delay mechanism, establishing the force and the parameters of an elastic-beam delay mechanism, deriving the mathematical model according to the rigid dynamic mechanics theory, establishing the finite element model by using Ls-dyna solver of the Ansys software, and carrying out the centrifugal test. Simulation and test results match theoretical results quite well. It is believed that the elastic-beam delaying mechanism is quite effective and useful to slow the speed of the movable part in MEMS devices.

Keywords: micro-electro-mechanical systems (MEMS); delaying mechanism; safety and arming system

1. Introduction

The function of the delaying mechanism is to delay the appropriate time to ensure that the mechanism completes the corresponding action. The traditional delaying mechanism mainly includes the non-returning clock mechanism, the gas or liquid damping mechanism, the quasi-fluid mechanism, and so on [1,2].

Micro-electro-mechanical systems (MEMS) are a relatively new and fast-growing field in microelectronics. Micro-electro-mechanical systems are commonly used as actuators, sensors, and radio frequency and microfluidic components, as well as bio-composites, with a wide variety of applications in health care, automotive, and military industries. It is expected that the market for MEMS will grow to over $30 B by 2050 [3–7].

With the rapid development of MEMS technology, the demand for miniaturization of delaying mechanisms has become more urgent. At present, there are a large number of MEMS delaying mechanism research, which are mainly divided into two categories. One category is the delaying mechanism based on electricity [8–16], which has high-control precision; Dalian University of Technology in China has studied the v-shaped micro-electric-thermal delaying mechanism based on MEMS technology [8], and the university of Toulouse in France has studied MEMS electro-powder delaying mechanism [9]. The other category is the mechanical delaying mechanism [17]; KAMAN INC. of the United States demonstrated a MEMS mechanical delaying mechanism and its delay action does not require electronic control, and it is free from electromagnetic interference and high reliability, and it is particularly suitable for battlefield use in complex electromagnetic environments.

In this paper, the MEMS Safety and Arming (S&A) system used in small-caliber ammunition is proposed innovatively, and the concrete structure as shown in Figure 1. The size of the MEMS S&A system is 10 mm × 13 mm × 0.5 mm. The MEMS S&A system is composed of threshold-value judging mechanism, lock-releasing mechanism, elastic-beam delaying mechanism, centrifugal lock, setback lock, sub-centrifugal slider, and main centrifugal slider. It has two functions [18–28]: one function is that after the ammunition goes out of the antiaircraft gun, it needs a certain time delay to ensure that the main centrifugal slider does not move to the designated position, so as to ensure that the ammunition does not explode at the muzzle or launch area, which can guarantee the safety of the ammunition. The other function is that when ammunition arrives at the designated area, it needs the main centrifugal slider to move to the designated position to ensure the reliable function of ammunition. The elastic-beam delaying mechanism plays an important role in ensuring the safety of ammunition. Figure 2 shows that the elastic-beam delaying mechanism is composed of baseplate, centrifugal slider, passive tooth, active tooth, and threshold elastic beam, which can slow the speed of the movable part to ensure a proper delay time and guarantee the safety of the ammunition.

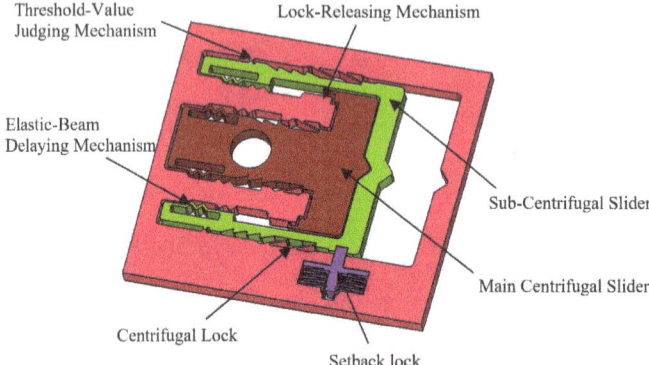

Figure 1. The mechanical MEMS S&A system.

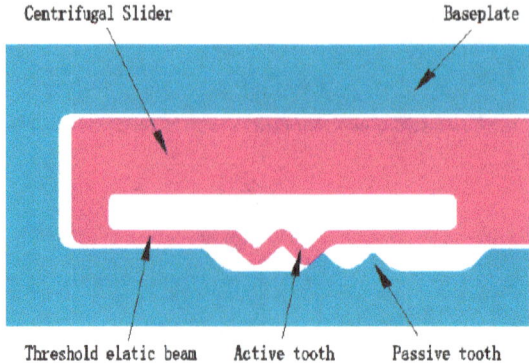

Figure 2. The elastic-beam delaying mechanism.

2. Model and Theoretical Analysis

2.1. Model

The center of gravity, rotation center, and stroke of the sub-centrifugal slider are shown in Figure 3. The parameters of elastic-beam delaying mechanism are shown in Figure 4. The force of elastic-beam

delaying mechanism are shown in Figure 5. Based on Figure 4, the main variables are parameterized, shown as Table 1.

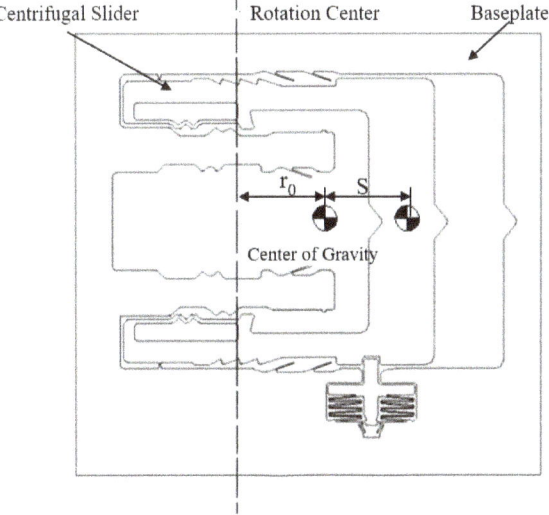

Figure 3. The center of gravity, rotation center, and stroke of the centrifugal slider.

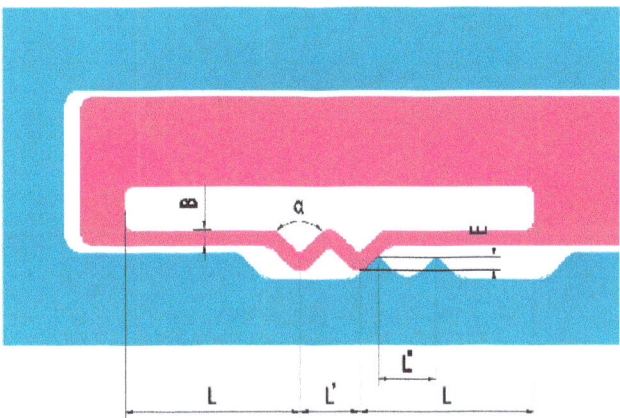

Figure 4. The parameters of elastic-beam delaying mechanism.

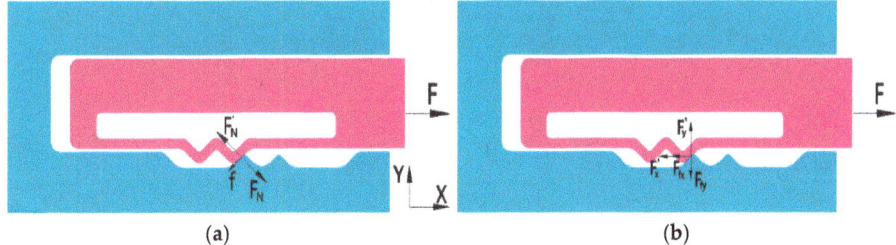

Figure 5. The force of elastic-beam delaying mechanism: (**a**) overall view, (**b**) decomposition view.

Table 1. The parameters of elastic-beam delaying mechanism.

Number	1	2	3	4	5
Width of the elastic-beam B/mm	0.1	0.08	0.1	0.1	0.1
Angle of active tooth α/°	90	90	70	90	90
Length of the elastic-beam L/mm	1.2	1.2	1.2	2.4	1.4
Gap between active and tooth and passive tooth E/mm	0.03	0.03	0.02	0.03	0.03
Number of the active tooth			2		1
Distance between two adjacent active teeth L'/mm			0.4		-
Distance between two adjacent passive teeth L''/mm			0.4		

2.2. Force Analysis

We have some hypotheses:

- Neglecting the factors such as friction and air resistance;
- The process of active tooth contact passive tooth movement is from uniform acceleration to uniform deceleration.

According to the Figures 4 and 5, we can get the centrifugal force of centrifugal slider

$$F = mr\omega^2 = mr\left(\frac{2\pi n}{60}\right)^2 \tag{1}$$

where, m is the mass of centrifugal slider; r is the eccentricity of centrifugal slider; n is the rotating speed of projectile.

According to the Figures 4 and 5b, we can get

$$\begin{cases} F_x' = F_N' \cos(\frac{\alpha}{2}) = F_N(w)\cos(\frac{\alpha}{2}) \\ F_y' = F_N' \sin(\frac{\alpha}{2}) = F_N(w)\sin(\frac{\alpha}{2}) \end{cases} \tag{2}$$

$$\begin{cases} F_{fx} = f\sin(\frac{\alpha}{2}) = \mu F_N(w)\sin(\frac{\alpha}{2}) \\ F_{fy} = f\cos(\frac{\alpha}{2}) = \mu F_N(w)\cos(\frac{\alpha}{2}) \end{cases} \tag{3}$$

where, $F_N(w)$ is the pressure of the active tooth to the passive tooth; μ is the friction coefficient of the active tooth to the passive tooth; f is the friction of the active tooth to the passive tooth; F_x', F_y', F_{fx}, F_{fy} is the decomposition for of $F_N(w)$ and f.

So, we can get the force of the centrifugal slider in the X and Y direction

$$\begin{cases} F_x = F - F_x' - F_{fx} = F - F_N(w)\left(\cos(\frac{\alpha}{2}) + \mu\sin(\frac{\alpha}{2})\right) \\ F_y = F_y' - F_{fy} = F_N(w)\left(\sin(\frac{\alpha}{2}) - \mu\cos(\frac{\alpha}{2})\right) \end{cases} \tag{4}$$

2.3. Deflection Calculation

According to the cantilever deflection equation of applied engineering mechanics, the Figure 6 can be obtained.

According to the Figure 6, we can get the deflection

$$w = \begin{cases} -\frac{Fb^2x^2}{6EIl}[3\frac{a}{l} - (1+2\frac{a}{l})\frac{x}{l}] & (0 \le x \le a) \\ -\frac{Fa^2(l-x)^2}{6EIl}[\frac{a}{l} - (1+2\frac{b}{l})\frac{(l-x)}{l}] & (a < x \le l) \end{cases} \tag{5}$$

According to Figures 4 and 6, we can get

$$F = F_y;\ b = L - \frac{L'}{4};\ a = L + \frac{5L'}{4};\ l = 2L + L';\ I = \frac{HB^3}{12} \tag{6}$$

where, H is the thickness of the elastic-beam.

If $F_x = 0$ in (4), the maximum value of $F_N(w)$ can be obtained

$$F_{Nmax}(w) = F\left[\cos(\tfrac{\alpha}{2}) + \mu\sin(\tfrac{\alpha}{2})\right]^{-1} \tag{7}$$

According to the (5)–(7), the maximum deflection at node C can be obtain.

$$w_{Cmax} = 4F\left(L - \tfrac{L'}{4}\right)^3\left(L + \tfrac{5L'}{4}\right)^3\left(\sin(\tfrac{\alpha}{2}) - \mu\cos(\tfrac{\alpha}{2})\right)\left[EHB^3(2L + L')^3\left(\cos(\tfrac{\alpha}{2}) + \mu\sin(\tfrac{\alpha}{2})\right)\right]^{-1} \tag{8}$$

Combined with Figure 4, the stroke of the centrifugal slider can be obtained

$$S = w_{Cmax}\tan(\tfrac{\alpha}{2}) \tag{9}$$

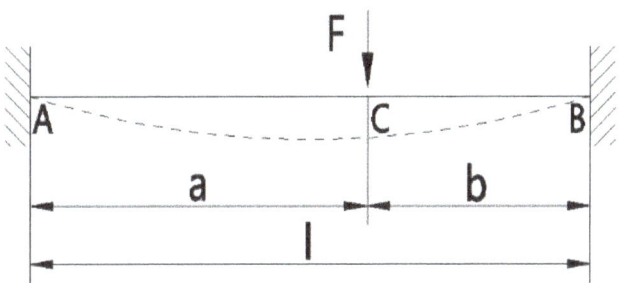

Figure 6. The schematic diagram of elastic-beam deflection analysis.

2.4. Kinematic Analysis and Time Estimating

The schematic diagram of kinematic analysis and time estimating are shown in Figure 7.

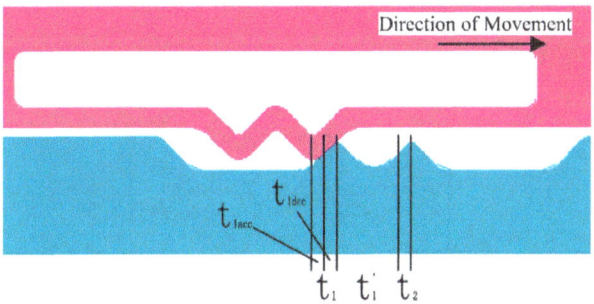

Figure 7. The schematic diagram of kinematic analysis and time estimating.

2.4.1. The First Contact of Active Tooth to Passive Tooth

The first contact is divided into two stages, one stage (t_{1acc}) is the uniformly acceleration process when the active tooth from the initial position to the first contact; the other stage (t_{1dec}) is the uniform deceleration process when the active tooth from the first contact to separation.

And

$$S = v_0 t + \tfrac{1}{2}at^2 \tag{10}$$

where, S is the displacement of the active tooth; v_0 is the initial velocity of the active tooth; a is the average acceleration of the active tooth; t is the time of the whole movement process.

- The uniformly acceleration process

We can get

$$\begin{cases} v_1 = 0 \\ a_1 = \frac{F}{m} \\ S_1 = E\tan(\frac{\alpha}{2}) \end{cases} \tag{11}$$

where, v_1 is the initial velocity of the active tooth at first contact; a_1 is the average acceleration of the active tooth at first contact; S_1 is the displacement of the active tooth from the initial position to the first contact.

So

$$t_{1acc} = \sqrt{\frac{2E\tan(\frac{\alpha}{2})}{r_0(\frac{2\pi n}{60})^2}} \tag{12}$$

where, r_0 is the initial eccentricity of centrifugal slider.

- The uniformly deceleration process

The uniform deceleration process is the inverse of the uniform acceleration process.

So

$$t_1 = t_{1acc} + t_{1dec} = 2t_{1acc} = 2\sqrt{\frac{2E\tan(\frac{\alpha}{2})}{r_0(\frac{2\pi n}{60})^2}} \tag{13}$$

2.4.2. The Second Contact of Active Tooth to Passive Tooth

The second contact is divided into two stages, one stage (t_1') is the uniformly acceleration process when the active tooth from separation to second contact; other stage (t_2) is the uniform deceleration process when the active tooth from the second contact to separation.

- The uniformly acceleration process

We can get

$$\begin{cases} v_1' = 0 \\ a_1' = \frac{(r_0+\frac{L'}{2})+(r_0+\frac{L'}{2}+L''-E\cdot\tan(\frac{\alpha}{2}))}{2}(\frac{2\pi n}{60})^2 \\ v_2 = \sqrt{2a_1'(L''-E\cdot\tan(\frac{\alpha}{2}))+v_1'^2} \end{cases} \tag{14}$$

where, v_1' is the velocity of the active tooth separation from the first passive tooth; a_1' is the average acceleration of the active tooth from separation to second contact; v_2 is the velocity of the active tooth at second contact

So

$$t_1' = \frac{v_2 - v_1'}{a_1'} \tag{15}$$

- The uniformly deceleration process

The uniform deceleration process is the inverse of the uniform acceleration process.

We can get

$$a_2 = \frac{(r_0+\frac{L'}{2}+L''-E\cdot\tan(\frac{\alpha}{2}))+(r_0+\frac{L'}{2}+L'')-2r_0}{2}(\frac{2\pi n}{60})^2 = \frac{L'+2L''-E\cdot\tan(\frac{\alpha}{2})}{2}(\frac{2\pi n}{60})^2 \tag{16}$$

where, a_2 is the average acceleration of the active tooth the second contact to separation.

So

$$E\cdot\tan(\frac{\alpha}{2}) = v_2 t_2 + \frac{1}{2}\bar{a}_2 t_2^2 \tag{17}$$

According to the (14), (16), and (17), we can get the time t_2.

According to the above theoretical analysis, the maximum deflection and time estimating of the different parameterized variable can be obtained, shown as Table 2.

Table 2. The theoretical result of maximum deflection and time estimating.

Number		1	2	3	4	5
Mass of centrifugal slider m/kg		2.88×10^{-6}	2.836×10^{-6}	2.866×10^{-6}	2.84×10^{-6}	2.868×10^{-6}
Maximum deflection w_{max}/mm		2.86×10^{-2}	5.50×10^{-2}	2.0×10^{-2}	0.20	3.23×10^{-2}
Time estimating/μs	t_1	27.6	28.3	23.1	73.0	29.3
	t_1'	49.7	48.0	50.3	36.5	49.5
	t_2	1.9	3.8	1.3	18.3	2.2
	t_{total}	79.2	80.1	74.7	129.8	81

3. Simulation Analysis

Base on the Figure 4 and Table 1, establishing the infinite model of elastic-beam delaying mechanism (Figure 8) by using HyperMesh, applying Material properties (Table 3), loading the acceleration of a constant 30,000 g, and making nonlinear dynamic mechanics simulation by using ANSYS/LS-DYNA, displacement-time curves (Figure 9a) and velocity-time curves (Figure 9b) are obtained.

Table 3. The material properties [21].

Name	Density ρ (kg/m^3)	Elasticity Modulus E (GPa)	Poisson's Ratio v
Si	2.3×10^3	180	0.3

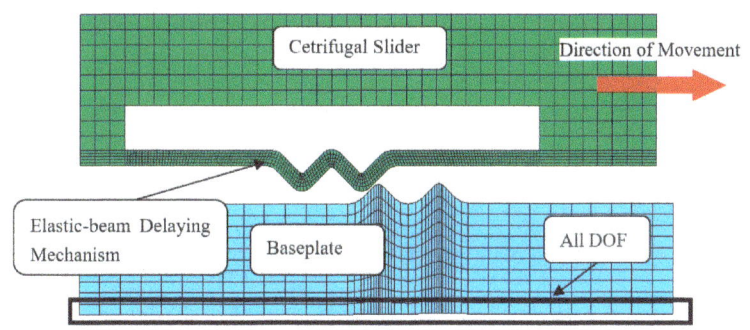

Figure 8. The infinite model of elastic-beam delaying mechanism.

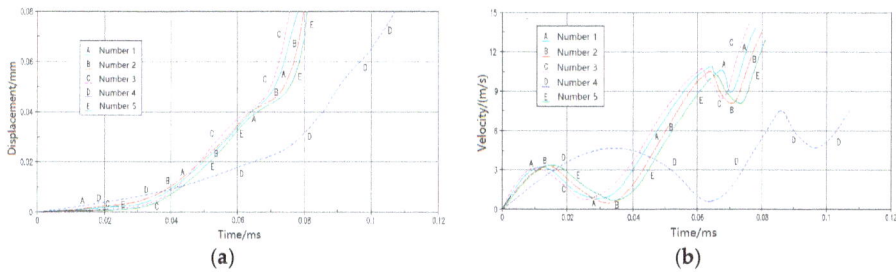

Figure 9. The simulation results of different structural parameters (**a**) displacement-time curves (**b**) velocity -time curves.

The Table 4 shown that the simulation results of the deflection are in good agreement with the theoretical results, and the maximum error is less than 6.5%, because of the residual velocity of separation between active teeth and passive when simulating.

Table 4. The maximum deformation and error between theoretical results and simulation results.

Number		1	2	3	4	5
Maximum deflection	Theoretical	2.86×10^{-2}	5.50×10^{-2}	2.0×10^{-2}	0.20	3.23×10^{-2}
w_{max}/mm	Simulation	2.92×10^{-2}	5.64×10^{-2}	2.1×10^{-2}	0.213	3.37×10^{-2}
Error (Simulation-Theoretical)/Theoretical		2.1%	2.5%	5.0%	6.5%	4.3%

According to Table 2, Table 4 and Figure 9, the following conclusions are obtained:

- In terms of movement trend, the simulation results are in good agreement with the theoretical calculation;
- In terms of the movement time for the same displacement, simulation results are shorter than the theoretical calculation. Because when active tooth is separated from the first passive tooth, the simulation results have residual velocity, and the theoretical value is 0;
- The gap between active tooth and passive tooth is the most important factor affecting the movement time.

4. Fabrication

The Figure 10 shows the micromachining process of silicon-based MEMS S&A system. It mainly includes the silicon-based MEMS S&A system in the middle and Benzocyclobutene (BCB) bonded glass on the upper and lower sides. As shown in Figure 11, the white part is the structure area, which is the main structure left by photolithography with bulk-micromachining technology, the blue part is the hollow area, the green part is the silicon-glass bonding area with BCB bonding, which enables the structure area to generate a certain gap, so that the key structure can move along the predetermined mode in the hollow area.

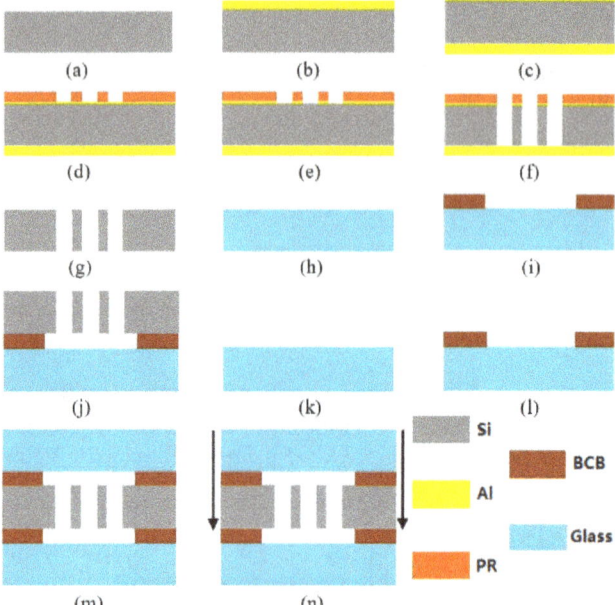

Figure 10. Micromachining process of silicon-based MEMS S&A system (**a**) Back-up; (**b**) Physical Vapor Deposition (PVD) Al 2um; (**c**) PVD Al 0.2um; (**d**) Structural lithography; (**e**) Corroded aluminum; (**f**) Deep Reactive Ion Etching (DRIE); (**g**) Degumming; (**h**) Back-up; (**i**) BCB Lithography; (**j**) BCB bond; (**k**) Buck-up; (**l**) BCB Lithography; (**m**) BCB bond; (**n**) Scribing.

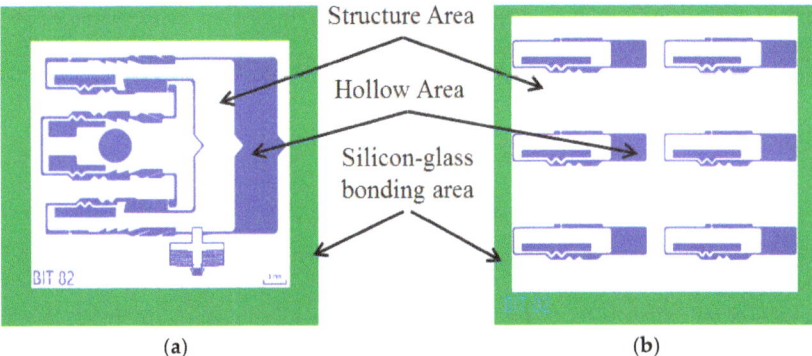

Figure 11. The processing layout (**a**) MEMS S&A system, (**b**) Elastic-beam delaying mechanism.

The micro-sample of the elastic-beam delaying mechanism can be obtained based on the digital microscope VHX-6000 series produced by Keyence Corporation, as shown in Figure 12a. And partial view can be obtained based on the Electron microscope, as shown in Figure 12b.

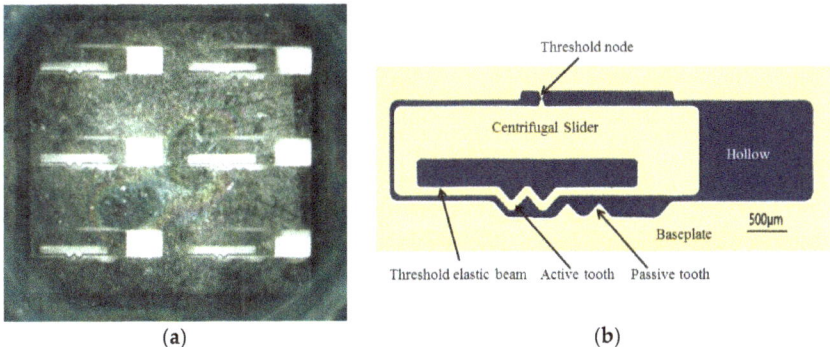

Figure 12. The micro-sample of the elastic-beam delaying mechanism (**a**) overall view, (**b**) partial view.

5. Test

There are two kinds of tests for delaying mechanism:

- The impact test. The aim is to break the threshold node so as to ensure the centrifugal test. The impact test platform is shown in Figure 13, and impact acceleration direction is shown in Figure 14.
- The centrifugal test. The aim is to obtain g-value of the centrifugal slider moving through the passive tooth of the delaying mechanism with different parameters. The centrifugal test platform is shown in Figure 15, and centrifugal acceleration direction is shown in Figure 16.

Four groups of centrifugal tests were carried out on the parameterized delaying mechanism, as shown in Table 1. The theoretical and simulation results obtained from Formula (8) were combined with centrifugal test to obtain Table 5 and Figure 17.

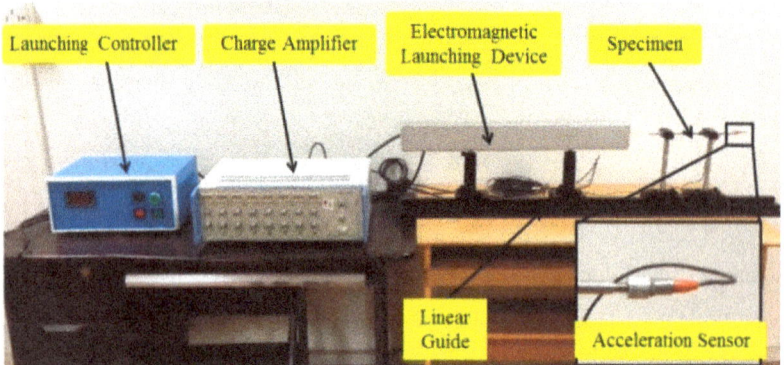

Figure 13. Impact test platform.

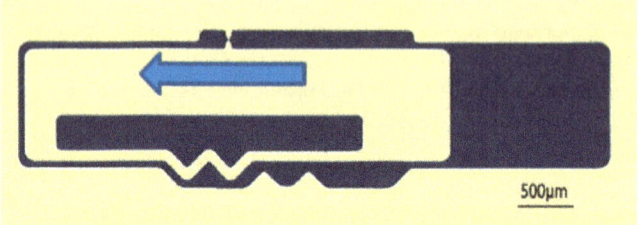

Figure 14. The direction of impact acceleration.

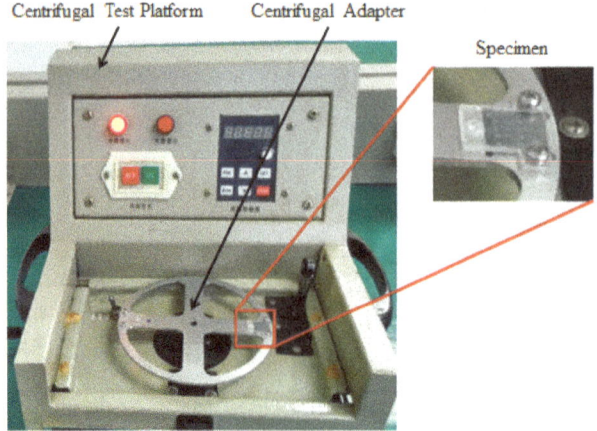

Figure 15. Centrifugal test platform and centrifugal adapter.

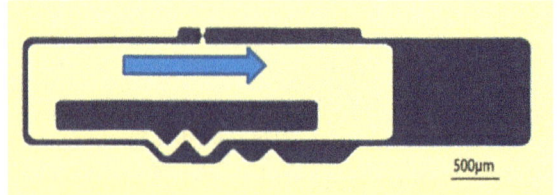

Figure 16. The direction of centrifugal acceleration.

Table 5. The g-value of the delaying mechanism with different parameters.

Number	Test Results/g					Theoretical Results/g	Simulation Results/g
	Group 1	Group 2	Group 3	Group 4	Average		
(1)	35,000	33,000	35,000	36,000	34,750	31,500	30,000
(2)	19,000	15,000	21000	22,000	19,250	16,370	15,000
(3)	33,000	31,000	34,000	36,000	33,500	30,120	28,000
(4)	9000	7000	10,000	9000	8750	4520	4100
(5)	24,000	22,000	24,000	26,000	24,000	20,500	19,000

According to Table 5 and Figure 17, it arrives at the conclusions as follows:

- The g-value of the delaying mechanism with the same processing batch and the same parameter is relatively discrete, indicating that the material properties of silicon have a certain degree of dispersion.
- The theoretical results are all higher than the simulation results, because the theoretical calculation is completely static, and the possible initial velocity is ignored. And the average value of the test results is higher than the theoretical results, because of the friction and gas resistance in the micro-sample.
- Theoretical results, simulation results and test results have a high degree of agreement, which can be used for initial optimization design.

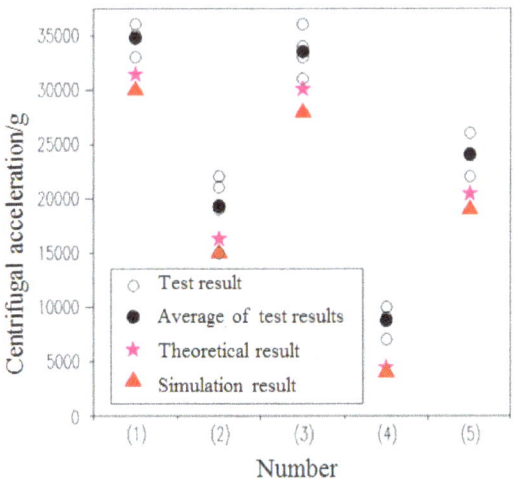

Figure 17. The g-value of the delaying mechanism with different parameters.

6. Conclusions

In this paper, the elastic-beam delaying mechanism has been proposed innovatively. Combining with the rigid dynamic mechanics theory, the mathematical model was established. Simulation and test results match theoretical results quite well. It is believed that the elastic-beam delaying mechanism is quite effective and useful to slow the speed of the movable part in MEMS devices.

Author Contributions: F.W. and L.Z. conceived the problem and designed the solution; F.W. and L.L. designed and performed the experiments; Z.Q., Q.C. and F.W. analyzed the data; F.W. wrote the paper.

Acknowledgments: The research was supported by Key Laboratory of Space Utilization, and sponsored by the National Project of China (Y7140211XN).

Conflicts of Interest: The authors declare no conflict of interest.

References

1. Ma, B.H. *Fuze Structure and Function*; National Defense Industry Press: Beijing, China, 1984.
2. Wang, S.Y. *Fuze System Analysis and Engineering Design Questions and Solutions*; Nanjing University of Science and Technology: Nanjing, China, 2005.
3. Shaeffer, D.K. MEMS inertial sensors: A tutorial overview. *IEEE Commun. Mag.* **2013**, *51*, 100–109. [CrossRef]
4. Allameh, S.M. An introduction to mechanical-properties-related issues in MEMS structures. *J. Mater. Sci.* **2003**, *38*, 4115–4123. [CrossRef]
5. Robinson, C.H.; Hoang, T.Q.; Gelak, M.R. *Materials, Fabrication and Assembly Technologies for Advanced MEMS-based Safety and Arming Mechanism for Projectile Munitions*; J. F. Rasmussen Axsun Technologies, Inc.: Billerica, MA, USA, 2006.
6. Zhou, X.; Shan, T.; Qi, X. Analysis and design of a high-power laser interrupter for MEMS based safety and arming systems. *Microsyst. Technol.* **2017**, *23*, 1–10. [CrossRef]
7. Wang, D.K.; Lou, W.Z.; Feng, Y.; Zhang, X.Z. Design of High-Reliability Micro Safety and Arming Devices for a Small Caliber Projectile. *Micromachines* **2017**, *8*, 234. [CrossRef]
8. Zhang, R.; Chu, J.K.; Wang, H.Y.; Chen, Z.P. SU-8 chevron electrothermal micro-actuator with three-layer structures. *Optics Precis. Eng.* **2012**, *7*, 1500–1508. [CrossRef]
9. Hélène, P.; Carole, R.; Marjorie, S.; Fabrice, M.; Xavier, D. Integration of a MEMS based safe arm and fire device. *Sensors Actuat. A Phys.* **2010**, *159*, 157–167.
10. Cope, R.D. *MEMS S-A Technology*; Naval Air Warfare Center, Weapons Division: China Lake, CA, USA, 1999.
11. Li, X.; Zhao, Y.; Hu, T. Design of a large displacement thermal actuator with a cascaded V-beam amplification for MEMS safety-and-arming devices. *Microsyst. Technol.* **2015**, *21*, 2367–2374. [CrossRef]
12. Zhao, Y.L.; Hu, T.J.; Li, X.Y. Design and characterization of a large displacement electro-thermal actuator for a new kind of safety-and-arming device. *Energy Harvest. Syst.* **2015**, *2*, 143–148. [CrossRef]
13. Yang, J.; Gao, J.Z.; Liu, Y.L.; Jiang, Z.D. Design and Fabrication of MEMS-Based Thermal Micro-Actuator. *Micronanoelectron. Technol.* **2005**, *4*, 175–179.
14. Fogel, O.; Winter, S.; Benjamin, E. 3D printing of functional metallic microstructures and its implementation in electrothermal actuators. *Addit. Manuf.* **2018**, *21*, 207–311. [CrossRef]
15. Kandula, P.; Dong, L. Robust Voltage Control for an Electrostatic Micro-Actuator. *J. Dyn. Sys. Meas. Control.* **2017**, *140*, 061012. [CrossRef]
16. Dong, L.; Kandula, P.; Gao, Z.; Wang, D. Active disturbance rejection control for an electric power assist steering system. *Int. J. Intell. Control Syst.* **2010**, *15*, 18–24.
17. Pezous, H.; Rossi, C.; Sanchez, M. Fabrication, assembly and tests of a MEMS-based safe, arm and fire device. *J. Phys. Chem. Solids* **2010**, *71*, 75–79. [CrossRef]
18. Robinson, C.H.; Wood, R.H. Ultra-Miniature Electro-Mechanical Safety and Arming Device. U.S. Patent 8,276,515, 2 October 2012.
19. Robert, R. MEMS Based Fuse Technology. In Proceedings of the 58th Annual NDIA Fuse Conference, Baltimore, MD, USA, 7–9 July 2015.
20. Wang, Y.; Lou, W.Z.; Feng, Y. High impact dynamic simulation of planar S-form micro-spring. *Key Eng. Mater.* **2013**, *562*, 1107–1110. [CrossRef]
21. Tabata, O.; Tsuchiya, T. *Reliability of MEMS*; Southeast University Press: Nanjing, Jiangsu, China, 2009.
22. Perrin, M. New Generation Naval Artillery Multi-Function Fuse. In Proceedings of the NDIA's 56th Annual Fuse Conference, Baltimore, MD, USA, 14–16 May 2012.
23. Zhou, Z.J.; Nie, W.R.; Wan, X.F. Study on parameters of MEMS planar zigzag slot for fuse. *Key Eng. Mater.* **2014**, *609*, 813–818.
24. Lyle, H.J. Precision Guidance Kit (PGK). In Proceedings of the NDIA 56th Annual Fuse Conference, Baltimore, MD, USA, 14–16 May 2012.
25. He, G. Micro-Mechanical Safety Mechanism Based on MEMS Technology Theory and Application. Ph.D. Thesis, Beijing Institute of Technology, Beijing, China, 2006.
26. Li, X.; Zhao, Y.; Hu, T. Design of a high-speed electrothermal linear micromotor for microelectromechanical systems safety-and-arming devices. *Micro Nano Lett.* **2016**, *11*, 692–696. [CrossRef]

27. Liu, J.K.; Qi, X.L.; Jia, J. Study on the reliability problem of MEMS fuse mechanism. *Adv. Mater. Res.* **2012**, *628*, 72–77. [CrossRef]

28. Renaud, L. Pyro-MEMS Technological breakthrough in fuse domain. In Proceedings of the NDIA 55th Annual Fuse Conference, Salt Lake City, UT, USA, 24–26 May 2011.

Article

A Novel, Hybrid-Integrated, High-Precision, Vacuum Microelectronic Accelerometer with Nano-Field Emission Tips

Haitao Liu [1,*], Kai Wei [1], Zhengzhou Li [1,*], Wengang Huang [2], Yi Xu [3] and Wei Cui [2]

1   School of Microelectronics and Communication Engineering of Chongqing University, Chongqing 400044, China; 201712131047@cqu.edu.cn
2   Chongqing acoustic optoelectronic Co. Ltd. of China Electronics Technology Group Corporation, Chongqing 400060, China; hwg_cq@163.com (W.H.); Cuiwei@analogfoundries.com (W.C.)
3   National Key Discipline Laboratory of Novel Micro/Nano Devices and Systems Technologies, Chongqing University, Chongqing 400044, China; xuyibbd@cqu.edu.cn
*   Correspondence: htliu@cqu.edu.cn (H.L.); lizhengzhou@cqu.edu.cn (Z.L.); Tel.: +86-23-6510-3544 (H.L. & Z.L.)

Received: 27 July 2018; Accepted: 16 September 2018; Published: 20 September 2018

Abstract: In this paper, a novel, hybrid-integrated, high-precision, vacuum microelectronic accelerometer is put forward, based on the theory of field emission; the accelerometer consists of a sensitive structure and an ASIC interface (application-specific integrated circuit). The sensitive structure has a cathode cone tip array, a folded beam, an emitter electrode, and a feedback electrode. The sensor is fabricated on a double-sided polished (1 0 0) N-type silicon wafer; the tip array of the cathode is shaped by wet etching with HNA (HNO_3, HF, and CH_3COOH) and metalized by TiW/Au thin film. The structure of the sensor is finally released by the ICP (inductively coupled plasma) process. The ASIC interface was designed and fabricated based on the P-JFET (Positive-Junction Field Effect Transistor) high-voltage bipolar process. The accelerometer was tested through a static field rollover test, and the test results show that the hybrid-integrated vacuum microelectronic accelerometer has good performance, with a sensitivity of 3.081 V/g, the non-linearity is 0.84% in the measuring range of -1 g~1 g, the average noise spectrum density value is 36.7 μV/Hz in the frequency range of 0–200 Hz, the resolution of the vacuum microelectronic accelerometer can reach 1.1×10^{-5} g, and the zero stability reaches 0.18 mg in 24 h.

Keywords: field emission; hybrid integrated; vacuum microelectronic; cathode tips array; interface ASIC

1. Introduction

There is an increasing demand for small-sized, lightweight, and low-powered sensing systems in micro-accelerometers. Especially, MEMS (micro-electromechanical system)-based accelerometers find great applications in navigation systems [1–4], inertial sensors [5–7], seismometers [8], space microgravity [9], military affairs [10], and optical devices [11]. Since the world's first field-based sensor was launched, field emission devices have been widely used due to their high accuracy, high sensitivity, and anti-radiation advantages.

The vacuum microelectronics accelerometer is based on the field emission principle [12]; while field emission has two distinct advantages over other accelerometers, due to the feature of cold cathode emission [13–15], the output current signal of the sensor changes exponentially with the acceleration, so that the sensitivity is very high, and the current output of the sensor makes its interface current relatively simple. In this paper, a novel, hybrid-integrated, high-precision vacuum microelectronic accelerometer is proposed. This paper presents the principle and structure, and the design and

fabrication of the sensor and interface circuits; then, it demonstrates the experimental results of the circuits and the accelerometer, and the conclusion is finally reported.

2. Principle and Structure

The vacuum microelectronics accelerometer is based on the metal field emission principle that metal field emission is a kind of electron emission phenomenon that relies on a strong external electric field to suppress the potential barrier of the metal surface, reduce the barrier, and narrow the barrier width. With the increasing strength of the applied electric field, the height of the surface barrier is not only reduced but the width is also narrowed. When the barrier width is narrow enough to be comparable to the electron wavelength, the electrons can escape through the potential barrier, thereby forming a field electron emission in the vacuum.

The metal field emission current can be calculated by the Fowler and Nordheim Formula (F–N) [16,17]. The simplified Fowler–Nordheim Formula is shown as Equation (1):

$$J = 1.5 \times 10^{-6} \frac{E^2}{\varnothing} \exp\left(\frac{10.4}{\varnothing^{1/2}}\right) \exp\left(-16.44 \times 10^7 \frac{\varnothing^{3/2}}{E}\right) \tag{1}$$

in which J is the field emission current density (A/cm^2), E is the surface electric field strength (V/cm), and $\varnothing$ is the work function (eV). The surface electric field strength of a tip is shown as Equation (2):

$$E = \frac{2}{r \ln\left(\frac{d}{r}\right)} V \tag{2}$$

in which r is the curvature radius of the cone tip, d is the distance between the cathode cone and the anode plate, and V is the voltage applied on the anode.

The distance between the emission current density and the cone tip radius of curvature, and the distance between the cathode cone and the anode plate, can be obtained by calculation, as shown in Figures 1 and 2. Figure 1 shows that the emission current density decreases exponentially, with the distance from the tip to the anode plate being at a certain emission voltage, and Figure 2 shows that the emission current density decreases with the radius of curvature of the cone tip. In order to obtain a larger emission current at a smaller operating voltage, the radius of curvature of the tip and the distance from the tip to the anode plate should be smaller.

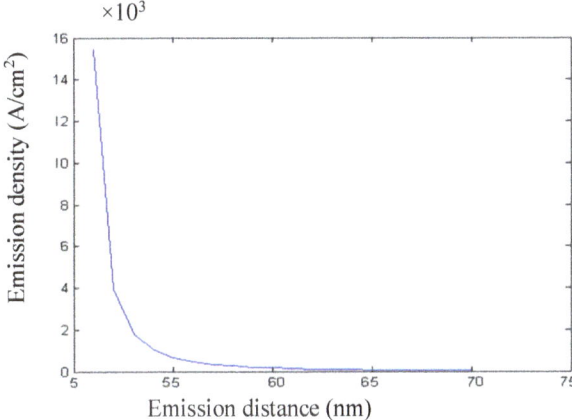

Figure 1. Emission current density vs. distance between the cathode tips and the anode.

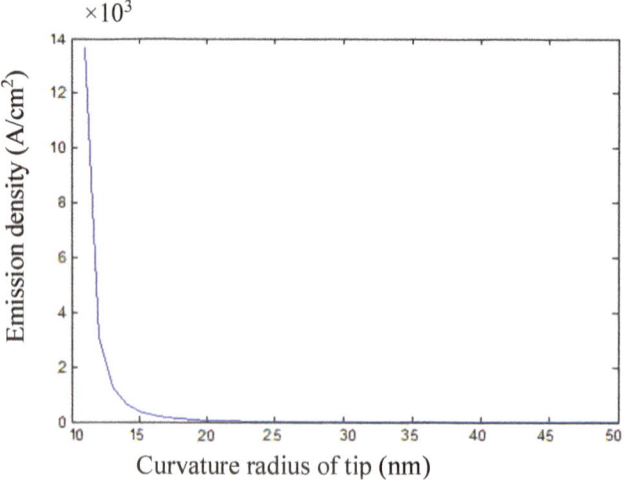

Figure 2. Emission current density vs. the curvature radius of the tip.

The structure of sensor consists of a proof mass with a field emission cathode tip array, a cantilever beam, an anode, and feedback electrodes, as shown in Figure 3. There are about 10,000 tip arrays on the cathode, which makes the emission current increase greatly. The external acceleration will lead to the proof mass moving, and this will cause the distance between the cathode and the anode tip array to change, causing the cathode field emission current to dramatically vary with an exponential relationship, so that the acceleration can be obtained by detecting the cathode emission current.

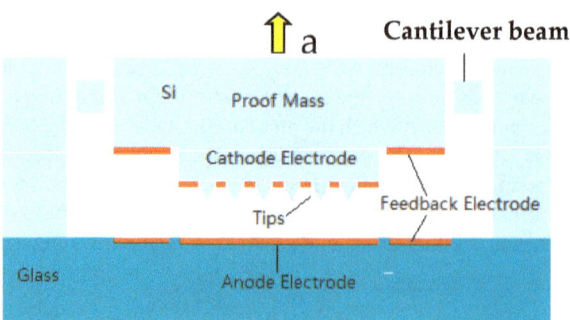

Figure 3. Structure of the vacuum microelectronic accelerometer.

3. Design and Simulations

3.1. The Sensor-Sensitive Unit

The sensitive element was designed by a cantilever-mass, and the cantilever was a novel folded beam structure, as shown in Figure 4. Based on the deformation energy analysis method, and mechanical and electric analyses, the rectangular folding beam elastic stiffness of the vacuum microelectronic accelerometer was calculated [18]. The design parameters of the structure are shown in Table 1.

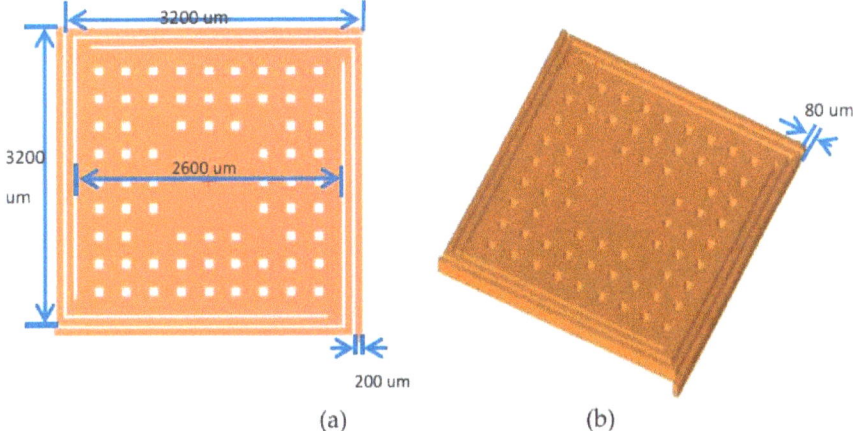

(a)   (b)

Figure 4. Structure of the sensitive structure of the vacuum microelectronic accelerometer. (**a**) Plan view of the sensitive structure; (**b**) 3D map view of the sensitive structure.

Table 1. Key structure parameters of the sensor.

Parameters	Design Value
Beam Length	3200 µm
Beam Width	200 µm
Beam Thickness	80 µm
Mass Side Length	2600 µm
Mass Thickness	80 µm

The resonant frequency and modal response of the sensor were analyzed by finite element simulation (FEM); the analysis results by ANSYS are shown in Figure 5 and Table 2. Figure 4a shows the first mode in the working mode of the accelerometer; the resonant frequency was 979.08 Hz. Figure 4b shows the second mode, in which the proof mass rotates around a horizontal axis, and the resonant frequency was 1774.5 Hz. The third mode was the same as the second mode, only that the rotation axis was different, and the resonant frequency was also 1774.7 Hz, as shown in Figure 4c. Figure 4d shows the fourth mode in which the support beams vibrate, and the resonant frequency was 3069.8 Hz. The frequency of the interfering modes was far from the operating mode.

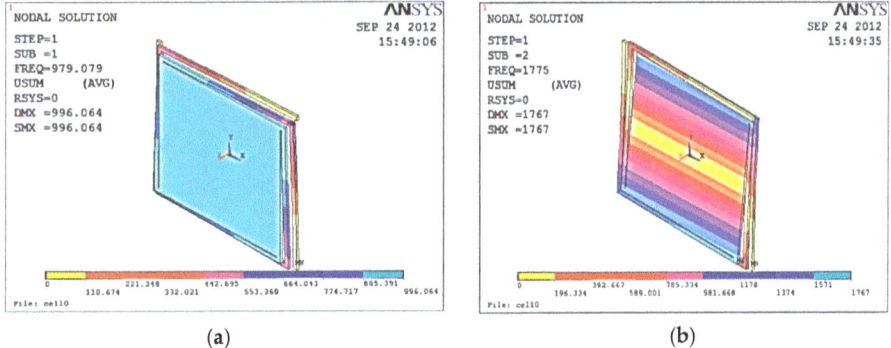

(a)   (b)

Figure 5. *Cont.*

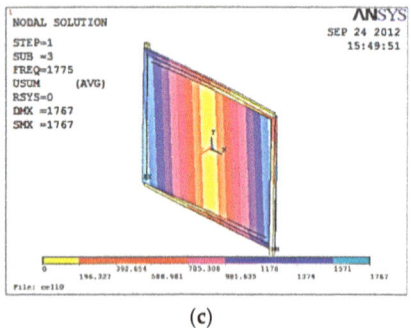

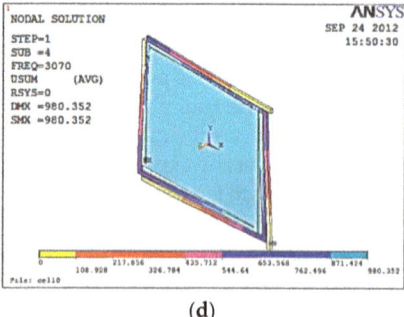

(c)   (d)

Figure 5. Modal graph of the vacuum microelectronic accelerometer. (**a**) The first mode, (**b**) the second mode, (**c**) the third mode, and (**d**) the fourth mode.

Table 2. Mode analysis results.

Set	Freq (Hz)	Loadstep	Substep	Cumulative
1	979.08	1	1	1
2	1774.5	1	2	2
3	1774.7	1	3	3
4	3069.8	1	4	4

The elastic stiffness K can be calculated by Equation (3):

$$K = \frac{ma}{x} \tag{3}$$

in which m is the quality of the proof mass, a is the acceleration, and x is the moving distance of the proof mass. Figure 6 shows the results of the static force analysis of the sensor; the displacement as represented by the color from the left to the right of the figure increases in turn. The simulation analysis results show that the displacement of the proof mass reached the maximum value of 0.13 μm when there was ± 1 g of acceleration along the Z axis; the quality of the proof mass was 4.72×10^{-6} kg; and the effect force applied on the proof mass was 4.72×10^{-6} N, so that the approximate elastic stiffness was 36 N/m by calculation.

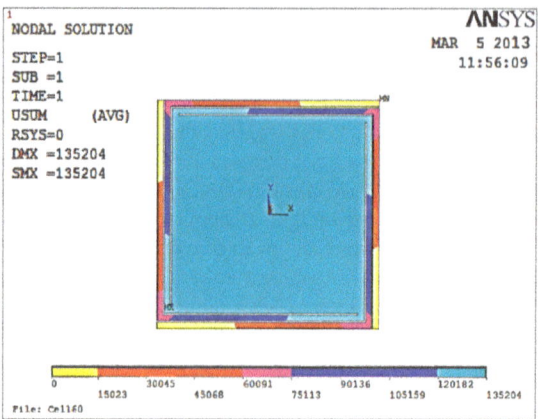

Figure 6. Displacement contour under +1 g acceleration.

81

3.2. Process and Fabrication

The vacuum microelectronic accelerometer is based on bulk silicon MEMS technology, which is based on the double-side polished N-type (1 0 0 direction) silicon wafer, which has a high tensile strength and low mechanical losses. The process of the fabrication of a vacuum microelectronic accelerometer includes the silicon process and the glass process, as shown in Figure 7. During the wafer process, etching groove windows are made to form a bonded anchor (a); etching groove windows are made to form the cone station (b), followed by corrosion cone and sharpening (c), and cone metallization after ion implantation (d); finally, ICP etching is performed to form the beam area (e). The glass processes form the electrodes on the glass surface (f). Finally, the silicon and glass are bonded (g), and the silicon is thinned by KOH etching (h); the ICP structures are then released to form the beam (i). After all these processes, the sensitive unit of the vacuum microelectronic accelerometer is formed.

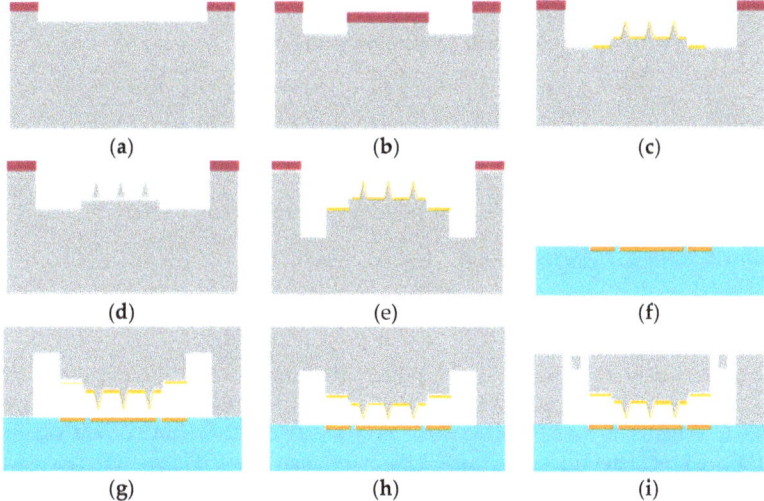

Figure 7. Process of the vacuum microelectronic accelerometer. (**a**) Etching groove windows to form a bonded anchor. (**b**) Etching groove windows to form the cone station. (**c**) Corrosion cone formation and sharpening. (**d**) After ion implantation, cone metallization is performed. (**e**) ICP etching of the front to form the beam area. (**f**) Growth of the electrodes on glass. (**g**) Bonding the silicon and glass. (**h**) The silicon is thinned by KOH etching. (**i**) ICP structure release to form the beam.

The tip shape is one of the main factors that affect the performance of the accelerometer, and the processes of the tip is also one of the key processes for the accelerometer. The silicon tip arrays are formed by wet etching with the HNA solution (HNO$_3$, HF, and CH$_3$COOH) and metalized by TiW/Au thin film; the morphology of the tip is an ideal pyramid, as shown in Figure 8.

The cathode cone tip surface of the vacuum microelectronic accelerometer is easily oxidized under a low vacuum, which leads to an emission current drop and instability. In order to improve the emission current stability of the vacuum microelectronic acceleration sensor, the surface was coated with a layer composite metal film to protect the cone tip, which can effectively improve the stability of the emission current.

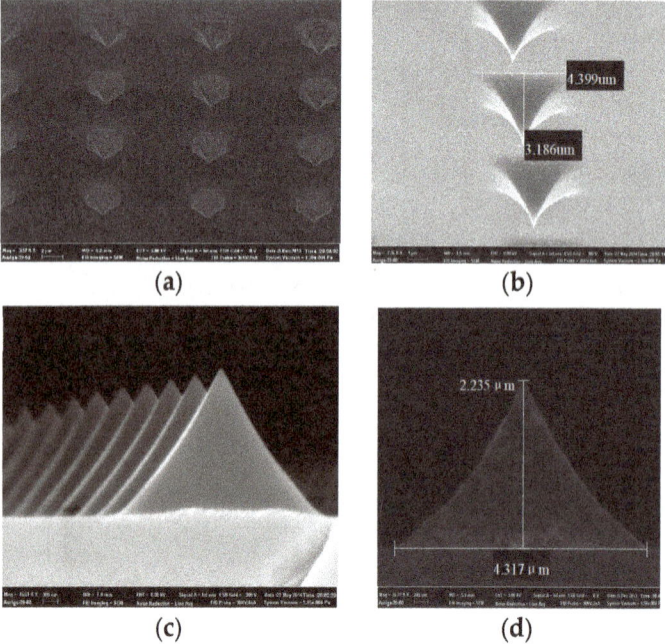

(a)   (b)

(c)   (d)

Figure 8. Scanning electron microscopy (SEM) of the cathode tip array. (**a**) Array front view, (**b**) partial front view, (**c**) array side view, and (**d**) partial side view.

3.3. Interface Circuit

The interface circuit includes a current–voltage conversion, a differential amplifier circuit, and electrostatic force feedback circuit modules, as shown in Figure 9. The mass will produce a slight displacement under acceleration, and it will lead to changes in the emission current; after the current is converted into voltage and amplified, the voltage signal V_o associated with the displacement of the mass is obtained.

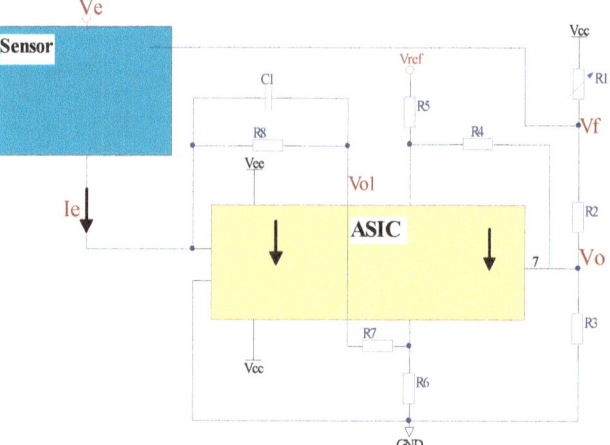

Figure 9. Schematic of the interface circuit of the vacuum microelectronic accelerometer.

The transmission current of the cone is exponentially related to the change of the displacement. When the displacement change is very small, the current can be approximated linearly. The voltage after the *I*–*V* conversion is calculated by Equation (4):

$$V_{o1} = R_8 \cdot I_e (1 - \alpha \Delta x) \tag{4}$$

When the resistance $R_5 = R_7$ and $R_4 = R_6$, the output voltage V_o after the differential amplification is calculated by Equation (5).

$$V_o = \frac{R_4}{R_5} \left(V_{o1} - V_{ref} \right) \tag{5}$$

In order to improve the output linearity and the dynamic response range of the vacuum microelectronic accelerometer, the linear feedback network was designed by the electrostatic force balance technique. The output voltage V_f of the feedback circuit is

$$V_f = \frac{R_2 + R_3}{R_1 + R_2 + R_3} \times V_{cc} + \frac{R_1}{R_1 + R_2} \times V_o \tag{6}$$

In order to improve the signal-to-noise ratio of the circuit, the two op amps, U1A and U1B in the circuit, were designed as application-specific integrated circuits (ASICs), which include an integrated dual operational amplifier. The basic structure of the interface ASIC is shown in Figures 10 and 11 is the schematic circuit of the interface ASIC.

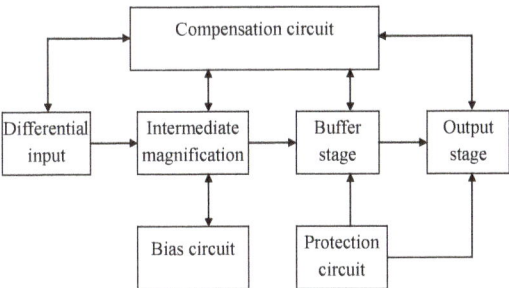

Figure 10. Basic structure of the interface ASIC circuit.

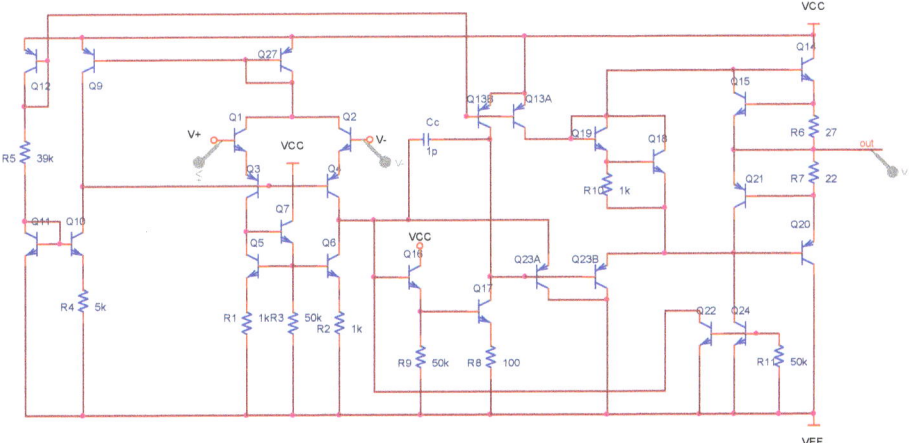

Figure 11. Schematic circuit of the ASIC.

Finally, the ASIC was fabricated based on the P-JFET high voltage bipolar process in the Chongqing acoustic optoelectronic Co. Ltd. of China Electronics Technology Group Corporation. Figure 12 shows the layout of the basic component units of the ASIC interface.

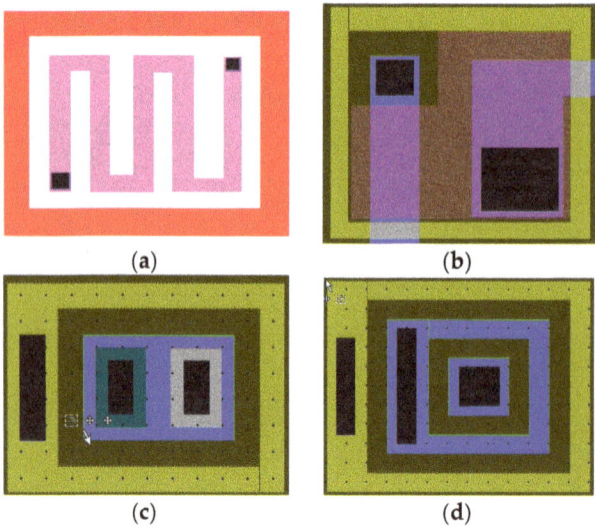

Figure 12. Layout of the basic component. (**a**) Resistor unit layout, (**b**) MOS (Metal Oxide Semiconductor) capacitor layout, (**c**) NPN (Negative Positive Negative) transistor layout, and (**d**) Substrate lateral PNP transistor layout.

4. Tests and Results

Finally, the sensitive structure unit and the interface ASIC chip were integrated on a PCB board, and the accelerometer was packed in the vacuum packaging machine, whose vacuum degree could reach 10^{-4} Pa; getter was added into the tube to maintain a high vacuum degree for a long time. The photo of hybrid integrated vacuum microelectronic accelerometer is shown in Figure 13. The working voltage of the accelerometer was a ±15 V power supply.

Figure 13. Photo of the hybrid integrated vacuum microelectronic accelerometer.

Table 3 shows the test results of main parameter of the interface ASIC. The offset voltage V_{os} was −1.29 mV; the offset current I_{os} was −0.5 nA; the common-mode rejection ratio CMRR was 81 dB; the voltage gain open-loop differential mode voltage gain, A_{vo}, was 103 dB; the power supply rejection ratio, PSRR, was 106 dB; and all these parameters showed that the interface ASIC has very good performance indicators, and that it can realize the signal amplification and the processing of the sensor.

Table 3. Main parameters of the ASIC.

Parameters	V_{os} (mV)	I_{b+} (nA)	I_{b-} (nA)	I_{os} (nA)	A_{vo} (dB)	CMRR (dB)	PSRR (dB)
Design specifications	<5	<150	<150	<30	>96	>70	>65
Test Results	−1.29	−16.3	−15.8	−0.5	103	81	106

A gravitational field static rollover test was carried out in order to test the sensitivity, linearity, and zero stability, to use the mirroring precision rotary indexing head. The testing data are shown in Figure 14 and Table 4. Test results showed that the measuring range was −1 g~1 g, and that the sensitivity of the accelerometer was 3.081 V/g; the least squares fitting correlation coefficient reached 0.99998, and the non-linearity was 0.84%.

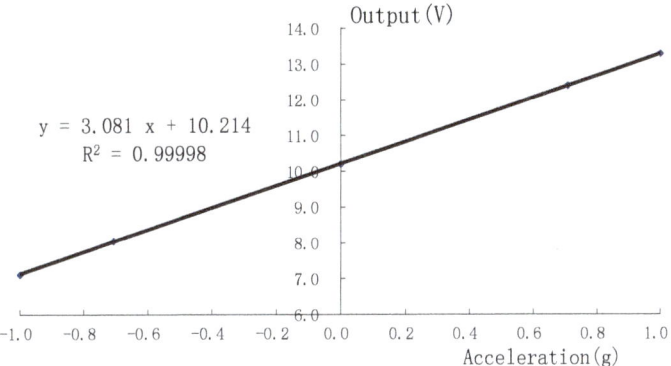

Figure 14. Curve of the output data vs. input data.

Table 4. Datum of the static gravitational field roll test.

Acceleration (g)	Measure Value (V)	Fitting Value (V)	Deviation Value (V)	Non-Linearity
+1.0	13.292	13.314	−0.022	−0.71%
+0.7	12.397	12.411	−0.014	−0.46%
0	10.206	10.230	−0.024	−0.76%
−0.7	8.053	8.049	0.004	0.12%
−1.0	7.120	7.146	−0.026	−0.84%

Figure 15 shows the spectrum density of output signal, in which the x-axis is the frequency and the y axis is the peak-to-peak spectrum density (μV/Hz), and the average noise spectrum density value is 36.7 μV/Hz in the frequency range of 0–200 Hz. Because the output sensitivity of the accelerometer is 3.081 V/g, the resolution of the vacuum microelectronic accelerometer can reach 1.1×10^{-5} g.

The accelerometer is placed at the zero acceleration position, and the output value is measured every 0.5 h. The output value of the accelerometer is recorded in 24 h to calculate its bias stability shown in Figure 16; the result shows that the zero stability reaches 0.18 mg in 24 h.

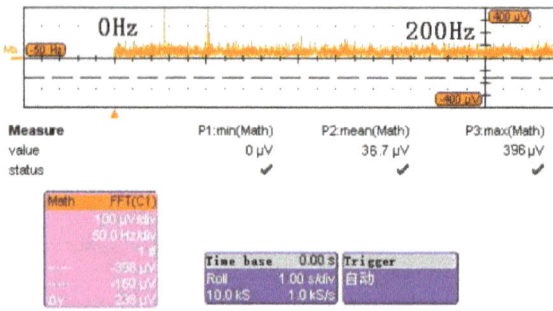

Figure 15. Noise spectrum density of the vacuum microelectronic accelerometer.

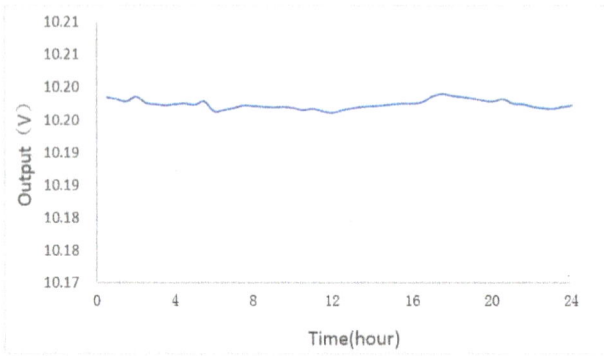

Figure 16. Zero stability of accelerometer in 24 h.

5. Conclusions

In this study, we presented a novel hybrid integrated vacuum microelectronic accelerometer. The structure and the working principles of the sensor were studied in detail, and the mechanistic characteristics of the sensitive structure were analyzed by finite element analysis. Furthermore, the fabrication process and the interface ASIC circuits were designed.

Because of the optimized design of the structural design and process while improving the cone tip production process, the most critical point of the vacuum microelectronic accelerometer is that the interface circuit was designed based on the application-specific integrated circuit, so that the system's signal-to-noise ratio was improved greatly. The test results of the vacuum microelectronic accelerometer show that the sensitivity is about 3.081 V/g, the nonlinearity is about 0.84% over a range of -1 g~1 g, the average noise spectrum density value is 36.7 μV/Hz in the frequency range of 0–200 Hz, the resolution of the vacuum microelectronic accelerometer can reach 1.1 × 10^{-5} g, and the zero stability reaches 0.18 mg in 24 h. This can be widely applied in high-precision inertial systems and other similar applications, such as those for acoustic measurement, navigation, earthquake monitoring, aerospace, and so on.

There are also some problems that need to be resolved; for example, performance should be greatly boosted if the sensitive structure and interface ASIC are monolithic integrated on one chip, along with chip-level vacuum packaging.

Author Contributions: H.L. and K.W. conceived and designed the sensor; Z.L. performed the experiments; W.H. designed the circuit; W.C. performed the process; and H.L. and Y.X. wrote the paper.

Acknowledgments: The work was supported by the Common Key technology innovation projects in the Key Industries of Chongqing under Grant cstc2016zdcy-ztzx0034.

Conflicts of Interest: The authors declare no conflict of interest.

References

1. Syed, Z.; Aggarwal, P.; Niu, X.; Sheimy, N. Economical and robust inertial sensor configuration for a portable navigation system. *Phys. Lett. A* **2007**, *365*, 263–267.
2. Mougenot, D.; Thorburn, N. MEMS-based 3D accelerometers for land seismic acquisition: Is it time? *Lead. Edge* **2004**, *23*, 246–250. [CrossRef]
3. Monajemi, P.; Ayazi, F. Design optimization and implementation of a micro-gravity capacitive HARPSS accelerometer. *IEEE Sens.* **2006**, *6*, 39–46. [CrossRef]
4. Dong, Y.G. *Microsensor*; Tsinghua University Press: Beijing, China, 2007.
5. King, K.; Yoon, S.W.; Perkins, N.C.; Najafi, K. Wireless MEMS inertial sensor system for golf swing dynamic. *Sens. Actuators A Phys.* **2008**, *141*, 619–630. [CrossRef]
6. Nakamura, S. MEMS inertial sensor toward higher accuracy & multi-axis sensing. *Sensors* **2005**, *4*, 939–942.
7. Sun, C.M.; Tsai, M.H.; Liu, Y.C.; Fang, W. Implementation of a monolithic single proof-mass tri-axis accelerometer using CMOS–MEMS technique. *IEEE Trans. Electron Devices* **2010**, *57*, 1670–1679. [CrossRef]
8. Ding, H.G. Advances trends and recommendations in micro-nano technology. *Nanotechnol. Precis. Eng.* **2006**, *4*, 249–255.
9. Tan, X.Y.; Liu, X.W. Development of Small satellite and Micro-satellite Speed up By MEMS Technology. *Chin. J. Sci. Instrum.* **2004**, *25*, 598–600.
10. Ruffin, P.B.; Burgetr, S.J. Recent progress in MEMS technology development for military application. *Proc. SPIE* **2001**, *4334*, 1–12.
11. Wu, M.C.; Solgaard, O.; Ford, J.E. Optical MEMS for lightwave communication. *J. Lightw. Technol.* **2006**, *24*, 4433–4454. [CrossRef]
12. Xia, S.H. Research and Development of Vacuum Microelectronic Sensors. *J. Mech. Strength* **2001**, *23*, 535–538.
13. Peng, S.C.; Wen, Z.Y.; Wen, Z.Q.; Pan, Y.S.; Li, X. Finite Element Analysis on Vacuum Microelectronic Acceleration Sensor. *Micronanoelectron. Technol.* **2003**, *7*, 292–301.
14. Paul, J.; Anthony, J.K. A Planar CMOS Field-Emission Vacuum Magnetic Sensor. *IEEE Trans. Electron Devices* **2009**, *56*, 692–695.
15. Xu, S.L. Study on Vacuum Microelectronic Pressure Sensor. Master's Thesis, Chongqing University, Chongqing, China, 2003.
16. Wang, B.P. *Vacuum Microelectronics and Its Application*; Southeast University Press: Nanjing, China, 2002.
17. Neamen, A.D. *Semiconductor Physics and Devices*, 3rd ed.; Electronic Industry Press: Beijing, China, 2005.
18. Liu, H.T.; Wen, Z.Y.; Shang, Z.G.; Chen, L. A new method to analyze the stiffness of MEMS accelerometer. *Key Eng. Mater.* **2014**, *609–610*, 710–714. [CrossRef]

Article

LSS-RM: Using Multi-Mounted Devices to Construct a Lightweight Site-Survey Radio Map for WiFi Positioning

Wei Yang, Chundi Xiu *, Jiarui Ye, Zhixing Lin, Haisong Wei, Dayu Yan and Dongkai Yang

School of Electronic and Information Engineering, Beihang University, Beijing 100083, China;
yangwei89@buaa.edu.cn (W.Y.); yejiarui@buaa.edu.cn (J.Y.); linzhixing@buaa.edu.cn (Z.L.);
sy1702122@buaa.edu.cn (H.W.); dyaxb@buaa.edu.cn (D.Y.); edkyang@buaa.edu.cn (D.Y.)
* Correspondence: xcd@buaa.edu.cn; Tel.: +86-10-8231-7222

Received: 31 July 2018; Accepted: 10 September 2018; Published: 12 September 2018

Abstract: A WiFi-received signal strength index (RSSI) fingerprinting-based indoor positioning system (WiFi-RSSI IPS) is widely studied due to advantages of low cost and high accuracy, especially in a complex indoor environment where performance of the ranging method is limited. The key drawback that limits the large-scale deployment of WiFi-RSSI IPS is time-consuming offline site surveys. To solve this problem, we developed a method using multi-mounted devices to construct a lightweight site-survey radio map (LSS-RM) for WiFi positioning. A smartphone was mounted on the foot (Phone-F) and another on the waist (Phone-W) to scan WiFi-RSSI and simultaneously sample microelectromechanical system inertial measurement-unit (MEMS-IMU) readings, including triaxial accelerometer, gyroscope, and magnetometer measurements. The offline site-survey phase in LSS-RM is a client–server model of a data collection and preprocessing process, and a post calibration process. Reference-point (RP) coordinates were estimated using the pedestrian dead-reckoning algorithm. The heading was calculated with a corner detected by Phone-W and the preassigned site-survey trajectory. Step number and stride length were estimated using Phone-F based on the stance-phase detection algorithm. Finally, the WiFi-RSSI radio map was constructed with the RP coordinates and timestamps of each stance phase. Experimental results show that our LSS-RM method can reduce the time consumption of constructing a WiFi-RSSI radio map from 54 min to 7.6 min compared with the manual site-survey method. The average positioning error was below 2.5 m with three rounds along the preassigned site-survey trajectory. LSS-RM aims to reduce offline site-survey time consumption, which would cut down on manpower. It can be used in the large-scale implementation of WiFi-RSSI IPS, such as shopping malls, hospitals, and parking lots.

Keywords: indoor positioning; WiFi-RSSI radio map; MEMS-IMU accelerometer; zero-velocity update; step detection; stride length estimation

---

1. Introduction

The positioning method is a basic component of location-based services (LBSs) such as navigating a customer to the nearest restaurant in a shopping mall, finding your car in an underground parking lot, guiding tourists in a museum, or aiding during a fire emergency [1]. As to positioning outdoors, global navigation satellite systems (GNSSs) can provide global services and users can get an accurate position, velocity, and time (PVT) in open air [2,3]. However, in an indoor environment and urban canyons, GNSS signal availability is limited, and indoor-positioning systems (IPSs) need to be studied [4].

IPSs can be classified roughly into two kinds: the infrastructure-free system and the infrastructure-based system [5]. A typical infrastructure-free system is the inertial navigation system (INS) [6]. Pedestrian dead-reckoning (PDR)-based INS (PDR-INS) utilizes the pedestrian kinetic

model to do step detection, stride-length estimation, and heading estimation [7–10]. Considering the drift of gyroscope readings and the fluctuation of indoor magnetic fields, an accurate heading estimation is difficult, and map information, including walls, corridors, and rooms, can be fused using a particle filter to get more accurate heading estimation [11]. The positioning accuracy of PDR-INS is easily influenced by the carry mode of devices, and the stride-length model needs parameters like height, leg length, or walking frequency, which should be tuned according to different users [12]. In some other pedestrian inertial navigation systems, especially for fire-emergency applications, microelectromechanical system inertial-measurement units (MEMS-IMU) are mounted on the foot. With the triaxial accelerometer and gyroscope readings, a zero-velocity update (ZUPT) algorithm is developed to measure velocity errors in the stance phase of a gait cycle [13,14]. ZUPT estimates pseudo measurements into the extended Kalman filter (EKF) navigation-error corrector, which allows the EKF to correct velocity errors during each gait cycle, breaking the cubic-in-time error growth and replacing it with an error accumulation that is linear with the number of steps [15–17]. This kind of foot-mounted inertial navigation system is called IEZ-INS and IEZ is short for the first letters of INS, EKF, and ZUPT. Foot-mounted MEMS-IMU can be used to get more accurate stride-length estimation [18–20]. There is no clear boundary between PDR-INS and IEZ-INS. The principle of method choice depends on the accuracy of MEMS-IMU and the specific application.

Unlike infrastructure-free IPSs, infrastructure-based ones need preinstalled transmitters such as WiFi [9,21,22], Bluetooth [23,24], near-field communication [25], RFID [26], ultrawide band (UWB) [27,28], LED [29,30], or ultrasound [31]. These methods can provide sufficient positioning accuracy for LBSs in different applications. Widely deployed private- or public-access points (APs) in large-scale buildings can provide free and dense signals for WiFi-based IPS. Furthermore, smartphones are embedded with WiFi chips that can easily obtain a received signal-strength index (RSSI). The methods used in WiFi-based IPS can be classified by time of arrival (TOA) [32], time difference of arrival (TDOA) [33], angle of arrival (AOA) [34], and RSSI fingerprinting [35], CSI-fingerprinting [36], and round-trip time (RTT) [37]. Each of these methods has shortcomings and limitations. TOA, TDOA, and AOA are easily influenced by indoor environments. Channel state information (CSI) fingerprinting needs an Intel 5300 wireless local area networks (WLAN) card that is not available for smartphones. Similar with CSI fingerprinting, the AP of the RTT method must support IEEE 802.11 mc, which is brand-new and not in the commercial market yet. Therefore, considering the complex multipath effect and the available hardware, RSSI fingerprinting is widely researched.

The WiFi-fingerprinting method consists of two phases, the offline site-survey phase and the online positioning phase [38]. In the offline phase, the WiFi-RSSI of selected APs is collected from each reference point (RP), and a radio map is built up. In the online phase, the collected WiFi-RSSI samples are compared with the radio map using matching algorithms such as k-nearest neighbor (KNN) to get the position estimation [39]. One of the most important reasons that limit the large-scale implementation of WiFi-based IPS is that site surveys are very time-consuming and labor-intensive [40]. For example, if we wanted to deploy the WiFi-IPS in a 10 m × 10 m room with a one-meter interval of each RP, and the WiFi-RSSI sampling time was 2 min of each RP, it would take 200 min to build the WiFi-RSSI radio map of this room in total. Time consumption would rise rapidly with the area of the place deploying WiFi-IPS [41].

To make WiFi-based IPS more practical, many researchers have focused on how to build the radio map in an energy-efficient way. The most common format of a WiFi-RSSI radio map is {RP coordinates, WiFi-RSSI vectors}. The offline WiFi-RSSI radio-training phase is very time-consuming because volunteers must stand still for a while to collect WiFi-RSSI from every RP [42]. Therefore, methods aiming to reduce the manpower of building a WiFi-RSSI radio map either research the model of WiFi-RSSI or add extra devices and sensors to help estimate RP coordinates [38,40]. According to Reference [43], methods that try to replace the construction of the radio map by using indoor radio-propagation models cannot capture all the details of the indoor structure and dynamics. These methods either achieve a very unsatisfactory performance or rectify model inaccuracies through

extensive calibration and exhaustive postprocessing to obtain a map that can achieve more acceptable localization accuracy. In Reference [44], a WiFi-enabled laptop and an attached GPS device were used to scan the WiFi-RSSI in the metropolitan-scale area. RP coordinates were provided by the GPS device recording the latitude–longitude coordinates of the war driver when the WiFi-RSSI scan was performed. However, this method is limited indoors, and large GPS positioning is not accurate enough for calculating RP coordinates. In Reference [45], nanoscale unmanned aerial vehicles (UAVs) were used to automate the WiFi-RSSI collection process. RP coordinates were calculated through a UWB-based localization subsystem. This method can build a 3D WiFi-RSSI radio map with low manpower. With the high accuracy of UWB-based IPS, the performance of an automatically built WiFi-RSSI radio map is similar with the manually built one. However, it needs extra equipment, like UAVs and UWB anchors that increase the cost.

With the rapid development of the electronic industry, a state-of-the-art smartphone is a good platform embedded with multiple sensors like accelerometers, gyroscopes, and magnetometers to measure pedestrian movement, and WiFi chips to scan WiFi-RSSI. In Reference [46], a method called crowdsourcing and multisource fusion-based fingerprint sensing (CMFS) was presented. Based on the floor plan, RSSI is collected uniformly by WiFi scanner at fixed time intervals. Simultaneously, stride length, step number, and heading direction of volunteers were estimated using PDR-INS method. The drawback of this method is the drawback of PDR-INS. Stride-length parameters like pedestrian height and step frequency must be tuned according to different volunteers. In addition, heading direction is calculated using magnetic-field strength that may have non-negligible bias indoors. In Reference [47], a WiFi-RSSI radio map was built using an inertial navigation solution from a Trusted Portable Navigator (T-PN) with handheld smartphones. The method of calculating RP coordinates was T-PN, which improves the accuracy of RP coordinates with absolute measurements like A-GPS, magnetometer, or barometer as filter updates. However, the accuracy of T-PN decreases indoors. In Reference [48], a zero-effort crowdsourcing method (Zee) was developed. Zee used inertial sensors of smartphones and detailed map information to count steps and estimate heading offset. An augmented particle filter was utilized to estimate stride length with map information. Then, WiFi-RSSI was recorded to the radio map with RP coordinates through the same timestamps. However, Zee needs detailed map information that is not always available.

In this paper, we proposed a method called LSS-RM, short for Lightweight Site-Survey Radio Map, to construct the WiFi-RSSI radio map. This method, which can scan WiFi-RSSI, can significantly reduce the time consumption of offline WiFi-RSSI radio-map construction. Similar with other WiFi fingerprinting-based systems, our method also consists of an offline site-survey phase and online-positioning phase. The offline phase of LSS-RM is divided into the data collection and preprocessing process, and the post calibration process. The site survey is a client–server model in which the WiFi-based IPS service provider hires some volunteers to participate in site-survey work. To attract more volunteers to a boring site survey, the incentive mechanisms developed in Reference [49] can be applied.

The remainder of this paper is organized as follows: Section 2 is the overview of our WiFi-RSSI LSS-RM. In Section 3, the data collection and preprocessing processes are described. In Section 4, the post calibration process is described. In Section 5, experiments were performed, and the performance of the WiFi-RSSI radio map built with our method was evaluated. Finally, Section 6 concludes this work and offers future research suggestions.

2. Overview of LSS-RM

The WiFi-RSSI radio map consists of RP coordinates and a WiFi-RSSI vector. In the traditional manual site survey, a floor plan with detailed RP coordinates is provided to trained persons who are familiar with the site-survey process, usually from the service-provider group. Then, site-survey participants stand on each RP to scan WiFi-RSSI using smartphones. The most time-consuming part of the site survey is the manual WiFi-RSSI scan process. In this paper, site-survey participants did not

need to stand still at each RP, instead walking along the preassigned site-survey trajectory, and the WiFi-RSSI radio map was automatically constructed.

The structure of LSS-RM is shown in Figure 1. The preassigned site-survey trajectory was told to volunteers, and then two smartphones were mounted on the waist (Phone-W) and the foot (Phone-F), respectively. The movement of the volunteer was recorded using a smartphone-embedded MEMS accelerometer, gyroscope, and magnetometer. WiFi-RSSI was also scanned by Phone-W and Phone-F at the same time. Volunteers did not need to take care of their walking frequency or step stride, but just took two smartphones to walk along the preassigned site-survey trajectory, and the WiFi-RSSI radio map was built up automatically. Phone-F could detect zero velocity and the IEZ-PDR algorithm was implemented to calculate the position of volunteers. The angular rate energy-detection algorithm using Phone-W motion data was applied to detect corners of the preassigned site-survey trajectory. Positioning results, stance phase estimation results, and WiFi-RSSI were transferred to the server in real time through the public 4G long term evolution (LTE) network for the post calibration process. In this process, a stance-phase detection-based algorithm is used to count steps and to estimate stride length. With accurate corner detection and preassigned site-survey trajectory information, we could calculate the accurate heading of the volunteers. Finally, with accurate step numbers, stride length, and heading, precise RP coordinates were calculated. The bridge between RP coordinates and WiFi-RSSI in LSS-RM is the start and end timestamp of each stance phase. After the two processes of the offline site-survey phase, a WiFi-RSSI radio map was built up. In the online-positioning phase, when a user sent a demand for the position service, real-time WiFi-RSSI was collected from the user's smartphone, and a matching algorithm like k-nearest neighbor (KNN) was used to obtain the user's position. Finally, the real-time localization results were sent back to the user's smartphone.

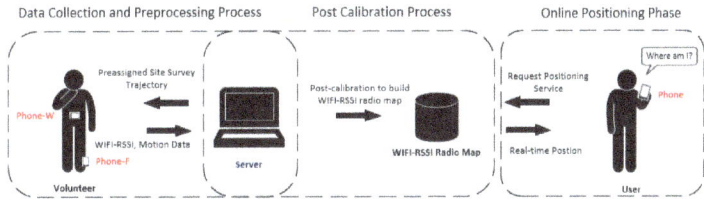

Figure 1. Structure of our WiFi-received signal strength index (WiFi-RSSI) radio map construction method with a lightweight site survey (LSS-RM). It consists of an offline site-survey phase and online-positioning phase. The offline site survey in LSS-RM has two processes: data collection and preprocessing process, and the post calibration processes. A WiFi-RSSI radio map is constructed in the offline phase. In the online-positioning phase, real-time WIF-RSSI is sent to the server to match the radio map and the user's position is then calculated.

The flowchart of LSS-RM is shown in Figure 2. It has the following nine steps. Steps (1) to (5) happen in the data collection and preprocessing process, and Steps (6) to (9) happen in the post calibration process.

(1) The volunteer is told they should walk along the preassigned site-survey. The server analyzes whether the volunteer walks in the right way.

(2) Two smartphones are mounted on the foot (Phone-F) and waist (Phone-W) of the volunteer, respectively. Motion data of the volunteer, such as accelerometer readings, gyroscope readings, and magnetometer readings are recorded in a format {Timestamp, Triaxial Accelerations, Triaxial Angular Rates, Triaxial Magnetic Field Strength}. Simultaneously, WiFi-RSSI data are recorded by both smartphones in a format {Timestamp, WiFi-RSSI Vector}. The timestamp can be used as a medium to connect the two different kinds of data.

(3) The timestamp difference of Phone-F and Phone-W is measured. Then, a timestamp-synchronization process is taken to align data from the two smartphones.

(4)   The position of the volunteer is calculated using the IEZ-INS method based on the accelerometer readings, gyroscope readings, and magnetometer readings of Phone-F. In this step, the stance-phase result of ZUPT is very important and will be used in the post calibration process.

(5)   The angular-rate energy detector (ARE) is used to detect the corner based on gyroscope readings of Phone-W. The corner-detection result can be used to calculate the heading with the preassigned site-survey trajectory in the post calibration process.

(6)   Step number and stride length are estimated based on stance-phase detection from Phone-F.

(7)   Heading of the volunteer is calculated based on preassigned site-survey estimation and corner-detection result from Phone-W.

(8)   RP coordinates are calculated using the post calibrated step number, stride length, and heading based on the PDR-INS method.

(9)   A radio map is built up with RP coordinates and WiFi-RSSI vectors in a traditional radio map format {RP coordinates, WiFi-RSSI vectors}. The bridge between RP coordinates and WiFi-RSSI vectors in the LSS-RM method is the start and end timestamp of each stance phase.

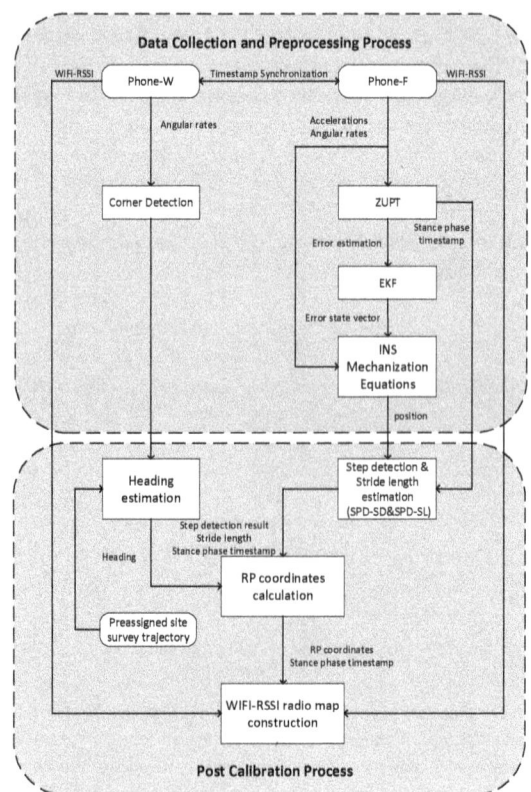

Figure 2. Flowchart of the WiFi-RSSI LSS-RM.

The offline site-survey phase of WiFi-fingerprinting-based IPS is a trade-off between manpower and radio-map accuracy. Like other methods, our LSS-RM method also aims to find a balance in the trade-off. Comparing with other methods, the advantage of LSS-RM is that no extra devices except for smartphones are needed, and volunteers don't need to take care of their step frequency or stride length to reduce the time-consumption of the offline site survey.

However, LSS-RM still has some drawbacks. Firstly, the system is more complex. Two smartphones are needed in LSS-RM. Despite the advantage of better RP coordinates and denser WiFi-RSSI, the system is more complicated than the traditional one. Secondly, the preassigned site-survey trajectory is needed, and the volunteer must walk along it. Thirdly, considering that the scanning time of LSS-RM on each RP is shorter than in manual site surveys, initial positioning accuracy will be lower. However, with more volunteers joining in the offline phase, positioning accuracy increases. Fourthly, the reliability of LSS-RM depends on the accuracy of RP-coordinate estimation, which can be influenced by many factors like the drift of the MEMS-IMU, the timestamp-alignment accuracy of the two smartphones, and the accuracy of the preassigned site-survey trajectory. With the post calibration process, reliability can be improved. In conclusion, there is still a long way to realize the complete site-survey free WiFi-RSSI radio-map construction method.

3. Data Collection and Preprocessing Process

3.1. Timestamp Alignment

Timestamp alignment is the first step of all multidevice-model systems. In our Android APP, timestamps are recorded along with the data. Each timestamp is an index to mark the time when data collected. The timestamp of Android platforms is Unix time that starts from 00:00:00 Coordinated Universal Time (UTC), Thursday, 1 January, 1970. Therefore, in principle, timestamps of different smartphones can be easily aligned because they are under the same time system. However, there are still differences between different smartphones. We performed an experiment, and the result is shown in Figure 3. To simplify X-axis of the figures in this paper, the timestamp sequence has been subtracted by the first timestamp. Two smartphones are tied together and move along the vertical direction. If their timestamps are synchronous, the first peak of acceleration waveform of Z-axis should has the same timestamp, but there exists a difference, Δt. Therefore, timestamps should be synchronized between different smartphones.

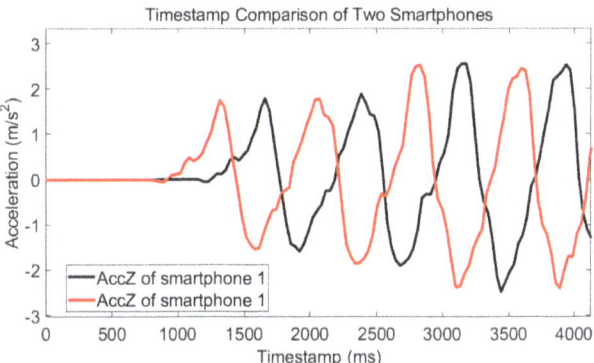

Figure 3. Timestamp comparison of the two smartphones. We tied together two smartphones and shook them along the vertical direction. If their timestamps were synchronous, the timestamp of the first peak of acceleration waveform of Z-axis should have been nearly the same, but there existed a timestamp difference Δt and the timestamp should have been synchronized between different smartphones.

The timestamp-synchronization algorithm is simple in LSS-RM. Firstly, the two-norm of the triaxial accelerations of Phone-F and Phone-W is calculated:

$$Acc_f(k) = \sqrt{Acc_f^x(k) + Acc_f^y(k) + Acc_f^z(k)} \tag{1}$$

$$Acc_w(k) = \sqrt{Acc_w^x(k) + Acc_w^y(k) + Acc_w^z(k)} \tag{2}$$

where k is the timestamp. Acc_f and Acc_w are the two-norm of triaxial accelerations of Phone-F and Phone-W, respectively. Acc_f^x, Acc_f^y, Acc_f^z, Acc_w^x, Acc_w^y, Acc_w^z, are X-axis, Y-axis, and Z-axis acceleration of Phone-F and Phone-W, respectively. The timestamp difference is calculated using the difference of the first acceleration peak between Phone-W and Phone-F:

$$\Delta t = timestamp_f^1 - timestamp_w^1 + \varepsilon_1 + \varepsilon_2 \tag{3}$$

where Δt is the timestamp difference between Phone-W and Phone-F; $timestamp_f^1$ is the timestamp of the 1st peak of Acc_f; and $timestamp_w^1$ is the timestamp of the 1st peak of Acc_w. ε_1 is the timestamp alignment error caused by the sampling process. ε_2 is the timestamp alignment error caused by the asynchronous motion of different parts of the body. Finally, timestamps of the two smartphones are aligned with a translation using Δt.

Three tests were performed to test the timestamp-alignment method. The results shown in Table 1 reveal that the Δt of different smartphones is not a constant. It can even reach 1189 ms, which means the positioning result of two smartphones could be over 2 m with a normal walking speed of 2 m/s. The timestamp alignment algorithm of LSS-RM can estimate the timestamp difference of Phone-W and Phone-F.

Table 1. Tests of timestamp difference of two smartphones.

Test Number	Timestamp of First Peak of Smartphone 1 (ms)	Timestamp of First Peak of Smartphone 2 (ms)	Timestamp Difference (ms)
1	1532703421657	1532703421312	345
2	1532705867200	1532705866011	1189
3	1532706440599	1532706439882	717

We want to discuss the influence of ε_1 and ε_2 in this section. The maximum of ε_1 is two times the sampling time. The sampling frequency of MEMS-IMU of our APP was set to 30 Hz, which means the timestamp alignment error was within (−66 ms, 66 ms). The positioning error caused by ε_1 was 13.2 cm if the walking speed were 2 m/s. It is quite a small error considering that the positioning accuracy of WiFi-fingerprinting-based IPS is 1–5 m.

Although Phone-F and Phone-W were mounted on different parts of the body, the motion of the body was almost coordinated. To reduce ε_2, the volunteer could have a jump at the beginning of the walk. A test was taken to show a time trace of the two mounted smartphones. The volunteer jumped at the beginning and stood still for several seconds before walking. The result is shown in Figure 4. It is obvious that the jump peaks are more distinguishable than the walk peaks. The timestamp difference calculated using jump peaks was 563 ms (regarded as the reference timestamp error) compared with 473 ms of walk peaks. ε_2 was 90 ms for this test. The positioning error caused by ε_2 was 18 cm if the walking speed were 2 m/s. Similar with ε_1, it is a small error and has little influence on LSS-RM.

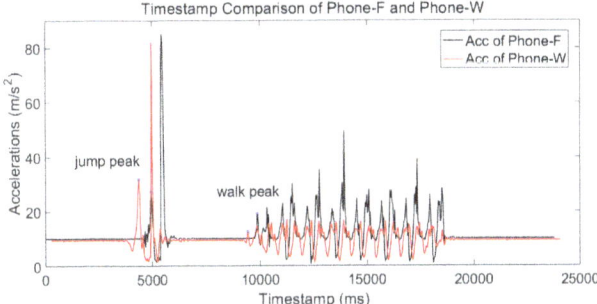

Figure 4. Timestamp comparison of Phone-F and Phone-W. The volunteer jumped at the beginning and stood still for several seconds before walking. The jump peaks are more distinguishable than the walk peaks.

3.2. Foot-Mounted Inertial Navigation Using Zero-Velocity Update-Aided Extended Kalman Filter (IEZ-INS)

The outputs of Phone-F-embedded MEMS-IMU are in the sensor body coordinate frame (b-frame) and should be transferred to the navigation coordinate frame (n-frame) using a rotation matrix C_b^n. The definition of b-frame and n-frame are shown in Figure 5.

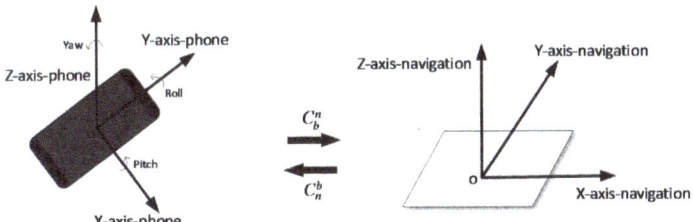

Figure 5. Sketch map of transformation between different coordinate frames. The coordinate frame of the smartphone (b-frame) is fixed, Y-axis directs to the head of the phone, Z-axis directs up perpendicular to the screen, and the X-axis was determined according to the right-hand screw rule. The navigation-coordinate frame (n-frame) used in our method is the east–north–up (ENU) coordinate system. C_b^n was used to transfer the data from b-frame to n-frame and C_n^b is from n-frame to b-frame.

The b-frame is determined by the smartphone and usually defined as a right-handed Cartesian coordinate system. *Y-axis*, directs to the head of the phone, *Z-axis*, directs up perpendicular to the screen, and the *X-axis*, was determined according to the right-hand screw rule. Considering the convenience of usage, the n-frame applied in our system is the local east–north–up (ENU) Cartesian coordinate system whose origin is the same as b-frame. The east was labelled *X-axis*, the north *Y-axis*, and the up *Z-axis*. The MEMS-IMU readings, including acceleration, angular rate, and magnetic-field strength are in b-frame and should be transferred to ENU to derive velocity and position. The two different coordinate systems are transferred through the rotation matrix C_b^n. Details of how to use accelerometer and magnetometer readings to calculate the rotation matrix can be found in Reference [17].

After MEMS-IMU readings transferred from the b-frame to the n-frame using C_b^n, the accelerometer, gyroscope, and magnetometer readings of Phone-F could be used in the INS mechanization equations to calculate the volunteer's position. Firstly, gravity should be subtracted from accelerometer readings in n-frame. Then, the position is calculated with the gravity-free acceleration value. At last, the orientation of the MEMS-IMU is updated with the gyroscope readings.

These equations take slightly different forms in different navigation frames. The basic equation utilizing accelerometers and gyroscopes to calculate position is [50]:

$$
\begin{bmatrix} v_k \\ p_k \\ q_k \end{bmatrix} = \begin{bmatrix} v_{k-1} + (q_{k-1}a_k q_{k-1}^{-1} - g)dt_k \\ p_{k-1} + v_{k-1}dt_k \\ \Omega(w_k dt_k)q_{k-1} \end{bmatrix}
\tag{4}
$$

where k is a timestamp, g is the gravity, v_k is the velocity of the pedestrian, a_k is the accelerometer readings, p_k is the position of the person, q_k is the quaternion describing the orientation frame, dt is the time differential, and $\Omega(\cdot)$ is the quaternion update matrix.

Considering the drift of the low-cost smartphone-embedded MEMS-IMU, accumulative error would rise rapidly only by using Equation (4). To solve this problem, the velocity of the stance phase was used as the measurement of the Extended Kalman Filter. This INS with the ZUPT-aided EKF method is called IEZ [16]. The error-state vector of this system is a 15-element vector, $\delta x = [\delta r, \delta v, \delta \varphi, \delta a, \delta w]$, where δr is the position error, δv is the velocity bias, $\delta \varphi$ is the attitude error, δa is the accelerometer bias, and δw is the gyroscope error. In addition, $\delta r, \delta v, \delta \varphi, \delta a, \delta w$ are all three-dimensional vectors. The state-transition matrix F is:

$$
F = \begin{bmatrix} I & I \cdot \Delta t & O & O & O \\ O & I & St \cdot \Delta t & C_b^n \cdot \Delta t & O \\ O & O & I & O & -C_b^n \cdot \Delta t \\ O & O & O & I & O \\ O & O & O & O & I \end{bmatrix}
\tag{5}
$$

where Δt is the sample interval, and O and I are the three-dimensional null matrix and unit matrix, respectively. $St(k)$ is the skew-symmetric matrix of acceleration:

$$
St(k) = \begin{bmatrix} 0 & -a_z(k) & a_y(k) \\ a_z(k) & 0 & -a_x(k) \\ -a_y(k) & a_x(k) & 0 \end{bmatrix}
\tag{6}
$$

The measurement model is:

$$
z(k) = H\delta x(k) + n(k)
\tag{7}
$$

where $z(k)$ is the measurement, δx is the error-state vector at timestamp k, $n(k)$ is the measurement-noise vector at the timestamp k, and H is the measurement matrix:

$$
H = \begin{bmatrix} O & I & O & O & O \end{bmatrix}
\tag{8}
$$

Steps of IEZ can be found in References [16,17]. An experiment was conducted to validate the IEZ-INS algorithm. A volunteer walked along a rectangular corridor and walked back to the start point. The experimental setup is summarised in Table 2. The MI6 smartphone was mounted on the left foot, and the sampling frequency was 30 Hz.

Table 2. Experimental setup.

Setup Content	Description
Experiment site	A rectangular corridor
Total length of the corridor	128 m
Mounting place of the smartphone	Left foot
Smartphone used	MI6 from Xiaomi
Sensors used	Triaxial accelerometer, gyroscope, and magnetometer
Sampling frequency	30 Hz

The positioning result is shown in Figure 6. It is obvious that the positioning result is badly influenced by the heading error. Therefore, the post calibration process was needed to get accurate RP coordinates.

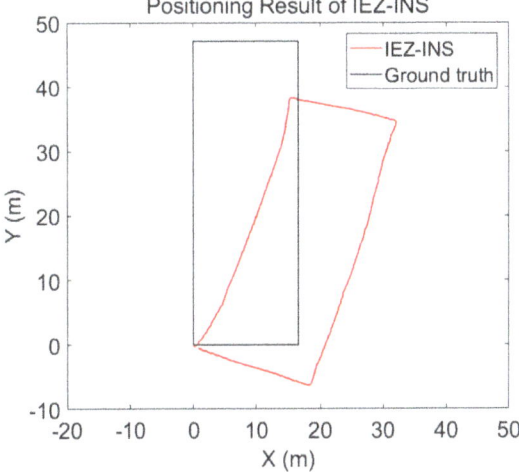

Figure 6. Positioning result of Foot-Mounted Inertial Navigation Using Zero-Velocity Update-Aided Extended Kalman Filter (IEZ-INS).

3.3. Stance-Phase Detection Using Phone-F-Embedded MEMS-IMU

The velocity error estimated during the stance phase is the measurement vector in IEZ. Considering that positioning errors will accumulate fast due to sensor drift, zero-velocity information is efficient in error correction. Furthermore, in our LSS-RM method, the stance phase is used to count steps and estimate stride length in the post calibration process.

To get more robust stance detection, three ZUPT detectors were fused; stance phase occurs when the results of all three detectors were in the stance phase. These three detectors used in this paper are from Reference [13]: the acceleration moving-variance detector (MV), the acceleration magnitude detector (MAG), and the ARE.

$$T_{mv}(k) = \frac{1}{W} \sum_{n=k}^{k+W-1} \frac{1}{\sigma_a^2} \|a(n) - \overline{a(k)}\|^2 \tag{9}$$

$$T_{mag}(k) = \frac{1}{W} \sum_{n=k}^{k+W-1} \frac{1}{\sigma_a^2} (\|a(n)\| - g)^2 \tag{10}$$

$$T_{are}(k) = \frac{1}{W} \sum_{n=k}^{k+W-1} \frac{1}{\sigma_\omega^2} \|\omega(n)\|^2 \tag{11}$$

where k is a time index. W is the window length. g is gravity. σ_a^2 and σ_ω^2 denote the accelerometer and gyroscope noise variance. $\|\cdot\|$ is the 2-norm calculation. T_{mv}, T_{mag}, and T_{are} are the test statics of the MV detector, MAG detector, and ARE detector, respectively. $\overline{a(k)}$ is the average of a during the average window W at time index k:

$$\overline{a(k)} = \frac{1}{W} \sum_{n=k}^{k+W-1} a(n) \tag{12}$$

T_{mv} and T_{mag} use accelerometer readings, while T_{are} uses gyroscope readings. The stance phase occurs when all these three ZUPT detectors are below their thresholds:

$$MV(k) = \begin{cases} 1 & if \quad T_{mv}(k) < \gamma_{mv} \\ 0 & if \quad T_{mv}(k) \geq \gamma_{mv} \end{cases}$$

$$MAG(k) = \begin{cases} 1 & if \quad T_{mag}(k) < \gamma_{mag} \\ 0 & if \quad T_{mag}(k) \geq \gamma_{mag} \end{cases} \tag{13}$$

$$ARE(k) = \begin{cases} 1 & if \quad T_{are}(k) < \gamma_{are} \\ 0 & if \quad T_{are}(k) \geq \gamma_{are} \end{cases}$$

$$MMA(k) = MV(k)\&MAG(k)\&ARE(k)$$

where $MV(k)$, $MAG(k)$, and $ARE(k)$ are stance-phase estimation results. γ_{mv}, γ_{mag}, and γ_{are} are the threshold of MV, MAG, and ARE, respectively. $MMA(k)$ is the combination of the previous three detectors and MMA is short for the first letters of MV, MAG, and ARE. An experiment was taken to verify the stance phase estimation result of different detectors. The results shown in Figure 7 depict that, in this experiment, stance-phase detectors using accelerometer readings have some errors and ARE using gyroscope readings perform better. This is not always right and in some other scenarios, like taking an elevator, MV and MAG may have a better performance. In any case, MMA will always have the best stance-phase estimation among MV, MAG, and ARE.

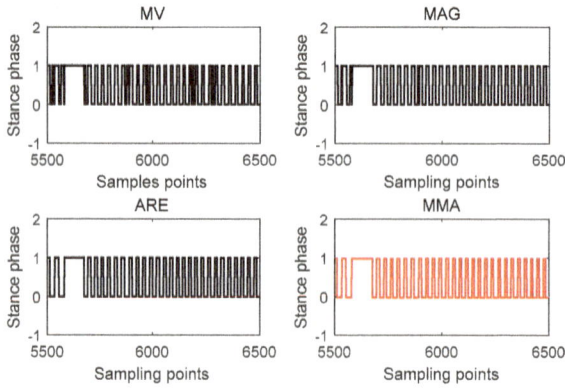

Figure 7. Stance-phase estimation results of acceleration moving-variance (MV), acceleration magnitude (MAG), angular-rate energy (ARE), and MMA detectors. 1 and 0 represent the pedestrian is in the stance phase and swing phase, respectively.

3.4. Corner Detection Using Phone-W-Embedded MEMS-IMU

The gyroscope readings of the waist-mounted smartphone can be used to detect the timestamp when a volunteer walks around a corner. The ARE method shown in Equation (11) is used:

$$Corner(k) = \begin{cases} Corner(k-1)+1 & if \quad \frac{1}{W}\sum_{n=k}^{k+W-1}\frac{1}{\sigma_\omega^2}\|\omega(n)\|^2 > \gamma_{corner} \\ Corner(k) & if \quad \frac{1}{W}\sum_{n=k}^{k+W-1}\frac{1}{\sigma_\omega^2}\|\omega(n)\|^2 \leq \gamma_{corner} \end{cases} \tag{14}$$

where $Corner(k)$ is the corner-detection result, and γ_{corner} is the threshold of corner detection. Considering that the turning speed of each site-survey process is different, the threshold γ_{corner} is not a fixed value. It is chosen as 10 times the average value of the whole ARE detector sequence:

$$\gamma_{corner} = 10 \times \frac{\sum\limits_{k=1}^{N} T_{are}(k)}{N} \tag{15}$$

where T_{are} is the ARE detector of Equation (11), and N is the length of T_{are}.

The ARE detector of Phone-W can clearly distinguish the difference of walking a straight line and tuning around a corner, which is shown in Figure 8.

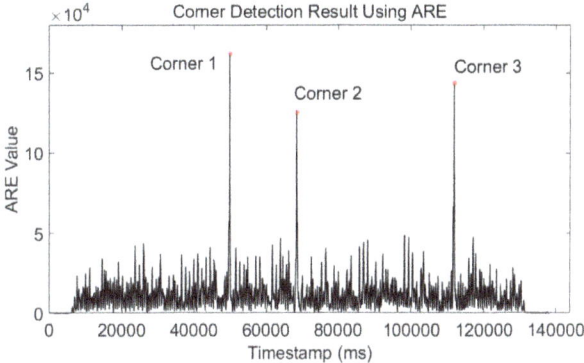

Figure 8. Corner-detection result using the ARE detector. This experiment was performed with the user walking around a square corridor and doing three turns. From the ARE values, we can clearly pick out the corners.

Three more experiments were conducted to verify the corner-detection algorithm. The volunteer walked around a square corridor for one, three, and six turns, respectively. The true and estimated number of corners of these three tests is summed in Table 3. The results show that the ARE-based corner-detection algorithm can accurately estimate turning movement and provide the timestamp when a turning movement occurs. We took a smartphone to record the video of the volunteer walking along the preassigned site-survey trajectory. From the video, we can take the average timestamp of the turning motion as the reference timestamp. All timestamp differences were smaller than 500 ms, which is smaller than the time duration of one step and has little influence on RP-coordinate estimation.

Table 3. Tests of ARE-based corner-detection algorithm using a waist-mounted smartphone.

Test Number	True Number of Corners	Estimated Number of Corners	Corner-Detection Error	Average Timestamp Error (ms)
1	3	3	0	324
2	11	11	0	426
3	23	23	0	233

This corner-detection method is not only applicable to 90-degree corners. A test was conducted to validate our method in other situations. The volunteer turned 45 degrees, 90 degrees, 135 degrees, and 180 degrees during a walk. The result is shown in Figure 9. The four typical degrees of corners can be detected correctly. We need to point out that, if the degree of the corner is too small, this corner-detection algorithm may have a wrong estimation. However, we can avoid this situation with a proper preassigned site-survey trajectory design.

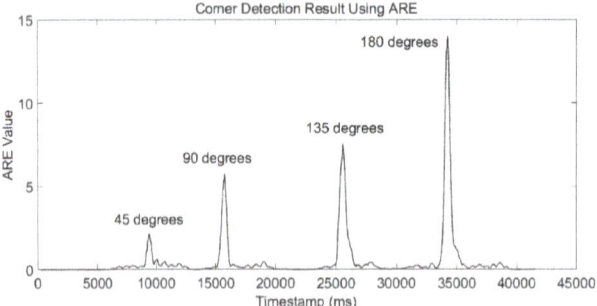

Figure 9. Corner-detection result using an ARE detector. The volunteer turned 45 degrees, 90 degrees, 135 degrees, and 180 degrees during a walk. The four corners were detected correctly.

4. Post Calibration Process

4.1. Stance-Phase Detection Based on Step Detection (SPD-SD)

One gait cycle consists of a stance phase and a swing phase. Five complete gait cycles are shown in Figure 10. It is very intuitive that we can count steps through counting stance phases. This method is SPD-SD. Furthermore, the start and end timestamp of the stance phase can be used to withdraw WiFi-RSSI from the complete WiFi-RSSI sequence and mark it with RP coordinates.

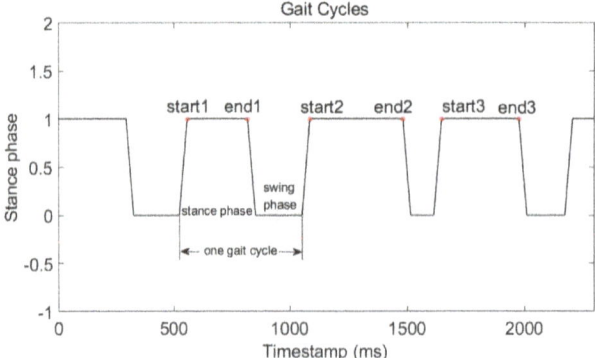

Figure 10. Five complete gait cycles during a walk. Each gait cycle consists of a stance phase and a swing phase. The start timestamp and the end timestamp of the stance phase can be detected from the rising edge and falling edge, respectively.

The equation of the SPD-SD method is:

$$step(k) = \begin{cases} step(k-1) + 0.5 & if \quad ARE(k) - ARE(k-1) = 1 \\ step(k-1) + 0.5 & if \quad ARE(k) - ARE(k-1) = -1 \\ step(k-1) & if \quad ARE(k) - ARE(k-1) = 0 \end{cases} \qquad (16)$$

where $step(k)$ is the step-detection result at timestamp k. The volunteer's movement always starts from the stance phase and ends with the stance phase, which means starting from a falling edge and ending with a rising edge of the stance-phase waveform. Therefore, the result of SPD-SD, which is equal to $step(end)$, must be an integer. We took a test walking 22 steps. Figure 11a shows the stance-phase waveform. Figure 11b shows the rising and falling edge of the stance-phase waveform. Figure 11c

shows the step-detection result using Equation (16). The step-detection result is 22 steps, which is the same as the true number.

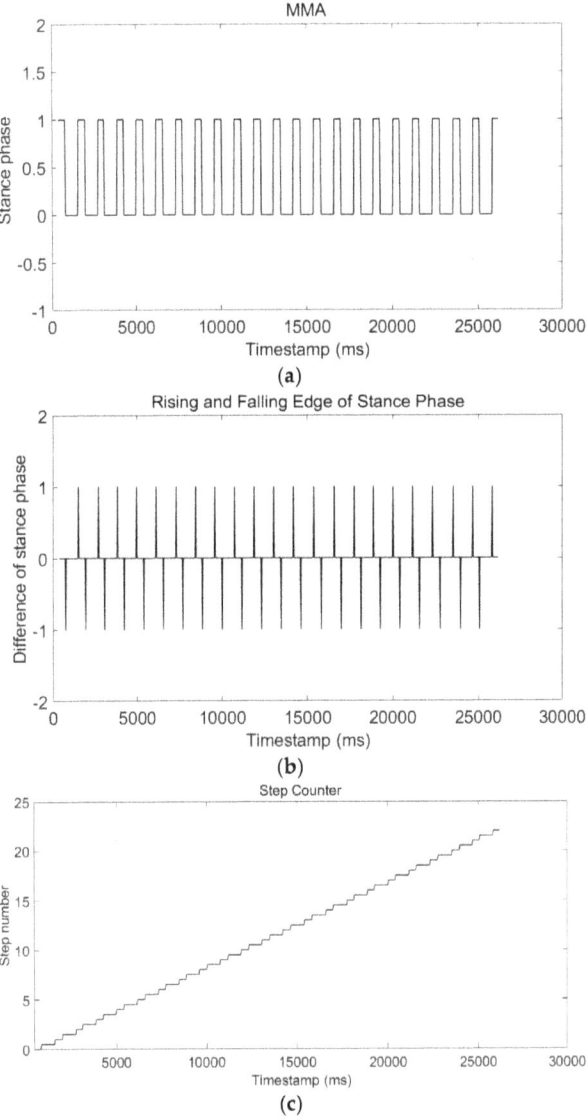

Figure 11. Results of the stance-phase detection based on step detection (SPD-SD) algorithm applied to a test of walking 22 steps. (**a**) Waveform of stance phase using the MMA algorithm. (**b**) Rising and falling edge of stance-phase waveform. 1 is the rising edge and −1 is the falling edge. (**c**) Step-detection result using Equation (16) and final step detection was 22 steps, which was the same as the true number.

To better verify the SPD-SD algorithm, we performed a much longer test, and the length of the trajectory was around 500 m. This test was repeated three times. A volunteer took a camera and recorded a video to count the true step number. The results are shown in Table 4. The step-detection results of the first two tests were the same as the true number. However, the step error of the

third test was 88. The reason is that in the third test the movement mode of the volunteer was running, not walking. The stance-phase detector with a fixed threshold had limited performance. In Reference [17], we developed a MAG-ZUPT method to estimate the stance phase of running using magnetic-field strength. In Reference [51], we developed an adaptive-threshold method of walking and running stance-phase detection. These two methods can solve the stance-phase detection problem but add extra sensors or increase computation complexity. Luckily, unlike the first responders, site-survey volunteers did not need to run, and the SPD-SD algorithm could provide good performance.

Table 4. Tests of SPD-SD algorithm using foot-mounted smartphone.

Test Number	True Number of Steps	Estimated Number of Steps	Error
1	426	426	0
2	437	437	0
3	413	325	88

4.2. Stance-Phase Detection-Based Stride-Length Estimation (SPD-SL)

The positioning result is the moving trail of Phone-F, which touches the ground only in the stance phase. Therefore, we can calculate the position in stance phase, and the distance between the neighbor stance-phase positions is stride length. This is the fundamental principle of the SPD-SL algorithm. To reduce fluctuation during stance phase, the position of each stance phase was averaged, with the window length calculated with the start and end timestamp of the stance phase:

$$spwl(k) = sp_k(end) - sp_k(start) \tag{17}$$

where $spwl(k)$ is the number of timestamps of the k-th stance phase, $sp_k(end)$ is the end timestamp of the k-th stance phase, and $sp_k(start)$ is the start timestamp of the k-th stance phase. The coordinates of the k-th stance phase are averaged with $spwl(k)$:

$$spc(k) = \frac{1}{spwl(k)} \times \sum_{i=sp_k(start)}^{sp_k(end)} traj(i) \tag{18}$$

where $spc(k)$ are the average coordinates of the k-th stance phase, and $traj$ are the coordinates of the whole trajectory. The stride length between the $(k-1)$-th stance phase and the k-th stance phase are calculated using:

$$sl(k) = \sqrt{(spc_x(k) - spc_x(k-1))^2 + (spc_y(k) - spc_y(k-1))^2} \tag{19}$$

where $sl(k)$ is the k-th step length, spc_x and spc_y are the x and y values of spc, respectively.

Considering that the smartphone was mounted on one foot, the stride length of SPD-SL was the stride length between the same foot, which was nearly two times longer than the stride length between the left and right foot. We took two tests to verify the SPD-SL algorithm. Test had a stride length of 1.2 m, and Test 2 of 0.6 m. The stride length of each step is shown in Figure 12. Comparison of true stride length and estimated stride length is shown in Table 5. The stride length estimation errors of Test 1 and Test 2 are 0.02 m and 0.03 m, respectively. It must be pointed out that it is difficult to have an accurate measurement of true stride length of each step. The volunteer tried the best to keep a constant stride length. However, considering that the positioning error of WiFi-fingerprinting based IPS is usually larger than 1 m, the stride length estimation accuracy using SPD-SL is enough to calculate RP coordinates.

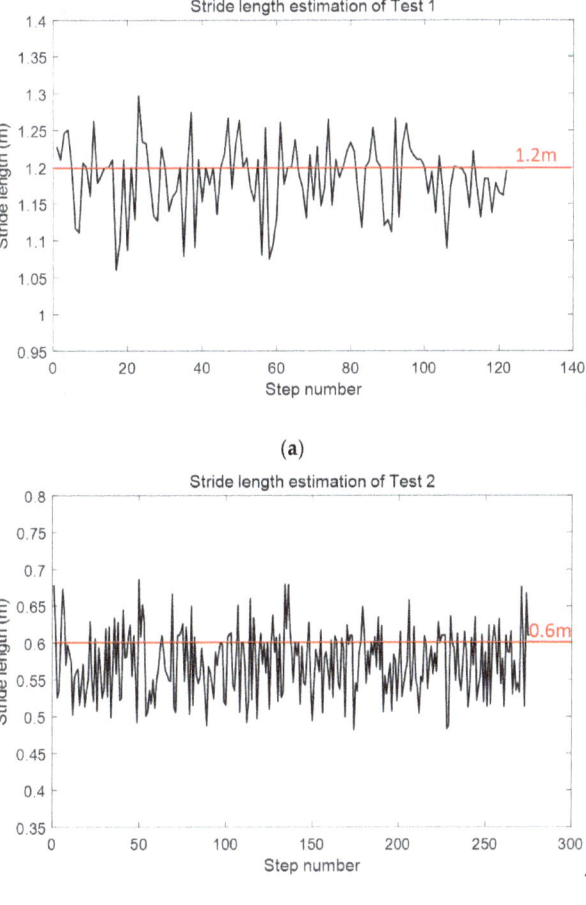

(a)

(b)

Figure 12. Stride-length estimation results of Test 1 and Test 2. (**a**) Result of Test 1. The volunteer walked 122 steps trying to keep stride length to 1.2 m. (**b**) Result of Test 2. The volunteer walked 276 steps trying to keep stride length to 0.6 m.

Table 5. Comparison of true stride length and estimated stride length.

Test Number	True Stride Length (m)	Average Estimated Stride Length (m)	Error (m)
1	1.2	1.18	0.02
2	0.6	0.57	0.03

4.3. Post Calibration with Preassigned Site-Survey Trajectory

Pedestrian dead-reckoning-based inertial navigation system (PDR-INS) integrates step lengths and heading estimations at each detected step to compute the position of the pedestrian [18]. The relationship between the k-th step and the $(k-1)$-th step is:

$$\begin{cases} x(k) = x(k-1) + l(k) * \cos(\varphi(k)) \\ y(k) = y(k-1) + l(k) * \sin(\varphi(k)) \end{cases} \quad (20)$$

where k is the step number calculated using the SPD-SD algorithm and can be mapped to the timestamp. $[x(k), y(k)]$ and $[x(k-1), y(k-1)]$ are position coordinates of k-th step and $(k-1)$-th step, respectively. l is the stride length calculated using the SPD-SL algorithm. φ is the heading calculated using corner-detection results and the preassigned site-survey trajectory. Our heading-estimation algorithm is very intuitive. With the knowledge that the volunteer cannot cross the wall, when they meet the corner they have to turn in the direction of the corridor. Therefore, the corner detected by Phone-W using gyroscope reading is an indicator of heading change and the value of the heading difference can be easily obtained using the preassigned site-survey trajectory.

Finally, with accurate RP coordinates and the timestamp duration of each stance phase, the WiFi-RSSI radio map is constructed. The WIFI-RSSI radio map can be built up with the correct RP coordinates and the corresponding timestamp. The time interval, shown in Equation (17), between the start and the end timestamp of each stance phase is the bridge to connect RP coordinates and WiFi-RSSI vectors.

An experiment was conducted to show the post calibration process. The test site was a square corridor. The step-detection result using the SPD-SD algorithm is shown in Figure 13, and the total step number was 108, which matched the true step number.

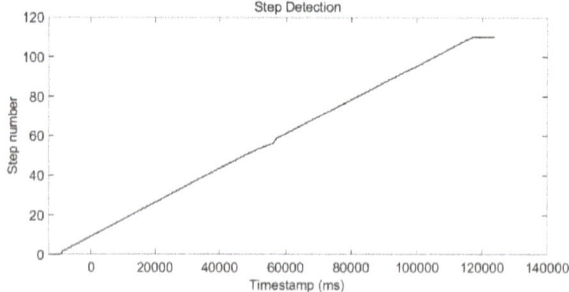

Figure 13. Step detection using the SPD-SD algorithm.

Stride length is estimated using SPD-SL, and the result of each step is shown in Figure 14. The estimated stride length of each step is around 1.2 m, which was the nearly the same as the true stride length.

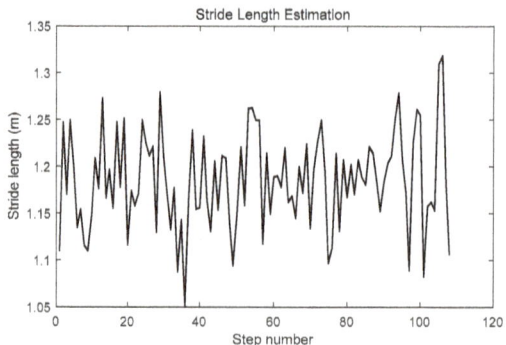

Figure 14. Stride length estimation using the stance-phase detection-based stride-length estimation (SPD-SL) algorithm.

The heading was estimated using the corner-detection result from Phone-W and the preassigned site-survey trajectory, which is shown in Figure 15. According to the direction of the corridor,

when a corner occurs the heading is plus or minus 90 degrees. RP coordinates calculated with the LSS-RM method, positioning results using Phone-F based on IEZ-INS, and the ground truth were compared. The positioning result of Phone-F based on IEZ-INS was badly influenced by the drift of the low-cost MEMS-IMU. However, a series of algorithms, such as SPD-DS, SPD-LS, and corner detection, can calibrate inaccurate IEZ-INS results into accurate RP coordinates.

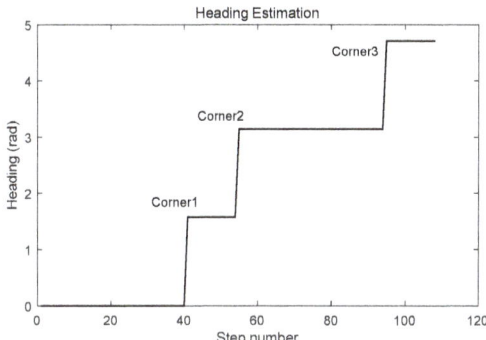

Figure 15. Heading estimation using corner-detection result from Phone-W and preassigned site-survey trajectory.

RP coordinates were calculated using Equation (20), and results are shown in Figure 16, even though the positioning accuracy of the smartphone-embedded low-cost MEMS-IMU is limited. We can still use a series of algorithms in LSS-RM to get the accurate RP coordinates that are a good match with the ground truth.

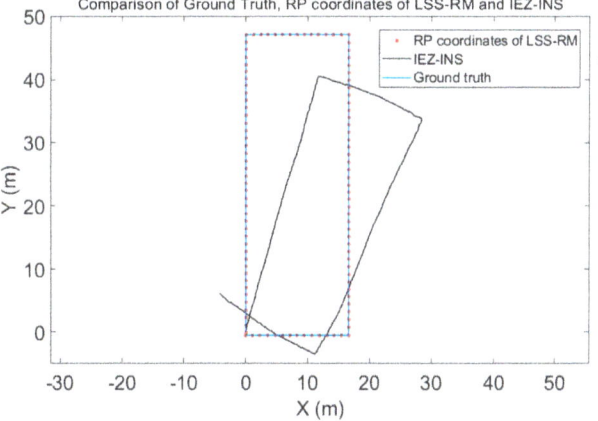

Figure 16. RP coordinates calculated with the LSS-RM method, positioning results using Phone-F based on IEZ-INS, and the ground truth are compared, although the positioning accuracy of the smartphone-embedded low-cost MEMS-IMU is limited. We can still use a series of algorithms in LSS-RM to get accurate RP coordinates.

5. Experimental Results

5.1. Summary of Submodule Tests in Previous Sections

Firstly, the experimental results of each submodule of our LSS-RM method have already been depicted in the corresponding sections. We would like to have a summary here:

(1) Timestamp alignment: Table 1 shows that our timestamp-alignment algorithm is very efficient. Figure 4 shows the timestamp comparison of Phone-F and Phone-W.

(2) IEZ-INS: Figure 6 shows that the positioning result of IEZ-INS using Phone-F is influenced by the heading error. The post calibration process is needed for accurate RP coordinates.

(3) Corner detection: Figures 8 and 9, and Table 3 show that we can detect corners correctly using ARE.

(4) SPD-SD: Figure 11 and Table 4 show that the SPD-SD algorithm can have accurate step detection.

(5) SPD-SL: Figure 12 and Table 5 show that the SPD-SL algorithm is accurate enough for calculating RP coordinates for WiFi-fingerprinting-based IPS.

(6) Post calibration: Figure 16 is the post calibration result. RP coordinates are matched with the ground truth.

Then, we conducted a comprehensive experiment to compare the WIFI-RSSI radio map built using our LSS-RSS with the one built with the traditional manual site survey method.

5.2. Comprehensive Experiment to Verify LSS-RM Method

This experiment was conducted in F6, New Main Building, Beihang University. The site is a square corridor, and the total length of the corridor is 128 m, shown in Figure 17a. Phone-F and Phone-W that were used in this experiment were MI6 (Xiaomi, Beijing, China), which contains a triaxial accelerometer, a triaxial gyroscope, and a triaxial magnetometer. The operating system of the two smartphones is MIUI based on Android 8.0, and we have developed an Android APP to sample accelerometer readings, gyroscope readings, magnetometer readings, and WiFi-RSSI. The sampling frequency of WiFi-RSSI and MEMS-IMU was 10 Hz and 30 Hz, respectively.

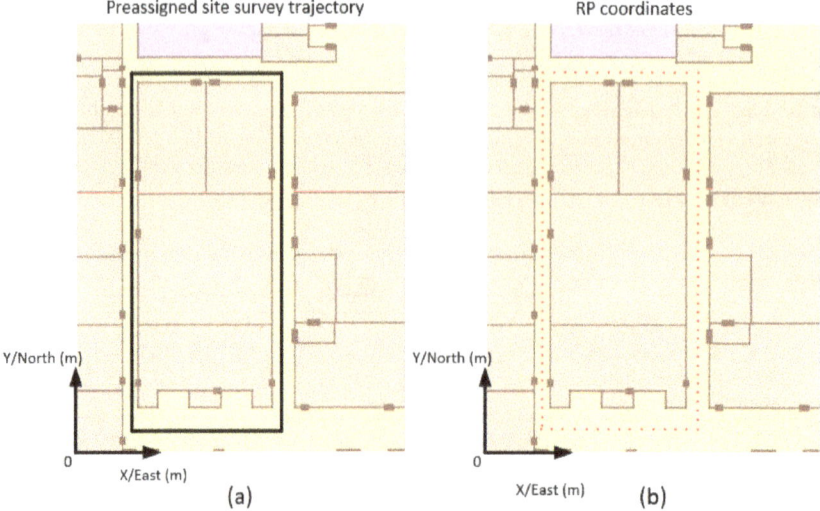

Figure 17. Preassigned site-survey trajectory and RP coordinates. (**a**) Preassigned site-survey trajectory along a square corridor. The volunteer was asked to walking along this trajectory. (**b**) Calculated RP coordinates using timestamp alignment, SPD-SD, SPD-SL, and heading estimation based on corner detection. There are 108 RPs.

Firstly, as a comparison, the manual site survey was conducted with 108 RPs. The distance between adjacent RP was 1.2 m and the sampling time on each RP was 0.5 min. A manual WiFi-RSSI radio map was built in 54 min. Then a volunteer mounted with Phone-W and Phone-F walked along the preassigned site survey trajectory. RP coordinates, shown in Figure 17b, were calculated using

a series of algorithms in LSS-RM like timestamp alignment, SPD-SD, SPD-SL, and heading estimation based on corner detection.

The radio map constructed by LSS-RM was compared with the traditional manual one. The number of test points was also 106 for both the manual radio map and the LSS-RM based radio map. The cumulative distribution function (CDF) plots of the WiFi-fingerprinting-based positioning using the manual radio map and the LSS-RM-based radio map are shown in Figure 18.

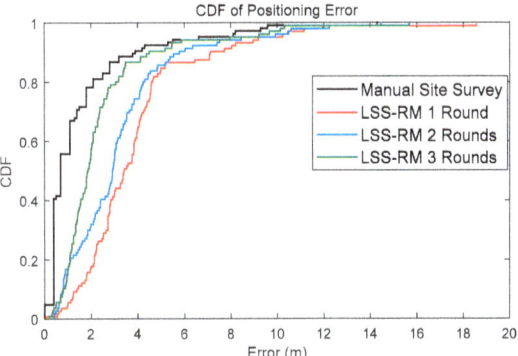

Figure 18. Comparison of CDFs of different radio maps. The black line is the WiFi positioning result using the manual radio map. The red line is the positioning result using our LSS-RM based radio map for one round along the preassigned site-survey trajectory. The blue line is using the LSS-RM-based radio map for two rounds, and green line for three rounds. Although the positioning accuracy of the manual site survey method is higher, its time consumption is nonnegligible. Furthermore, with more volunteers walking more times, positioning accuracy would rise remarkably.

The time consumption and average-positioning error of different site-survey methods are summarized in Table 6. The results show that the manual site survey method has the best positioning accuracy, but time consumption is 54 min, which is several times longer than of the LSS-RM method. As for LSS-RM, positioning accuracy rises with more walking, which means more WiFi-RSSI can make the radio map more robust.

Table 6. Comparison of time consumption and average positioning error.

Test Type	Time-Consumption (Minute)	Average Positioning Error (m)
LSS-RM of one round	2.6	3.91
LSS-RM of two rounds	5.1	3.25
LSS-RM of three rounds	7.8	2.47
Manual site survey	54	1.61

The most important advantage of LSS-RM is that it can conspicuously reduce the time consumption of offline site surveys. This characteristic helps the large-scale commercial deployment of WIFI-RSSI indoor positioning systems. Taking the shopping market application as an example, LSS-RM helps to build the WiFi-RSSI radio map in a much shorter time compared with the traditional manual site-survey method. Then, customers can retrieve their positions by matching the WiFi-RSSI collected in real time with the WiFi-RSSI radio map built using LSS-RM. With the position information and the indoor map of the shopping mall, customers can navigate to the nearest shoe shop or exit. Furthermore, shopping-mall managers can use the position information of customers to optimize the arrangement of the stores.

6. Conclusions

LSS-RM is developed in this paper to reduce the time consumption of offline site-survey processes. The offline phase of LSS-RM consists of data collection and preprocessing, and post calibration. The use of MEMS accelerometer and gyroscope readings of Phone-F can easily detect the stance phase of the volunteer. Furthermore, stance-phase information can be used to count steps (SPD-SD) and estimate stride length (SPD-SL). Using MEMS gyroscope readings of Phone-W can detect the corner of preassigned site-survey trajectories and accurate headings can be estimated in the post calibration process. The pedestrian dead-reckoning algorithm is used to calculate RP coordinates. A radio map is built with the RP coordinates and WiFi-RSSI vectors in a traditional radio-map format {RP coordinates, WiFi-RSSI vectors}. The bridge between RP coordinates and WiFi-RSSI vectors in the LSS-RM method is the start and end timestamp of each stance phase.

Several experiments were conducted to evaluate the submodules of the LSS-RM method. The results show that timestamp alignment, corner detection using ARE, step detection using SPD-SD, stride estimation using SPD-SL, heading estimation using corner information, preassigned site-survey trajectory, and RP-coordinate calculation all performed well. Finally, a comprehensive experiment was conducted to compare the performance of the traditional manual site survey and the LSS-RM method. The result shows that the manual radio map has better positioning accuracy, while time consumption is 54 min compared with the 2.6–7.8 min of the LSS-RM method. Furthermore, positioning accuracy of the LSS-RM method can be improved by more volunteers joining in the site-survey work. In our future work, we will research a site-survey-free method to construct the WIFI-RSSI radio map.

Author Contributions: W.Y., C.X. and J.Y. conceived the system and designed the experiments; W.Y., J.Y., Z.L., H.W. and D.Y. (Dayu Yan) performed the experiments; the indoor map was built by Z.L.; data-collection APP was developed by J.Y. and H.W.; D.Y. (Dayu Yan) developed the communication module through a public 4G LTE network; and W.Y. wrote the paper. All the work in this paper was done under the supervision of C.X. and D.Y. (Dongkai Yang).

Funding: This work was supported by the Beihang Beidou Technology Industrialization Funding program (Grant No.: BARI1701).

Acknowledgments: We appreciate Beihang University and Jinhua Science and Innovation Park for providing the experiment sites. We also appreciate Yanzhao Wang and Shangyin Liang for the data-collection work.

Conflicts of Interest: The authors declare no conflict of interest.

References

1. Zou, D.; Meng, W.; Han, S.; He, K.; Zhang, Z. Toward Ubiquitous LBS: Multi-Radio Localization and Seamless Positioning. *IEEE Wirel. Commun.* **2016**, *23*, 107–113. [CrossRef]
2. Kaplan, E.D.; Hegarty, C. *Understanding GPS: Principles and Applications*; Artech House: Norwood, MA, USA, 2005; pp. 598–599.
3. Zhang, T.; Liu, H.; Chen, Q.; Zhang, H.; Niu, X. Improvement of GNSS Carrier Phase Accuracy Using MEMS Accelerometer-Aided Phase-Locked Loops for Earthquake Monitoring. *Micromachines* **2017**, *8*, 191. [CrossRef]
4. Mautz, R. Indoor Positioning Technologies. Ph.D. Thesis, ETH Zurich, Zurich, Switzerland, 2012.
5. Nilsson, J.O. Infrastructure-Free Pedestrian Localization. Ph.D. Thesis, KTH Royal Institute of Technology, Stockholm, Sweden, 2013.
6. Zhang, R.; Hoflinger, F.; Reindl, L. Inertial Sensor Based Indoor Localization and Monitoring System for Emergency Responders. *Sens. J. IEEE* **2013**, *13*, 838–848. [CrossRef]
7. Kang, W.; Han, Y. SmartPDR: Smartphone-based pedestrian dead reckoning for indoor localization. *IEEE Sens. J* **2015**, *15*, 2906–2916. [CrossRef]
8. Zhuang, Y.; Lan, H.; Li, Y.; El-Sheimy, N. PDR/INS/WiFi Integration Based on Handheld Devices for Indoor Pedestrian Navigation. *Micromachines* **2015**, *6*, 793–812. [CrossRef]
9. Deng, Z.; Hu, Y.; Yu, J.; Na, Z. Extended Kalman Filter for Real Time Indoor Localization by Fusing WiFi and Smartphone Inertial Sensors. *Micromachines* **2015**, *6*, 523–543. [CrossRef]

10. Tian, Q.; Salcic, Z.; Wang, K.I.; Pan, Y. A Multi-Mode Dead Reckoning System for Pedestrian Tracking Using Smartphones. *IEEE Sens. J.* **2016**, *16*, 2079–2093. [CrossRef]

11. Lee, M.S.; Ju, H.; Park, C.G. Map assisted PDR/Wi-Fi fusion for indoor positioning using smartphone. *Int J Control Autom. Syst.* **2017**, *15*, 627–639. [CrossRef]

12. Ho, N.; Truong, P.; Jeong, G. Step-Detection and Adaptive Step-Length Estimation for Pedestrian Dead-Reckoning at Various Walking Speeds Using a Smartphone. *Sensors* **2016**, *16*, 1423. [CrossRef] [PubMed]

13. Skog, I.; Nilsson, J.O.; Händel, P. Evaluation of zero-velocity detectors for foot-mounted inertial navigation systems. In Proceedings of the International Conference on Indoor Positioning and Indoor Navigation, Zurich, Switzerland, 15–17 September 2010; pp. 1–6.

14. Skog, I.; Handel, P.; Nilsson, J.O.; Rantakokko, J. Zero-Velocity Detection—An Algorithm Evaluation. *IEEE Trans. Biomed. Eng.* **2010**, *57*, 2657–2666. [CrossRef] [PubMed]

15. Foxlin, E. Pedestrian Tracking with Shoe-Mounted Inertial Sensors. *IEEE Comput. Graph. Appl.* **2005**, *25*, 38–46. [CrossRef] [PubMed]

16. Jiménez, A.R.; Seco, F.; Prieto, J.C.; Guevara, J. Indoor pedestrian navigation using an INS/EKF framework for yaw drift reduction and a foot-mounted IMU. In Proceedings of the Positioning Navigation and Communication, Dresden, Germany, 11–12 March 2010; pp. 135–143.

17. Yang, W.; Xiu, C.; Zhang, J.; Yang, D. A Novel 3D Pedestrian Navigation Method for a Multiple Sensors-Based Foot-Mounted Inertial System. *Sensors* **2017**, *17*, 2695. [CrossRef] [PubMed]

18. Jimenez, A.R.; Seco, F.; Prieto, C.; Guevara, J. A comparison of Pedestrian Dead-Reckoning algorithms using a low-cost MEMS IMU. In Proceedings of the IEEE International Symposium on Intelligent Signal Processing, Budapest, Hungary, 26–28 August 2009; pp. 37–42.

19. Beauregard, S. Omnidirectional Pedestrian Navigation for First Responders. In Proceedings of the 4th IEEE Workshop on Positioning, Navigation and Communication, Hannover, Germany, 22 March 2007; pp. 33–36.

20. Feliz Alonso, R.; Zalama Casanova, E.; Gómez García-Bermejo, J. Pedestrian tracking using inertial sensors. *J. Phys. Agents (JoPha)* **2009**, *3*, 35–43. [CrossRef]

21. Du, Y.; Yang, D.; Yang, H.; Xiu, C. Flexible indoor localization and tracking system based on mobile phone. *J. Netw. Comput. Appl.* **2016**, *69*, 107–116.

22. Youssef, M.A.; Agrawala, A.; Shankar, A.U. WLAN Location Determination via Clustering and Probability Distributions. In Proceedings of the IEEE International Conference on Pervasive Computing and Communications, Fort Worth, TX, USA, 23–26 March 2003; pp. 143–150.

23. Diaz, J.J.M.; Maues, R.D.A.; Soares, R.B.; Nakamura, E.F.; Figueiredo, C.M.S. Bluepass: An indoor Bluetooth-based localization system for mobile applications. In Proceedings of the IEEE Symposium on Computers and Communications, Riccione, Italy, 22–25 June 2010; pp. 778–783.

24. Kriz, P.; Maly, F.; Kozel, T. Improving Indoor Localization Using Bluetooth Low Energy Beacons. *Mob. Inf. Syst.* **2016**, *2016*, 1–11. [CrossRef]

25. Ozdenizci, B.; Coskun, V.; Ok, K. NFC Internal: An Indoor Navigation System. *Sensors* **2015**, *15*, 7571–7595. [CrossRef] [PubMed]

26. Shirehjini, A.A.; Yassine, A.; Shirmohammadi, S. Equipment location in hospitals using RFID-based positioning system. *IEEE Trans. Inf. Technol. Biomed.* **2012**, *16*, 1058–1069. [CrossRef] [PubMed]

27. Jourdan, D.; Dardari, D.; Win, M.Z. Position error bound for UWB localization in dense cluttered environments. *Aerosp. Electronic Syst. IEEE Trans.* **2008**, *44*, 613–628. [CrossRef]

28. Zhang, C.; Kuhn, M.J.; Merkl, B.C.; Fathy, A.E. Real-Time Noncoherent UWB Positioning Radar with Millimeter Range Accuracy: Theory and Experiment. *IEEE Trans. Microw. Theory Tech.* **2010**, *58*, 9–20. [CrossRef]

29. Yoshino, M.; Haruyama, S.; Nakagawa, M. High-accuracy positioning system using visible LED lights and image sensor. In Proceedings of the Radio and Wireless Symposium, Orlando, FL, USA, 22–24 January 2008; pp. 439–442.

30. Ajmani, M.; Sinanović, S.; Boutaleb, T. Optimal beam radius for LED-based indoor positioning algorithm. In Proceedings of the Students on Applied Engineering, Newcastle Upon Tyne, UK, 20–21 October 2016; pp. 357–361.

31. Holm, S. Ultrasound positioning based on time-of-flight and signal strength. In Proceedings of the International Conference on Indoor Positioning and Indoor Navigation, Montbéliard-Belfort, France, 28–31 October 2013; pp. 1–6.

32. Ciurana, M.; Giustiniano, D.; Neira, A.; Barcelo-Arroyo, F.; Martin-Escalona, I. Performance stability of software ToA-based ranging in WLAN. In Proceedings of the International Conference on Indoor Positioning and Indoor Navigation, Zurich, Switzerland, 15–17 September 2010; pp. 1–8.

33. Keunecke, K.; Scholl, G. IEEE 802.11 n-based TDOA performance evaluation in an indoor multipath environment. In Proceedings of the European Conference on Antennas and Propagation, The Hague, The Netherlands, 6–11 April 2014; pp. 2131–2135.

34. Ng, B.P. *Robust Methods for AOA Geo-Location in a Real-Time Indoor WiFi System*; Taylor & Francis, Inc.: Oxford, UK, 2008; pp. 112–121.

35. Lemic, F.; Handziski, V.; Caso, G.; Nardis, L.D. Enriched Training Database for improving the WiFi RSSI-based indoor fingerprinting performance. In Proceedings of the IEEE Consumer Communications & NETWORKING Conference, Las Vegas, NV, USA, 6–13 January 2016; pp. 875–881.

36. Gui, L.; Yang, M.; Yu, H.; Li, J.; Shu, F.; Xiao, F. A Cramer-Rao Lower Bound of CSI-based Indoor Localization. *IEEE Trans. Veh. Technol.* **2017**, *67*, 2814–2818. [CrossRef]

37. Lee, K.H.; Yoo, J.; Kang, Y.M.; Kim, C.K. 802.11mc: Using Packet Collision as an Opportunity in Heterogeneous MIMO-Based Wi-Fi Networks. *IEEE Trans. Veh. Technol.* **2015**, *64*, 287–302. [CrossRef]

38. Hossain, A.K.M.M.; Soh, W. A survey of calibration-free indoor positioning systems. *Comput. Commun.* **2015**, *66*, 1–13. [CrossRef]

39. Tian, Z.; Fang, X.; Zhou, M.; Li, L. Smartphone-Based Indoor Integrated WiFi/MEMS Positioning Algorithm in a Multi-Floor Environment. *Micromachines* **2015**, *6*, 347–363. [CrossRef]

40. He, S.; Chan, S.H.G. Wi-Fi Fingerprint-Based Indoor Positioning: Recent Advances and Comparisons. *IEEE Commun. Surv. Tutor.* **2016**, *18*, 466–490. [CrossRef]

41. Du, Y.; Yang, D.; Xiu, C. A Novel Method for Constructing a WIFI Positioning System with Efficient Manpower. *Sensors* **2015**, *15*, 8358–8381. [CrossRef] [PubMed]

42. Kim, Y.; Shin, H.; Chon, Y.; Cha, H. Crowdsensing-based Wi-Fi radio map management using a lightweight site survey. *Comput. Commun.* **2015**, *60*, 86–96. [CrossRef]

43. Sorour, S.; Lostanlen, Y.; Valaee, S.; Majeed, K. Joint Indoor Localization and Radio Map Construction with Limited Deployment Load. *IEEE Trans. Mob. Comput.* **2015**, *14*, 1031–1043. [CrossRef]

44. Cheng, Y.; Chawathe, Y.; LaMarca, A.; Krumm, J. Accuracy characterization for metropolitan-scale Wi-Fi localization. In Proceedings of the 3rd ACM International Conference on Mobile Systems, Applications, and Services 2005, Seattle, WA, USA, 6–8 June 2005; pp. 233–245.

45. Chekuri, A.; Won, M. Automating WiFi Fingerprinting Based on Nano-Scale Unmanned Aerial Vehicles. In Proceedings of the 2017 IEEE 85th Vehicular Technology Conference (VTC Spring), Sydney, Australia, 4–7 June 2017; pp. 1–5.

46. Zhao, W.; Han, S.; Hu, R.Q.; Meng, W.; Jia, Z. Crowdsourcing and Multisource Fusion-Based Fingerprint Sensing in Smartphone Localization. *IEEE Sens. J.* **2018**, *18*, 3236–3247. [CrossRef]

47. Zhuang, Y.; Syed, Z.; Li, Y.; El-Sheimy, N. Evaluation of Two WiFi Positioning Systems Based on Autonomous Crowdsourcing of Handheld Devices for Indoor Navigation. *IEEE Trans. Mob. Comput.* **2016**, *15*, 1982–1995. [CrossRef]

48. Rai, A.; Chintalapudi, K.; Padmanabhan, V.; Sen, R. Zee: Zero-effort crowdsourcing for indoor localization. In Proceedings of the 18th Annual International Conference on Mobile Computing and Networking, Istanbul, Turkey, 22–26 August 2012; pp. 293–304.

49. Yang, D.; Xue, G.; Fang, X.; Tang, J. Incentive Mechanisms for Crowdsensing: Crowdsourcing with Smartphones. *IEEE/ACM Trans. Netw.* **2016**, *24*, 1732–1744. [CrossRef]

50. Nilsson, J.O.; Gupta, A.K.; Handel, P. Foot-mounted inertial navigation made easy. In Proceedings of the International Conference on Indoor Positioning and Indoor Navigation, Busan, Korea, 27–30 October 2014; pp. 24–29.

51. Zhang, J.; Xiu, C.; Yang, W.; Yang, D. Adaptive threshold zero-velocity update algorithm under multi-movement patterns. *J. Beijing Univ. Aeronaut. Astronaut.* **2018**, 636–644. (In Chinese) [CrossRef]

Article

Activity Monitoring with a Wrist-Worn, Accelerometer-Based Device

Wen-Yen Lin [1,2,*], Vijay Kumar Verma [1], Ming-Yih Lee [2,3] and Chao-Sung Lai [1,4,5]

[1]   Department of Electrical Engineering, Center for Biomedical Engineering, Chang Gung University,
     Tao-Yuan 33302, Taiwan; d0421006@stmail.cgu.edu.tw (V.K.V.); cslai@mail.cgu.edu.tw (C.-S.L.)
[2]   Division of Cardiology, Department of Internal Medicine, Chang Gung Memorial Hospital,
     Tao-Yuan 33305, Taiwan; leemiy@mail.cgu.edu.tw
[3]   Graduate Institute of Medical Mechatronics, Center for Biomedical Engineering, Chang Gung University,
     Tao-Yuan 33302, Taiwan
[4]   Department of Nephrology, Chang Gung Memorial Hospital, Linkou, Tao-Yuan 33305, Taiwan
[5]   Department of Materials Engineering, Ming Chi University of Technology, New Taipei 24301, Taiwan
*   Correspondence: wylin@mail.cgu.edu.tw; Tel.: +886-3-2118800 (ext. 3675)

Received: 2 August 2018; Accepted: 8 September 2018; Published: 10 September 2018

Abstract: This study condenses huge amount of raw data measured from a MEMS accelerometer-based, wrist-worn device on different levels of physical activities (PAs) for subjects wearing the device 24 h a day continuously. In this study, we have employed the device to build up assessment models for quantifying activities, to develop an algorithm for sleep duration detection and to assess the regularity of activity of daily living (ADL) quantitatively. A new parameter, the activity index (AI), has been proposed to represent the quantity of activities and can be used to categorize different PAs into 5 levels, namely, rest/sleep, sedentary, light, moderate, and vigorous activity states. Another new parameter, the regularity index (RI), was calculated to represent the degree of regularity for ADL. The methods proposed in this study have been used to monitor a subject's daily PA status and to access sleep quality, along with the quantitative assessment of the regularity of activity of daily living (ADL) with the 24-h continuously recorded data over several months to develop activity-based evaluation models for different medical-care applications. This work provides simple models for activity monitoring based on the accelerometer-based, wrist-worn device without trying to identify the details of types of activity and that are suitable for further applications combined with cloud computing services.

Keywords: accelerometer; activity monitoring; regularity of activity; sleep time duration detection

1. Introduction

Acceleration due to the human activities is proven to be of great importance in epidemiological research and physical activity-based health assessment. MEMS accelerometer-based, wrist-worn devices, such as smart watches and smart wrist-bands, are becoming more and more popular for PA monitoring. These devices are not only being used to perform sleep assessment [1–3] but also for activity monitoring. Most of them provide number of steps and calorie consumption during the wearing period, and may even try to identify types of activities performed from the acceleration data measured by the accelerometer inside the devices. However, there are still unanswered questions such as: How accurate are these for the identified types of activities and calories burned, due to the fact the information is derived indirectly from acceleration information?

Having a good physically active life style can reflect one's health condition and could be used to predict whether subjects might suffer from some diseases or not [4]. More importantly, having a regular life style and good sleep quality in the long run will even impact the subject's health condition.

However, there is no quantitative way to assess the regularity of daily life. Inadequate PA may result in many health problems and diseases related to the lungs, heart, etc. [5–7]. Tracking PA on daily basis provides valuable information related to the human body [8,9]. A number of platforms have been designed, implemented, and tested successfully in order to track subjects' PA based on the wearable MEMS accelerometer [10,11]. Recent growth of IoT [12,13] and the capability of smart-phones in the past few years made PA recognition a dynamic field of study [14–16]. Bender et al. conducted an empirical study of various fitness devices such as Fitbit Flex, Fitbit Charge HR, Garmin Vívoactive, and Apple Watch to compare PA recognition accuracy and device performance [17].

Several previous works also proposed methods for activity identification and PA pattern analysis using external software, such as those executed on PCs or smart phones. For instance, Guillermo R. Oviedo et al. [18] have used GT3X ActiGraph accelerometer (Firmware 4.4.0, ActiGraph™, FortWalton Beach, FL, USA), and data were downloaded with the ActiLife 6 Software (v.6.12.0., ActiGraph™, Fort Walton Beach, FL, USA) for certain types of activity identification through activity patterns on PCs. Chelsea Dobbins and Reza Rawassizadeh [19] used principal component analysis feature selection (PCAFs) and correlation feature selection (CFs) on PCs to refine clustering of raw accelerometer data that had a positive effect on the computational burden that is associated with processing large sets of data, as energy efficiency and resource use is decreased, because less data is processed by the clustering algorithms, but a tremendous amount of raw data from devices is still required. A. K. Chowdhury et al. [20] have proposed the use of posterior-adapted, class-based weighted decision fusion to effectively combine data from multiple accelerometer-based devices for improving physical activity recognition. Nan Zeng et al. [21] have used NL-1000 pedometer and ActiGraph GT3X accelerometer for assessing the reliability of using motion sensors to measure children's PA Levels in Exergaming. Matin Kheirkhahan et al. [22] have developed machine learning methods for identifying activity types and computing energy expenditures using standard statistical methods, as well as the bag-of-words (BoW) approach. However, this research poorly describes further applications for the results that have been analyzed.

In summary, these results are not found to be accurate enough or to have a close association with the timing information so that they could be considered for medical-care systems and clinical applications. Therefore, this research aims to bridge this gap and to develop simple mathematical models and algorithms for achieving parameters and results that are closely associated with actual on-set timing information. Moreover, the models and algorithms are simple enough to be easily deployed inside wrist-worn, accelerometer-based devices, and the results obtained are accurate and precise enough to be considered for medical-care systems and clinical applications.

Therefore, in this study we have proposed a methodology for the quantification of physical activities performed the daily life, i.e., activity index (AI), which is closely associated with human body acceleration [23]. These results can be used to categorize activities into 5 different levels, i.e., rest/sleep, sedentary, light, moderate, and vigorous activity states. Based on the AIs, a sleep duration detection algorithm has also been developed. Furthermore, by calculating the correlation coefficient for AIs on a daily or weekly basis, a quantitative method to express the regularity of daily life, i.e., the regularity index (RI), has been proposed in this study. By combining all these quantitative indices, such as activity, sleep duration, and regularity of ADL, this index can be used for many activity monitoring-based medical-care applications. Moreover, these results could be synced to smart phones for IoT applications with a greater capability of sensing human activity ubiquitously and unobtrusively through advancements in miniaturization and sensing abilities. This could unleash a systematic methodology in mobile-health for the prevention and early detection of chronic diseases.

2. Materials and Methods

This study was reviewed and approved by the institutional review board (IRB) of the Chang Gung Memorial Hospital, Taiwan, R.O.C. An accelerometer-based, wrist-worn device (GeneActiv, Activinsights Ltd., Huntingdon, UK, as in Figure 1) has been used to record wrist movement

acceleration in three orthogonal directions along with timestamp and body temperature. This device is equipped with MEMS-based accelerometer, which was set for a sensing range of ±8 g at a 12-bit digital resolution (i.e., 3.9 mg resolution, 1 mg = 1/1000 g), in which 'g' is the gravitational force. Even though the sampling frequency of this device was configurable, the sampling frequency was set to 20.0 Hz, as it was used to monitor ADL and also to prevent too much data were generated due to high sampling rate. Ten normal subjects were chosen and have been asked to wear the device for 24 h a day during their normal routine. After a maximum of 30-days, data were downloaded at the data-server station, and the subjects were given another device immediately for continuous data recording.

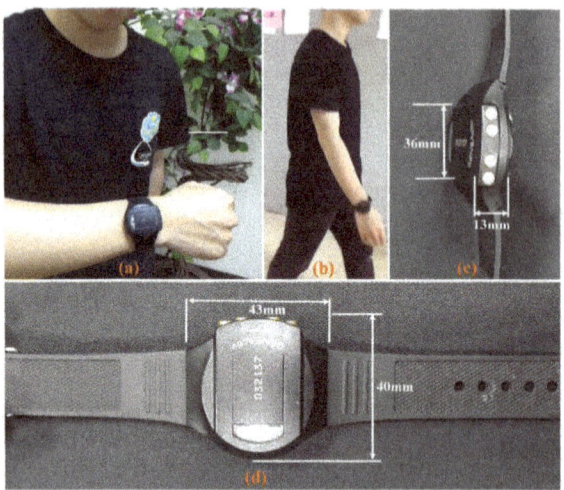

Figure 1. (**a**) Accelerometer-based GENEActiv device, Activinsights Ltd., Huntingdon, UK. (**b**) Walking subject with GENEActiv device on right-wrist. (**c**) Physical dimensions and side view of the GENEActiv device. (**d**) Top view of the device.

Downloaded data were then further analyzed and interpreted. A mathematical model has been proposed and implemented for the data processing and parameters for quantifying and assessing PAs that were defined, quantified, and evaluated. A brief description about data analytical models are as given below.

2.1. Data Analytical Models

With the fact that the kinematic energy associated with human activities can be given as (1),

$$E = \frac{1}{2}mv^2 \tag{1}$$

in which 'm' is the human's body mass moving with a velocity of 'v'. Since $v = a \cdot \Delta t + v_0$, in which 'a' is the moving acceleration of the body in the time interval of 'Δt', and 'v_0' is the initial velocity. If the movement starts from rest position, then the initial velocity, 'v_0', can be taken as 0, and $v = a \cdot \Delta t$. Now, the kinematic energy in (1) will be proportional to the square of acceleration, as in (2), assuming that the subject's mass is constant over the period Δt.

$$E \propto a^2 \tag{2}$$

The magnitude of acceleration, A_j, at the data instance j could be calculated as in (3), and since only the magnitude is calculated, the impact of the orientation of accelerometer will be eliminated.

$$A_j = \sqrt{\left(a_{x\_j}^2 + a_{y\_j}^2 + a_{z\_j}^2\right)} \tag{3}$$

In (3), a_{x_j}, a_{y_j}, and a_{z_j} are the raw data from accelerometer along the X-, Y-, and Z-axis sampled at the time instance j. The standard deviation, σ, of the acceleration magnitude within a pre-defined epoch period can be calculated as in (4).

$$\sigma = \sqrt{\frac{1}{N}\sum_{j=1}^{N}(A_j - \mu)^2}, \text{ where } \mu = \frac{1}{N}(A_1 + A_2 + A_3 + \cdots\cdots + A_N) \tag{4}$$

In (4), N is total number of acceleration data points measured in the epoch period, and μ is the mean value of the total acceleration within that period. Since the acceleration measured by the accelerometer also contains gravity force all the time; therefore, the average of total acceleration within a short period can be considered as the gravity. Hence, deducting the acceleration's mean value 'μ' from the magnitude of acceleration A_j measured by accelerometer will give the net acceleration of the body movement. Therefore, the square of standard deviation, σ, calculated in (4) can be considered as the average of the square of total net acceleration within the epoch period, such that the human kinematic energy will also be proportional to the square of standard deviation, σ, as in (5).

$$E \propto a^2 \propto \sigma^2 \tag{5}$$

Since the square of standard deviation, σ^2, over the epoch period is very small and usually less than 1, so the standard deviation, σ, is taken into consideration in the model proposed in this study for the larger value represented. Regarding the results, an activity index (AI) is proposed as the summation of the standard deviation, σ, over a desired time interval, as shown in (6).

$$AI = \sum_{k=1}^{M}\sigma_k \tag{6}$$

in which M is the total number of epoch periods within the time interval and σ_k represents the standard deviation of acceleration in the k-th epoch period within that time interval. In the implementation of the data analytical models described here, a 5-s epoch period is considered. The choice of 5-s epoch period typically will not cover more than two activities within that epoch period and will still generate manageable amounts of data in the computation. Since the sampling frequency of the accelerometer in the device was set to 20 Hz, within the 5-s epoch period, there will be 100 measured data points, i.e., $N = 100$ as in (4). To generate a minutely-wise AIs, i.e., choosing one minute as a basic time interval, then there will be 12 epoch periods in the time interval, and hence $M = 12$ as in (6).

2.2. Categorization for Different Levels of Activities

With the AI introduced in Section 2.1, testing of 14 different activities of daily life (ADL) and each activity lasted for 3 min was conducted with 10 normal subjects wearing the device. These activities included sleeping, sitting and watching TV, sitting and reading newspaper, sitting and web browsing, housekeeping, driving, walking—no hand-swing, walking—with hand-swing, upstairs—no hand-swing, upstairs—with hand-swing, downstairs—no hand-swing, downstairs—with hand-swing, jogging—no hand-swing, and jogging—with hand-swing. Among these 10 different normal subjects under test, only the types of activities to perform for testing were instructed; no detailed constraints, such as numbers of body turns during the sleep, moving (swing) frequency of the arms, walking speed, etc., were asked to be followed. Figure 2a shows a normal subject walking with hand-swing during the test, and Figure 2b depicts the magnitude of acceleration recorded over an epoch period, 5 s, for the activity, and a sample calculation for the standard deviation of acceleration recorded within that period is shown as Figure 2b. Figure 2b shows that the 12 standard deviations, σ's, calculated in a one minute time interval, and by accumulating such 12 σ's, a minute-wise activity index, AI, for the walking with hand-swing is obtained.

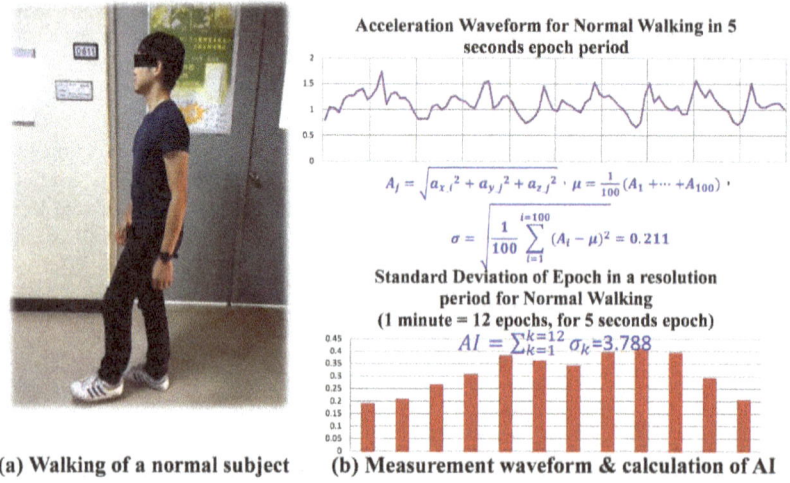

$$A_j = \sqrt{a_{x\,j}^2 + a_{y\,j}^2 + a_{z\,j}^2} \cdot \mu = \frac{1}{100}(A_1 + \cdots + A_{100}),$$

$$\sigma = \sqrt{\frac{1}{100}\sum_{i=1}^{i=100}(A_i - \mu)^2} = 0.211$$

Standard Deviation of Epoch in a resolution period for Normal Walking
(1 minute = 12 epochs, for 5 seconds epoch)

$$AI = \sum_{k=1}^{k=12}\sigma_k = 3.788$$

(a) Walking of a normal subject **(b) Measurement waveform & calculation of AI**

Figure 2. (**a**) Photo of walking test and (**b**) acceleration waveform of measured activity and calculation of AI.

Among these 14 ADL tested, sleeping is considered as a typical rest/sleep level of activity; the activities sitting and watching TV, sitting and reading newspaper, and sitting and web browsing, which were all performed in sitting position, are sedentary level activities; housekeeping, driving, and walking without hand-swing are classified as light level of activities; walking with hand-swing and up and down stairs, with or without hand-swing, are all moderate levels of activities; finally, jogging with or without hand-swing are considered vigorous levels of activity.

2.3. Sleep Duration Detection

Figure 3 shows the 24-h AI pattern of a subject. These AIs were calculated minute-wise, and hence there are 1440 AI values within 24-h period of daily life. The 24-h period started from 12:00:00.000 on the day and end at 11:59:59.950 on the next day, so that the typical sleep duration during the night can be covered in a 24-h minute-wise AI pattern completely. From this pattern, it is very clear that around 14:07, the level of AI is very low, as the subject was taking a nap, and the duration between 23:13:00 to 05:52:00 of the next day is also low, as the subject was sleeping.

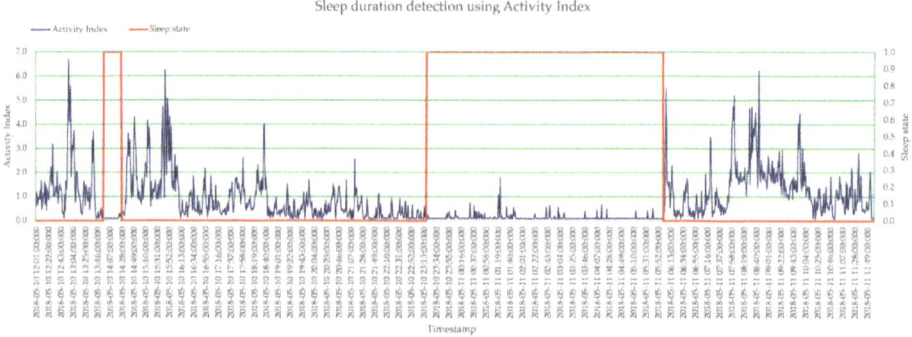

Figure 3. 24 h minute AI pattern with sleep state identified.

Although detecting sleep duration with accelerometer has been practiced for a long time, neither methods use the proposed AI for recognition nor results are precise and accurate enough so that they

can be used for clinical applications. Sleep duration detection is of utmost importance in the diagnosis of diseases like insomnia, drowsiness, and other sleep-related disorders. A smart sleep duration detection algorithm has been developed and is introduced in more detail below. The algorithm is based on the AI models that have been proposed in previous section.

The flow chart of the sleep duration detection algorithm is shown in Figure 4. The proposed algorithm consists of two phases. Phase I is to judge whether the subject in current time interval is in sleeping or awake state. In this phase, the proposed algorithm will thoroughly process all of the AIs corresponding to all basic time intervals. If the previous time interval is in awake state, the algorithm will judge if the subject falls asleep in the current time interval; otherwise, it will see if the subject wakes up during this time interval.

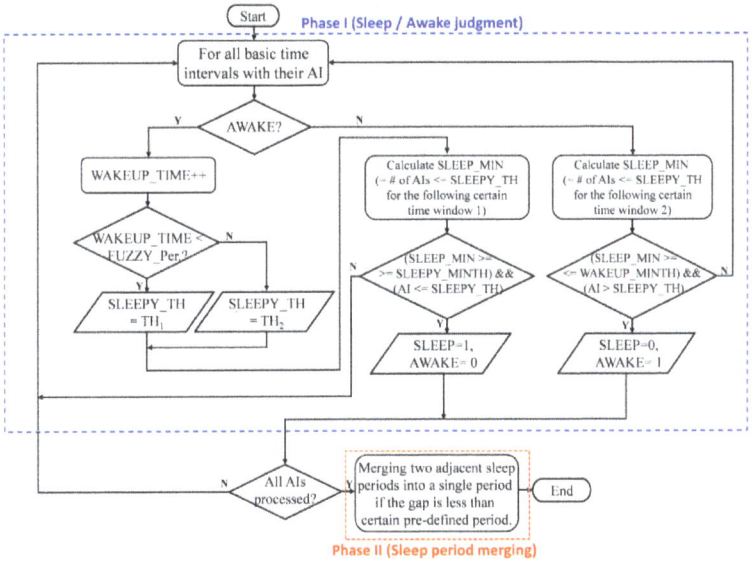

Figure 4. Flow chart of the proposed smart sleep duration detection algorithm based on AI model.

A subject is said to be in a fuzzy period when the subject just woke up within certain period of time within the sleep duration, and there are two different criteria to judge if the subject falls asleep in current time interval or not based on if the subject is in a fuzzy period. This reflects the fact that a subject will be most likely to fall asleep again when the awaking duration is not long enough. As a result, relaxed criteria (higher threshold value considered for current AI) could be used to judge if the subject falls into asleep or not when the subject is in a fuzzy state. Otherwise, stricter criteria (lower threshold value could be considered for current AI) is used to detect if the subject falls asleep. Besides the threshold value used, the criteria for a subject to fall asleep in current time interval requires the number of AIs, i.e., SLEEP_MIN, that is lower than the threshold value, SLEEP_TH, within the following certain time window 1 to be more than some pre-defined value, SLEEP_MINTH, and also that the AI of current time interval be lower than the threshold value.

Similarly, judging if the subject wakes up in current time interval requires meeting wake-up criteria. It is defined as the number of AIs, i.e., SLEEP_MIN, that are lower than the threshold value within the following certain time window 2 to be less than some pre-defined value, WAKEUP_MINTH, and for the AI of current time interval to be greater than the threshold value.

After all the AIs of the all basic time intervals have been processed thoroughly, all the sleeping periods are detected. Then, the algorithm goes into Phase II. Phase II is to merge any two adjacent sleeping periods into a single period if the duration is less than certain pre-defined time. This phase

could eliminate the results of fragmented sleep durations detected in Phase I that falsely indicate bad sleep quality. Typically, people will consider they sleep for a whole period of time but often wake up during sleep instead of having fragmented sleep durations. Phase II considers this situation and, as a result, will have more matched and accurate sleep duration with the subject's intuitive cognition.

As per the assessment of AI under rest/sleep level of activities in previous section, the AI is less than 0.1 within the 1-min time interval under rest/sleep status. Hence, AI value of 0.1 is used in the very beginning of the sleep duration detection algorithm to judge if that time interval is in AWAKE state or not.

2.4. Quantification of Regularity of ADL

In the IRB testing, a typical situation was that a single subject with un-regular living patterns in two consecutive days has been observed; this is shown in Figure 5. Indeed, the subject went to the emergency room (ER) for urgent health situation on the next day. Apparently, as in top of Figure 5, the subject's sleep duration was between 21:21:01 on the day to 04:41:01 of the next day. However, according to the AI pattern of the next day, as in the bottom of Figure 5, the subject went to bed around 04:00:01 in the morning and woke up at 07:21:01 with only very short sleep duration and also in extremely irregular time slot. This situation suggested that the subject's life style on these two consecutive days was extremely irregular, which might result in his visiting ER the next day. This observation motivated the need for a quantification assessment of regularity of ADL.

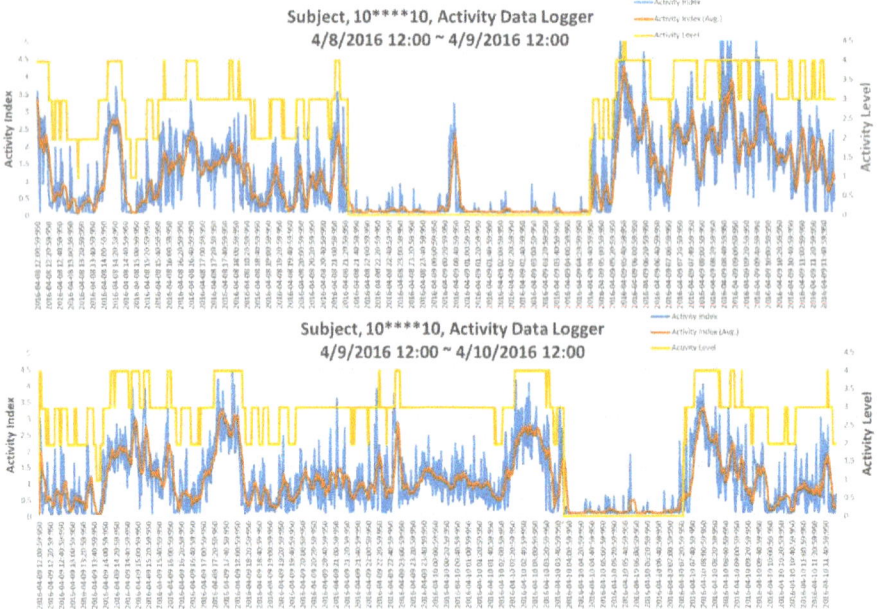

Figure 5. The AI pattern plots of a single subject over two consecutive days.

As shown in Figure 6a, there are 1440 data points of minute-wise AIs in a 24-h day, which raises significant complexities for further processing. Therefore, hour-based AI patterns were generated by taking the cumulative values of 60 min-wise AIs patterns into the total AI within an hour. As a result, hourly AI patterns of the day are generated, as shown in Figure 6b. By finding the correlation coefficient between the hourly AI patterns of day $i-1$ and day i, as shown in Figure 6c, the result can indicate the regularity of ADL of the day i with respect to the previous day $i-1$, namely, day-to-day Regularity Index (RI). Similarly, if the hourly AI patterns of day i are compared with the patterns of

the day one week before, i.e., day $i-7$, then it is called week-to-week Regularity Index. Since the range of the correlation coefficient is in between -1 and $+1$, the result of $+1$ means the ADL pattern of day i is the same as day $i-1$. 0 stands for totally uncorrelated and -1 stands for totally inversely correlated. With the continuous monitoring of days for human's ADL, the trend for the regularity of that subject's living style can be plotted as in Figure 6d.

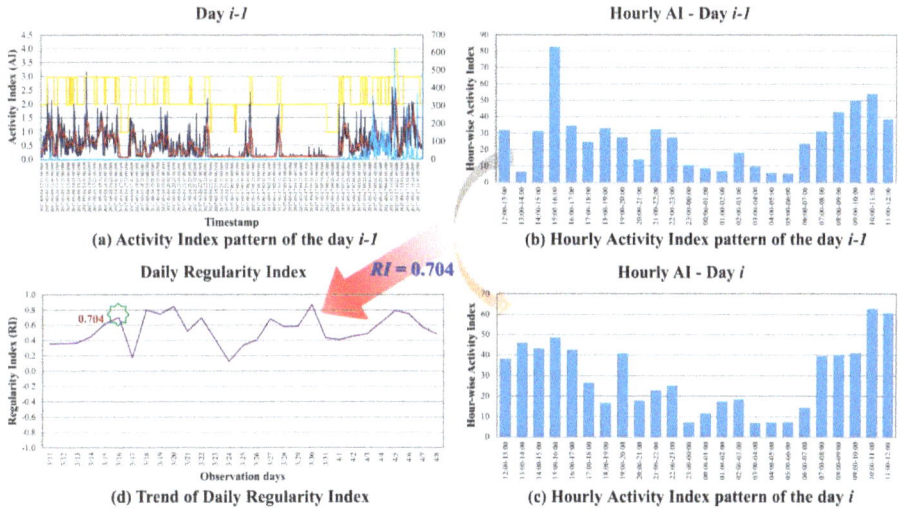

Figure 6. (**a**) Minute-wise AI pattern of day $i-1$; (**b**) hourly AI pattern of day $i-1$; (**c**) hourly AI pattern of day i; and (**d**): RI trend.

3. Results

As described in Section 2.2, 14 different types of daily physical activities (PAs) were performed by 10 normal subjects with GeneActiv devices worn on their wrists. Maximum and minimum values of AI for a specific PA have been identified; furthermore, the mean value of AI for that type of PA was calculated. As there were no detailed constraints, such as numbers of body turns during sleep, moving (swing) frequency of the arms, walking speed, etc., for any specific type of activity that were followed in the test, the range of maximum and minimum AIs for some types of activities were quite large. However, they can still be categorized into different levels of activity based upon their mean AI values; furthermore, the AI value falls between the specific maximum and minimum ranges, as shown in Table 1. For instance, mean AI value less than 0.1 indicates that the subject under observation was sleeping or resting, similarly, mean AI value between 0.1 and 0.5 will be considered as subject was performing sedentary types of PAs. For light PAs, the mean AI falls between 0.5 and 2.0. Moderate PAs could be considered as having mean AI in the range of 2.0~4.0. Additionally, all PAs having mean AI greater than 4.0 could be considered as performing the vigorous activities. The level of activities that have been categorized are shown as the yellow line in Figure 5, with rest/sleep leveled at 0 and vigorous activities leveled at 4 respectively.

With the same acceleration data, a 24-h minute-wise AI pattern starting from the noon of a day to the noon of next day were analyzed by the proposed smart sleep duration detection algorithm to detect the sleep duration with a precision up to 1 min. The red line in Figure 3 shows the identified sleep states, and it represents 'awake' state by '0' and 'sleep' state by '1'. As stated, through the phase II—sleep period merging, no more fragmented sleep periods were generated due to the bad sleep quality. The sleep quality can also be quantified by calculating the average minute-wised AI during the sleep duration. Larger value means there were more activities during the sleep, and hence it stands

for bad sleep quality. Smaller values mean few activities occurred during the sleep, which indicates good sleep quality.

Table 1. Classification of different PAs based upon AI.

Daily Life Activities	Level of Activities	Maximum	Minimum	Mean
Rest/Sleeping	REST/SLEEP	0.096975	0.082482	0.088523
Sit-Watching TV	SEDENTARY	0.345286	0.092849	0.186028
Sit-Reading News paper	SEDENTARY	0.466089	0.090514	0.301787
Sit-Web browsing	SEDENTARY	0.12198	0.100102	0.111355
Housekeeping	LIGHT	1.604074	0.829076	1.222838
Driving	LIGHT	1.378413	0.976756	1.171664
Walking-no Hand-Swing	LIGHT	2.334001	0.648251	1.59335
Walking-w/Hand-Swing	MODERATE	3.973975	2.048219	2.541614
DownStairs-no Hand-Swing	MODERATE	8.124398	1.673799	3.890505
DownStairs-w/Hand-Swing	MODERATE	3.40835	1.243286	2.415157
UpStairs-no Hand-Swing	MODERATE	2.602795	2.352814	2.515802
UpStairs-w/Hand-Swing	MODERATE	2.464628	2.308799	2.393241
Jogging-no Hand-Swing	VIGOROUS	12.95352	4.437786	8.742815
Jogging-w/Hand-Swing	VIGOROUS	7.27138	4.832407	6.033422

Regularity Index (RI) of ADL, which is the correlation coefficient between the hourly AI patterns of day $i-1$ and day i, effectively represents the regularity of PAs on hourly-basis on the day i and is in the range of ± 1. Table 2 lists an example of one-week data for a subject, arranged from noon to noon of next day; hourly AIs are listed in different columns with respect to the dates. RI value of +1 means the ADL pattern of the day i is the same as of day $i-1$; 0 stands for totally uncorrelated, and -1 stands for inversely correlated.

Table 2. A sample of one-week data, illustrating assessment of RI.

Hour\Date	9 May 2018	10 May 2018	11 May 2018	12 May 2018	13 May 2018	14 May 2018	15 May 2018
12:00:00	91.6985	7.9372	52.1803	39.1756	16.1481	120.7643	7.4088
13:00:00	68.0396	155.7581	104.2664	20.2958	190.7752	68.5014	59.4915
14:00:00	60.9835	132.4570	209.2683	47.8527	165.0358	24.0670	79.0501
15:00:00	127.9568	97.1954	152.4025	63.6552	72.7410	17.4707	110.0526
16:00:00	50.0016	37.5953	5.2637	20.1774	34.5341	125.6139	55.4181
17:00:00	52.6829	63.8394	5.0618	40.1613	60.5130	50.6716	35.2950
18:00:00	58.4110	44.1087	15.8595	39.5041	35.8857	16.3036	43.0587
19:00:00	30.3024	11.9919	23.8395	20.0936	34.3655	24.9894	19.2359
20:00:00	32.9313	44.0399	19.5678	19.0029	49.1668	21.5088	42.3071
21:00:00	25.9622	20.2555	32.2234	16.1118	45.0717	18.5637	26.7250
22:00:00	13.7704	16.6188	12.7696	9.2744	26.0573	9.6644	16.1690
23:00:00	8.8460	11.4903	13.3283	9.6019	16.6092	8.3738	10.8001
00:00:00	9.1765	8.0131	11.8799	9.2327	12.3928	7.2355	7.0924
01:00:00	11.7781	13.9831	9.7203	7.5271	18.6474	12.4477	7.9296
02:00:00	7.1743	25.7008	6.9460	10.2525	9.9166	7.7463	15.6859
03:00:00	6.9395	16.4114	8.0269	7.3836	13.6776	7.3144	8.6616
04:00:00	6.9992	18.5755	10.9319	6.9433	13.1182	7.3127	11.0475
05:00:00	6.6685	66.1852	33.5049	10.2133	87.0340	81.8432	69.0056
06:00:00	51.9434	60.5493	53.4563	45.3398	28.0925	201.7091	43.4253
07:00:00	60.3498	146.4305	152.9999	199.2066	72.4182	161.1731	138.3872
08:00:00	146.0198	185.8036	184.6546	132.9641	95.0854	98.3405	239.4251
09:00:00	109.5714	126.9927	105.3715	178.1697	16.5196	96.5270	150.3620
10:00:00	50.1309	97.9686	48.7204	84.0562	42.8646	131.4003	77.6424
11:00:00	53.7380	95.5183	34.1327	74.0714	37.0004	53.0020	78.3121
RI	-	0.7078	0.8386	0.6484	0.1430	0.1105	0.4395

4. Discussions and Conclusions

With the proposed analytical model representing the quantified PAs and different levels of PAs identified in this study, several parameters can be used to evaluate the status of subject with the activity-based information. As shown in Table 3, besides the AI/min. (minute-wised activity index) and day-to-day RI introduced previously, a simple total summation of the AIs over the whole day,

i.e., T_AI, can simply indicate the total amount of activities within the day performed by the subject. The SL_T and SL_Q representing the sleep hours and sleep quality can indicate the amount of time that the subject was asleep or in rest during the day and the quality of the sleep/rest, respectively. T_[level of activity] stands for the total number of hours that a subject can perform certain level of activities. For example, if the information of the duration that a subject performed activities above certain level (including the specific level) is required, then it can be calculated with T_[Level] + T_[Level+1] + ... + T_[4]. The last parameters, A[x]_T_[level of activity], which are of interest, constitute the time duration that a subject performed certain level of activities within certain time window from when the subject was awoke. This may indicate the capability of that subject performing certain level activities within the time window from getting enough rest, i.e., period from when they awoke. For example, in some clinical applications, the capability (evaluated by the time duration) that people can perform activity levels greater than or equal to 3 (moderate level) within 3 h when they awake may be a strong indication representing their lung condition.

Table 3. Parameters used for activity-based monitoring.

Parameters	Meanings
AI/min.	Activity Index/min.—minute-wised activity index
T_AI	Total Activity Index of a Day—summation of all the minute-wised AIs within a day
D_RI	D-to-D Regularity Index—regularity index between the day and the day before
SL_T	Sleep hours—number of hours in sleeping/resting
SL_Q	Sleep quality—average minute-wised AI in sleep duration
T_[Level of Activity]	Hours of Activity [Level]—total time duration performing certain level of activities in a day
A[x]_T_[Level of Activity]	Duration of Activity [Level] within [x] hours after awake—time duration of performing certain level of activities within [x] hours after awake

With these parameters or even by combining these, they could be strong indicators for many medical-care applications by analyzing either the changes of the long-term trends of these parameter or some machine learning algorithms. In this study, analytical models and methods were proposed to perform activity-based monitoring with accelerometer-based wearable devices. A new parameter, activity index (AI), has been introduced and used to categorize different PAs into 5 levels. Another new parameter, regularity index (RI), has been proposed to represent the degree of regularity of ADL. It provides a quantitative measurement of the regularity of living and can be used for many quantified risk assessments of certain diseases. The proposed models and calculations are simple enough to have them implemented into existing accelerometer-based wearable devices. Hence, they are extremely suitable for further applications combining cloud computing services and IoT-based online health monitoring platform, or for monitoring the health condition of a patient discharged from the hospital and predicting their next re-hospitalization by observing varying patterns in ADL.

Author Contributions: Conceptualization—W.L., M.L. and C.L.; Formal analysis—W.L. and V.K.V.; Funding acquisition—C.L.; Investigation—W.L. and M.L.; Methodology—M.L.; Supervision—C.L.; Writing original draft—V.K.V.; Writing review & editing—W.L.

Funding: This work was funded in part by the Chang-Gung Medical Research Project under Grant No. CIRPG5E0012 & CIRPD5E0012, and in part by the Ministry of Science and Technology, Taiwan, R.O.C. under Grant No. MOST 106-2627-E-182-001, and MOST 106-2221-E-182-034, as well as CGU fund BMRPC50 and BMRP138.

Acknowledgments: The authors are thankful to the Biomedical Engineering Research Center (BMERC) of Chang Gung University, Taiwan, R.O.C. for providing the necessary resources required for this work.

Conflicts of Interest: The authors declare no conflict of interest. The founding sponsors had no role in the design of the study; in the collection, analyses, or interpretation of data; in the writing of the manuscript; or in the decision to publish the results.

References

1. Van Hees, V.T.; Sabia, S.; Anderson, K.N.; Denton, S.J.; Oliver, J.; Catt, M.; Abell, J.G.; Kivimäki, M.; Trenell, M.I.; Singh-Manoux, A. A Novel, Open Access Method to Assess Sleep Duration Using a Wrist-Worn Accelerometer. *PLoS ONE* **2015**, *10*, e0142533. [CrossRef] [PubMed]

2. Cole, R.J.; Kripke, D.F.; Gruen, W.; Mullaney, D.J.; Gillin, J.C. Automatic sleep/wake identification from wrist activity. *Am. Sleep Disord. Assoc. Sleep Res. Soc.* **1992**, *15*, 461–469. [CrossRef]

3. Sadeh, A.; Sharkey, M.; Carskadon, M.A. Activity-Based Sleep-Wake Identification: An Empirical Test of Methodological Issues. *Am. Sleep Disord. Assoc. Sleep Res. Soc.* **1994**, *17*, 201–207. [CrossRef]

4. Osman, I.M.; Godden, D.J.; Friend, J.A.; Legge, J.S.; Douglas, J.G. Quality of life and hospital re-admission in patients with chronic obstructive pulmonary disease. *Thorax BM J.* **1997**, *52*, 67–71. [CrossRef]

5. Trost, S.G.; Pate, R.R.; Freedson, P.S.; Sallis, J.F.; Taylor, W.C. Using objective physical activity measures with youth: How many days of monitoring are needed? *Med. Sci. Sports Exerc.* **2000**, *32*, 426. [CrossRef] [PubMed]

6. Steele, B.G.; Holt, L.; Belza, B.; Ferris, S.; Lakshminaryan, S.; Buchner, D.M. Quantitating physical activity in copd using a triaxial accelerometer. *CHEST J.* **2000**, *117*, 1359–1367. [CrossRef]

7. Yang, C.-C.; Hsu, Y.-L. A review of accelerometry-based wearable motion detectors for physical activity monitoring. *Sensors* **2010**, *10*, 7772–7788. [CrossRef] [PubMed]

8. Arif, M.; Kattan, A. Physical activities monitoring using wearable acceleration sensors attached to the body. *PLoS ONE* **2015**, *10*, e0130851. [CrossRef] [PubMed]

9. Lester, J.; Choudhury, T.; Kern, N.; Borriello, G.; Hannaford, B. A hybrid discriminative/generative approach for modeling human activities. In Proceedings of the 19th International Joint Conference on Artificial Intelligence, Edinburgh, UK, 30 July–5 August 2005; pp. 766–772.

10. Aminian, K.; Robert, P.; Buchser, E.; Rutschmann, B.; Hayoz, D.; Depairon, M. Physical activity monitoring based on accelerometry: Validation and comparison with video observation. *Med. Biol. Eng. Comput.* **1999**, *37*, 304–308. [CrossRef] [PubMed]

11. Najafi, B.; Aminian, K.; Paraschiv-Ionescu, A.; Loew, F.; Bula, C.J.; Robert, P. Ambulatory system for human motion analysis using a kinematic sensor: Monitoring of daily physical activity in the elderly. *IEEE Trans. Biomed. Eng.* **2003**, *50*, 711–723. [CrossRef] [PubMed]

12. Gao, L.; Bourke, A.K.; Nelson, J. Evaluation of accelerometer based multi-sensor versus single-sensor activity recognition systems. *Med. Eng. Phys.* **2014**, *36*, 779–785. [CrossRef] [PubMed]

13. Zhu, C.; Sheng, W. Human Daily Activity Recognition in Robotassisted Living Using Multi-sensor Fusion. In Proceedings of the 2009 IEEE International Conference on Robotics and Automation, Kobe, Japan, 12–17 May 2009.

14. Amroun, H.; Ouarti, N.; Ammi, M. Recognition of human activity using Internet of Things in a non-controlled environment. In Proceedings of the 2016 14th International Conference on Control, Automation, Robotics and Vision (ICARCV), Phuket, Thailand, 13–15 November 2016.

15. San-Segundo, R.; Echeverry-Correa, J.D.; Salamea, C.; Pardo, J.M. Human activity monitoring based on hidden Markov models using a smartphone. *IEEE Instrum. Meas. Mag.* **2016**, *19*, 27–31. [CrossRef]

16. Anguita, D.; Ghio, A.; Oneto, L. Energy Efficient Smartphone-Based Activity Recognition using Fixed-Point Arithmetic. *J. Univ. Comput. Sci.* **2013**, *19*, 1295–1314.

17. Bender, C.; Hoffstot, J.C.; Combs, B.T.; Hooshangi, S.; Cappos, J. Measuring the fitness of fitness trackers. In Proceedings of the 2017 IEEE Sensors Applications Symposium (SAS), Glassboro, NJ, USA, 13–15 March 2017.

18. Oviedo, G.R.; Travier, N.; Guerra-Balic, M. Sedentary and Physical Activity Patterns in Adults with Intellectual Disability. *Int. J. Environ. Res. Public Health* **2017**, *14*, 1027. [CrossRef] [PubMed]

19. Dobbins, C.; Rawassizadeh, R. Towards Clustering of Mobile and Smartwatch Accelerometer Data for Physical Activity Recognition. *J. Inform.* **2018**, *5*, 29. [CrossRef]

20. Chowdhury, A.K.; Tjondronegoro, D.; Chandran, V.; Trost, S.G. Physical Activity Recognition Using Posterior-Adapted Class-Based Fusion of Multiaccelerometer Data. *IEEE J. Biomed. Health Inform.* **2018**, *22*, 678–685. [CrossRef] [PubMed]

21. Zeng, N.; Gao, X.; Liu, Y.; Lee, J.E.; Gao, Z. Reliability of Using Motion Sensors to Measure Children's Physical Activity Levels in Exergaming. *J. Clin. Med.* **2018**, *7*, 100. [CrossRef] [PubMed]

22. Kheirkhahan, M.; Mehta, S.; Nath, M.; Wanigatunga, A.A.; Corbett, D.B.; Manini, T.M.; Ranka, S. A Bag-of-Words Approach for Assessing Activities of Daily Living Using Wrist Accelerometer Data. In Proceedings of the IEEE International Conference Bioinformatics and Biomedicine (BIBM), Kansas City, MO, USA, 13–16 November 2017; pp. 678–685.

23. Verma, V.K.; Lin, W.Y.; Lee, M.Y.; Lai, C.S. Levels of Activity Identification & Sleep Duration Detection with a Wrist-Worn Accelerometer-Based Device. In Proceedings of the 2017 39th Annual International Conference of the IEEE Engineering in Medicine and Biology Society (EMBC), Seogwipo, Korea, 11–15 July 2017.

Article

A ΣΔ Closed-Loop Interface for a MEMS Accelerometer with Digital Built-In Self-Test Function

Dongliang Chen [1,2], Xiaowei Liu [1,2,3,*], Liang Yin [1,2], Yinhang Wang [1,2], Zhaohe Shi [1,2] and Guorui Zhang [1,2]

[1] MEMS Center, Harbin Institute of Technology, Harbin 150001, China; zoom_chen@126.com (D.C.); yinliang2003@126.com (L.Y.); 17B921023@stu.hit.edu.cn (Y.W.); smooth_nic@163.com (Z.S.); zgrhit@hotmail.com (G.Z.)
[2] Key Laboratory of Micro-Systems and Micro-Structures Manufacturing, Harbin Institute of Technology, Harbin 150001, China
[3] State Key Laboratory of Urban Water Resource & Environment, Harbin Institute of Technology, Harbin 150001, China
* Correspondence: lxw@hit.edu.cn

Received: 9 August 2018; Accepted: 31 August 2018; Published: 6 September 2018

Abstract: Sigma-delta (ΣΔ) closed-loop operation is the best candidate for realizing the interface circuit of MEMS accelerometers. However, stability and reliability problems are still the main obstacles hindering its further development for high-end applications. In situ self-testing and calibration is an alternative way to solve these problems in the current process condition, and thus, has received a lot of attention in recent years. However, circuit methods for self-testing of ΣΔ closed-loop accelerometers are rarely reported. In this paper, we propose a fifth-order ΣΔ closed-loop interface for a capacitive MEMS accelerometer. The nonlinearity problem of the system is detailed discussed, the source of it is analyzed, and the solutions are given. Furthermore, a built-in self-test (BIST) unit is integrated on-chip for in situ self-testing of the loop distortion. In BIST mode, a digital electrostatic excitation is generated by an on-chip digital resonator, which is also ΣΔ modulated. By single-bit ΣΔ-modulation, the noise and linearity of excitation is effectively improved, and a higher detection level for distortion is easily achieved, as opposed to the physical excitation generated by the motion of laboratory equipment.

Keywords: MEMS accelerometer; electromechanical delta-sigma; built-in self-test; in situ self-testing; digital resonator

1. Introduction

In recent years, electromechanical sigma-delta (EM-ΣΔ) closed-loop MEMS accelerometer has been an active research field, due to its high-performance, inherent digital output, and convenience for post-processing. As the research on the EM-ΣΔ accelerometer has gone in-depth, it has demonstrated competitive performance compared to traditional macro-scale devices [1–9]. Previous research mainly focuses on the enhancement of noise performance [2–4] and the realization of requisite high-order ΣΔ architecture [5–8]. Such an accelerometer, with a noise floor as low as 200 ng/$\sqrt{\text{Hz}}$ and a fifth-order EM-ΣΔ architecture, has already been reported [9].

Although the MEMS accelerometer has rapidly occupied the low-end commercial market with its high cost-performance, there are obstacles that limit its further development toward high-end applications (such as in aerospace and the military). In most of these situations, the accelerometer is often required to work in a harsh environment for a long period of time as a safety critical device. For these applications, the working reliability and performance stability are primary considerations [10]. Since any malfunction will induce disastrous consequences, any drift will be twice augmented by the integration, especially when performing long-term high-speed inertial navigation. However, due to

the relatively large micromechanical manufacturing error, stress variation in material, surface effects of planar process, and fatigue of material [11,12], the MEMS-sensing element has a relatively poor long-term stability with respect to traditional macro devices.

Besides waiting for the evolution of MEMS process, the in situ self-test and self-calibration will provide a promising new point of view on these problems [13] and, thus, attract extensive research attention worldwide [13–28]. Although use of built-in self-test (BIST) units has been a routine technique in most mixed-signal system-on-chip (SoC) design flows [14,15], obstacles are encountered when implanting to an EM-$\Sigma\Delta$ system. This is due to the fact that the measurand of these systems is essentially physical, which creates difficulty with respect to precision using an electrical-only stimulus.

Direct implementation of electrostatic stimulus is only valid in a basic functional test, which aims to diagnose the defective dies in functional test or malfunction in practical usage. Many researchers have proposed diverse functional BIST methods, by incorporating the MEMS structure into a phase-lock loop (PLL) [16], resonator [17], or charge-pump [18] circuit, then, the working state of the circuit will be an indicator of malfunction. A more precise functional BIST is static symmetry testing, which can identify the location of defects by applying an electrostatic force on symmetrically distributed testing electrodes, and observing the output response [19]. However, the efficacy of these methods is limited, and the implementations are too dedicated to be widely adopted.

In order to alleviate the difficulty in ensuring the precision and consistency of direct measurement, some researchers have resorted to indirect methods to realize the performance BIST. A widely adopted indirect test method is a so-called "alternate test", which is first proposed by the engineers from TI Inc. for enhancing the test efficiency of analog ICs [20]. Recently, many researchers have worked on using this method to predict the key performance of MEMS sensors [21–23]. It is based on the principle that the mechanical performance undergoes the same environmental variations as the electrical performance; if the relationship between them can be precisely established, then the mechanical performance can be predicted by electrical test only. However, the process of establishing a precise mapping relationship needs a lot of repetitive work on sample collection and statistical analysis, which is time-consuming and can only be realized in a factory.

Recently, some researchers have proposed a purely algorithmic method for calibrating the output of a 3-axis accelerometer [24–28]. It is based on the principle that in a static state, the vector sum of the 3-axis output should always be equal to the earth's gravity. Based on this principle, a series of uncorrelated static measurements are performed. Then, the problem is shifted to the solving of a series of nonlinear multivariable equations. However, in this method, only linear drift error is taken into consideration, and the process of calibration does not utilize the cooperation of on-chip circuit, and the error inside the sensor still exists.

The EM-$\Sigma\Delta$ technique has many advantages compared to open-loop and analog closed-loop implementation [12,29].

Open-loop is a simple and cost-effective implementation which has been adopted in early designs. However, several drawbacks have been identified: since it is necessary to provide sufficient damping, more Brownian noise is introduced. Furthermore, the design latitude is constrained by the contradictory trade-offs introduced by the sensing element [12,29]. Since each element in the signal chain will add a distortion in final performance, the linearity of it is, thus, relatively poor.

As the research goes in depth, an analog closed-loop architecture has been proposed. As opposed to the open-loop system, the proof mass is well controlled at the equilibrium position by electrostatic feedback force. Thus, the linearity of the system is greatly improved. Since the electrical damping effect is introduced by the feedback force, the design trade-offs in the sensing element are released, and vacuum packaging technique can be used, resulting in a significant reduction in Brownian noise. Moreover, the bandwidth of the system has been significantly expanded by the feedback loop.

Recently, an EM-$\Sigma\Delta$ closed-loop architecture has been proposed, which incorporates the sensing element into a $\Sigma\Delta$ modulation loop. This configuration inherits the merits of analog closed-loop architecture, and has an additional advantage of direct digital output and compact architecture. Due to

the time averaging effect of ΣΔ servo loop, the electrostatic feedback force is linearized, resulting in an improved linearity performance [1–9].

Thus, the EM-ΣΔ closed-loop accelerometer is the most advanced technique for building an interface circuit for accelerometers. However, the aforementioned methods either treat the sensing element as an individual device or treat the whole system as a black box. The BIST method, which is dedicated for this type of system, is rarely reported.

This paper has proposed a fifth-order EM-ΣΔ accelerometer with a digital built-in self-test function. The digital BIST circuitry is dedicated for the in situ dynamic distortion test of the EM-ΣΔ accelerometer. The traditional dynamic test method relies on the sophisticated vibration or shock machine [30], the inherent vibration distortion of which is usually large and limits the precision of the distortion test. The proposed BIST circuitry makes use of the ΣΔ modulated characteristic of the interface circuit. An on-chip 1-bit ΣΔ digital resonator is used to generate electrical excitation. Due to the noise reshaping nature of the ΣΔ loop, the in-band noise and distortion are well suppressed. The 1-bit signal has an inherently good linearity, and alleviates the need for a multi-bit multiplier. Thus, an area-efficient digital excitation source can be easily implemented on chip, as opposed to the difficulty in generation of analog or physical excitation.

This paper is organized as follows: Section 2 describes the system architecture and gives the theoretical analysis of the source of nonlinearity and the trade-offs between performance and stability. Section 3 gives the theoretical analysis and system level design of the BIST function. Section 4 gives implementation details and practical consideration of the proposed system. The experimental results are presented and discussed in Section 5, and the paper ends with conclusions in Section 6.

2. System Description and Topology Analysis

The block diagram of proposed interface system is shown in Figure 1. A capacitive MEMS accelerometer is incorporated in a ΣΔ modulation loop with a third-order electrical integrator, constituting a fifth-order EM-ΣΔ system [5,31]. By time multiplexing technique, capacitance sensing and force feedback could be performed through the same sensing electrode, alternatively. This collocated sensing mechanism will simplify the design of sensing element and reduce higher-order resonance phenomenon [32]. The sensing element is configured as a balanced capacitive bridge with a pair of reference capacitors, thus, a fully differential architecture can be established. In the feed-forward path, there is a charge amplifier with a correlated-double sampling (CDS) function to realize the capacitance detection and reduce the low frequency noise. A phase compensator is inserted between the charge amplifier and electrical loop filter, in order to add some phase lead compensation to insure the loop stability. The BIST function is realized by an on-chip ΣΔ digital resonator. In BIST mode, a single-bit ΣΔ modulated sinusoidal wave will be injected into the digital part of the system under the control of external pin. As will be described in detail next, the output will be a reflection of the total harmonic distortion in the whole loop.

As evident, the electromechanical interface is a hybrid nonlinear feedback system comprised of components in different domains: physical, analog, and digital. Thus, comprehensive consideration should be taken at the beginning of the design stage.

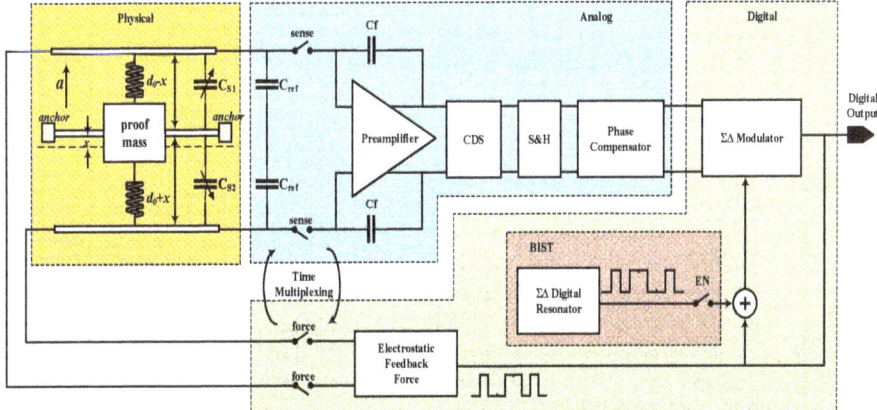

Figure 1. The block diagram of the whole electromechanical ΣΔ interface with built-in self-test (BIST) function.

2.1. Sensing Element

The sensing element is a critical part of the system which affects the stability, determines the sensitivity, and contributes to a major part of noise. A capacitive MEMS accelerometer is chosen as the front-end sensing element for its high output signal, low temperature sensitivity, and ease of applying electrostatic force to establish closed-loop control [33]. As shown in Figure 2, it generally consists of a proof mass suspended by cantilever beams anchored to a fixed frame and accompanied by a couple of fixed plates located on each side.

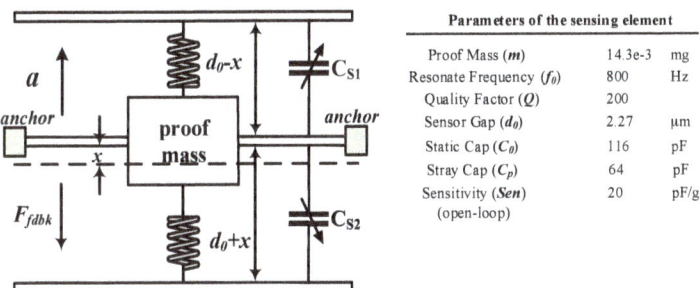

Parameters of the sensing element		
Proof Mass (*m*)	14.3e-3	mg
Resonate Frequency (*f₀*)	800	Hz
Quality Factor (*Q*)	200	
Sensor Gap (*d₀*)	2.27	μm
Static Cap (*C₀*)	116	pF
Stray Cap (*Cₚ*)	64	pF
Sensitivity (*Sen*)	20	pF/g
(open-loop)		

Figure 2. The mechanical model of the sensing element.

By using Newton's second law, the mechanical transfer function can be obtained:

$$H_{ms} = \frac{x}{a} = \frac{1}{s^2 + \frac{b}{m}s + \frac{k}{m}} = \frac{1}{s^2 + \frac{\omega_0}{Q}s + \omega_0^2}, \tag{1}$$

where m is the proof mass, b is the damping factor, k is the spring constant, $\omega_0 = \sqrt{k/m}$ is the resonate frequency, $Q = \sqrt{km}/b$ is the quality factor. In closed-loop configuration, the response of the system is a comprehensive result of sensing element and interface circuit, and thus, the above parameters no longer dominate the system bandwidth, sensitivity, and resonance characteristic, therefore, more latitude can be obtained in choosing the mechanical parameters.

The Brownian noise caused by the motion of gas molecules and suspension beam is the major noise source in the mechanical part [12]. Since it is directly added into the front-end without any

suppression, it will put a fundamental limit on the noise floor achievable and, thus, needs to be carefully treated. The Brownian noise equivalent acceleration (BNEA) can be expressed as:

$$\text{BNEA} = \frac{\sqrt{4k_B T b}}{9.8m} \quad [g/\sqrt{Hz}] , \tag{2}$$

where k_B is the Boltzmann constant, and T is the temperature in Kelvin. From the expression, we can find that a larger m and smaller b will help in reducing the intrinsic noise floor. Thus, most of the high-end MEMS accelerometers, including our design, use bulk micromachining technology and vacuum packaging technique to realize a larger proof mass and a smaller damping factor. The Brownian noise floor, in our design, is 17.4 ng/$\sqrt{Hz}$, calculated by the mechanical parameters used, thus, it is no longer a dominant noise source. However, the price paid here is that the mechanical part exhibits a highly underdamped response with a quality factor as high as 200, which makes insuring stability of the closed-loop a more challenging task. In our design, a phase-lead proportional differentiation (PD) controller is used as a phase compensator, but there are still trade-offs between stability and loop gain, which will be discussed next.

2.2. Electrostatic Feedback Force

For closed-loop operations, the system performance is mainly determined by the feedback path, as long as the loop gain is sufficiently large. In our system, the electrostatic feedback force is the only section in the feedback path, and the electrical BIST stimulus is converted to physical actuation force. Therefore, it has a significant effect on the key parameters of the system, including sensitivity, linearity, and dynamic range.

Consider, when a voltage drop V is imposed on either pair of sensing plates, there will be an attractive electrostatic force between them, which is given by

$$F_{elec} = \frac{C_0 d_0 V^2}{2(d_0 + x)^2}, \tag{3}$$

where C_0 is the static capacitance, d_0 is the initial distance between the plates, and x is the displacement of proof mass at that time. There are two problems faced by force feedback:

- Nonlinearity: the electrostatic force is second-order related to voltage, and is modulated by the displacement x.
- Applying mechanism: the electrostatic force is always attractive, and there is normally no extra electrode for applying it, due to structural limitations, and the existence of high-order resonance mode [32,34].

One way to solve this problem is getting the linear result by subtracting a pair of balanced preload forces which are differentially changed on both sides, and realizing the frequency domain separation by modulating the measurand to high frequency [35–37]. However, there are interactions between sensing and force feedback mode, and if the preload force is above a certain limit, its polarity will reverse, resulting in instability [33,37]. Thus, in our system, we resort to oversampled $\Sigma\Delta$ force feedback to realize the linearization, and time-multiplexing technique to realize the separation of sensing and force feedback mode in the time domain. The applying mechanism of $\Sigma\Delta$ time-multiplexed feedback force is shown in Figure 3.

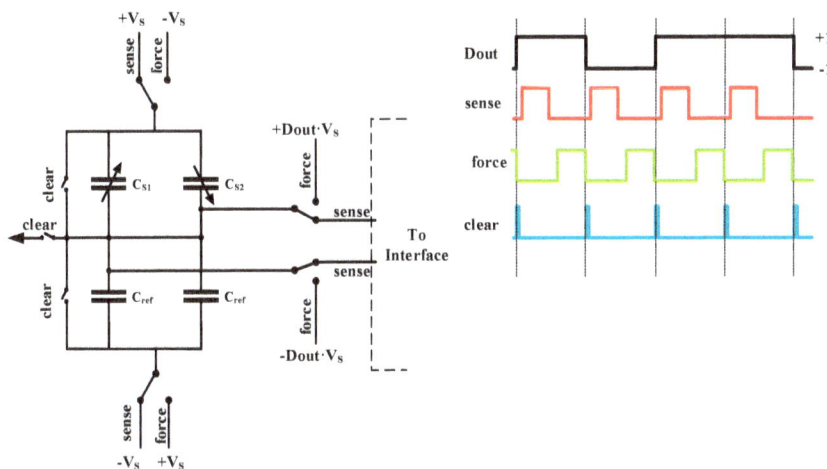

Figure 3. The diagram of ΣΔ time-multiplexed feedback mechanism.

As shown, in each cycle, the front-end capacitive bridge is switching between two working phases: sense and force. In the sense phase, the bridge is biased, with supply voltage $\pm V_S$ and connected to back-end interface circuit. In the force phase, the proof mass is biased to negative supply, and the fixed electrodes are biased to either of the supply rails determined by the digital output signal D_{out}. The two phases are working independently, and a clear phase is inserted to diminish the residue effect from the previous cycle, thus, the interaction effect is minimized. Since the oversampling frequency is well above the bandwidth of sensing element, it can be considered that the two phases are adding together at the same time.

The composite electrostatic feedback force subjected by the proof mass can be expressed as:

$$F_{Feedback} = F_P - F_N = \frac{1}{2}\frac{C_0 d_0(-V_S - D_{out}V_S)^2}{(d_0 + x)^2} - \frac{1}{2}\frac{C_0 d_0(-V_S + D_{out}V_S)^2}{(d_0 - x)^2}, \tag{4}$$

where D_{out} is the digital output of the system which is quantized to ±1, and x is the residue displacement which is filtered by the second-order low pass characteristic of the sensing element. F_P and F_N are the electrostatic force imposed on positive side and negative side, respectively. It should be pointed out that Equation (4) is a general expression of composite electrostatic force which is effective at each sampling cycle. After a rearrangement, Equation (4) can be expressed as

$$F_{Feedback} = \frac{\frac{1}{2}C_0 d_0 V_s^2[-4d_0 x(1 + D_{out}^2) + 4D_{out}(d_0^2 + x^2)]}{(d_0^2 - x^2)^2}. \tag{5}$$

Note that the digital output D_{out} is either 1 or −1, therefore, D_{out}^2 is always equal to 1 at each time point. Moreover, as the filtering characteristic of sensing element, only the low-frequency in-band part of the feedback force is effective, thus, consider the averaged feedback force:

$$\overline{F_{Feedback}} = \frac{\frac{1}{2}C_0 d_0 V_s^2[-8d_0 x + 4\overline{D_{out}}(d_0^2 + x^2)]}{(d_0^2 - x^2)^2} = \frac{F_0[-\frac{2x}{d_0} + \overline{D_{out}}(1 + \frac{x^2}{d_0^2})]}{(1 - \frac{x^2}{d_0^2})^2}, \tag{6}$$

where

$$F_0 = \frac{1}{2}\frac{C_0(2V_s)^2}{d_0}. \tag{7}$$

Equation (6) reveals that the average feedback force $\overline{F_{Feedback}}$ is first-order-related to the in-band output signal $\overline{D_{out}}$, and therefore, the relationship is linearized by a $\Sigma\Delta$ oversampling mechanism. On the other hand, it has to be said that there are first and second order modulation effects of displacement x, which will add additional distortion. If the loop gain is reasonably large, the condition of $x^2 \ll d_0^2$ can hold, and Equation (6) can be rewritten as

$$\overline{F_{Feedback}} \approx \overline{D_{out}} F_0 - \frac{2x}{d_0} F_0. \tag{8}$$

There are only two first order terms: an ideal feedback force term with a linear displacement modulation term. Although the second term will add a feed forward path from displacement to feedback force introducing a gain reduction effect, the resulting system is still a first order system, and therefore, the nonlinear effect is diminished, in principle. Further increasing the loop gain, if the condition $x \ll d_0$ holds, an ideal linear feedback system can be obtained. Thus, increasing the loop gain is the key to solving the distortion problem, but it will be obstructed by the stability problem, especially in a high-order $\Sigma\Delta$ system.

Besides introducing nonlinearity in feedback force, the displacement modulation effect will cause a more detrimental "pull-in" effect.

Consider when an electrostatic voltage is applied on one pair of parallel plates of the sensor, the equilibrium position of the proof mass can be found from the force-balance equation:

$$\frac{C_0 d_0 V^2}{2(d_0 + x)^2} = kx - f_{ext}, \tag{9}$$

where f_{ext} is the external force, which will introduce a zero-voltage gap of $x_0 = d_0 - f_{ext}/k$. When the applied voltage is smaller than the pull-in voltage $V_{pi} = \sqrt{\frac{8kx_0^3}{27C_0d_0}}$ [38], the above function will have two solutions. As the voltage amplitude increases, when it exceeds V_{pi}, no solution will be found and the system will collapse. Thus, the critical value of applied electrostatic voltage is V_{pi}, which corresponds to an equilibrium position of $2x_0/3$. This effect comes from the fact that the electrostatic force is inversely proportional to the squared displacement x, while the elastic force is linearly proportional to the displacement x. As the displacement increases, the electrostatic force will increase faster than the elastic force introduced by cantilever beam, and after a certain limit, the equilibrium state will never establish, and the proof mass will collapse onto one of the static plates. This phenomenon will not only limit the usable range; once it happens, it may cause irreversible structure damage.

The stability problem introduced by pull-in effect is a rather complicated problem, and it should be discussed with regard to different states [39–41].

For static state, the use of the proof mass is servo-controlled at the balanced place, as long as the input amplitude is within the representable range of the $\Sigma\Delta$ system. Thus, the static pull-in is well solved by the closed-loop control mechanism.

As for the dynamic "pull-in", when the voltage changes quickly, the quasi-static regime does not apply. Both the damping forces and mass inertia need to be included in the model [42]. Furthermore, the case of the pull-in due to a step input and the pull-in due to modulated voltage is different, and warrants different treatment [43]. Our system is the modulated voltage case, in which the pull-in trigger point should be calculated from the accumulation effect of a series bits [43]. The dynamic pull-in is hard to be modeled, due to its strong nonlinear characteristics and because of the multiple solutions of the system state [39,44,45]. The modeling of the sensing element and stability analysis technique for multistate nonlinear systems need to be further researched, which is outside the scope of this paper.

2.3. ΣΔ Closed-Loop Interface

As mentioned, incorporating the sensing element in a ΣΔ closed-loop will release the design trade-offs faced by it, but the problem is shifted to the design stage of the back-end interface. Since the filtering ability of the sensing element is always insufficient for suppressing the quantization noise, the use of a high-order electrical filter in the succeeding interface circuit is a must, but will impair the stabilization of the closed-loop further as the quality factor of the front-end is made rather high, in consideration of Brownian noise. This section will be devoted to the design consideration and trade-offs with respect to these issues in the back-end interface circuit.

2.3.1. Performance

The mathematical model of the whole system can be abstracted, as shown in Figure 4. Each element in Figure 1 is expressed by its mathematical function.

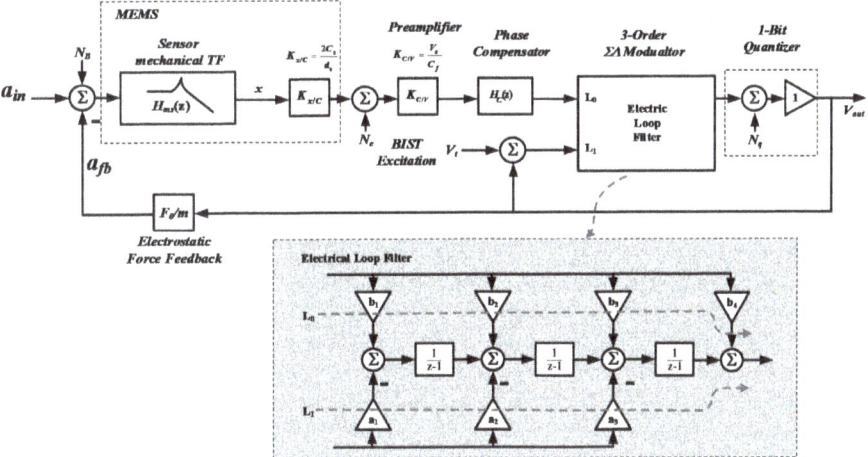

Figure 4. The mathematical abstracted model of the whole system.

The sensing element is expressed with its transfer function $H_{ms}(z)$ in z domain, followed by a gain stage $K_{x/C}$ to translate it to capacitance change. The transformation from $H_{ms}(s)$ in s domain to $H_{ms}(z)$ in z domain, taking time-multiplexed effect into consideration, is derived as detailed in paper [46], and thus, is not discussed in this paper. The only conclusion needed to be cited here is that the use of time-multiplexing only introduces a gain loss and a duty-cycle-related time delay. Besides that $H_{ms}(z)$ still exhibits a second order filtering characteristic. The front-end pre-amplifier is abstracted as a gain stage $K_{C/V}$, and the phase compensator is expressed as its z-form expression $Hc(z)$. To provide a direct and concise viewpoint, the 3-order ΣΔ modulator is expressed by its feed-forward path transfer function L_0, and feedback path transfer function L_1 [47]. For the distributed feedback topology we take, L_0 and L_1 is given by

$$L_0(z) = \sum_{i=1}^{4} b_i \left(\frac{1}{z-1}\right)^{4-i}, \tag{10}$$

$$L_1(z) = -\sum_{i=1}^{3} a_i \left(\frac{1}{z-1}\right)^{3-i}. \tag{11}$$

Then, the signal transfer function $STF_{3-order}$ and noise transfer function $NTF_{3-order}$ of the 3-order electrical $\Sigma\Delta$ modulator can be expressed by

$$STF_{3-order}(z) = \frac{L_0(z)}{1 - L_1(z)}, \tag{12}$$

$$NTF_{3-order}(z) = \frac{1}{1 - L_1(z)}. \tag{13}$$

Before discussing the influence factors of the performance of EM-$\Sigma\Delta$ loop, it should be noted here that the two major sources of performance degradation, quantization noise and residue displacement, undergo different loop processing; thus, they should be discussed separately.

First, consider the quantization noise. After sufficiently suppressing Brownian noise, the quantization noise will be the major limitation of the noise floor achievable. It is added into the loop at the quantizer node, and the noise transfer function NTF_{EM} from quantization noise to output can be derived by analyzing the mathematical model shown in Figure 4:

$$NTF_{EM}(z) = \frac{1}{1 - L_1(z) - L_0(z)G(z)}, \tag{14}$$

$$G(z) = \frac{F_0 K_{x/C} K_{C/V} H_{ms}(z) H_C(z)}{m}, \tag{15}$$

where $G(z)$ is the overall transfer function of the mechanical branch. The denominator of Equation (10) indicates that the quantizer noise is suppressed by a composite filtering effect provided by a 3-order term L_1 and a 5-order term L_0G. However, it should be noted that the overall noise performance is inferior to a 5-order electrical $\Sigma\Delta$ modulator. This is due to the fact that, as opposed to a 2-order electrical integer, the in-band gain of the mechanical branch is flat and limited, moreover, it will be weakened by the use of time-multiplexing technique and phase compensator. Thus, the efficacy of quantization noise suppression is mainly determined by the inside 3-order electrical integer, as shown in Figure 5. Therefore, for our system, in order to achieve sub-$\mu g/\sqrt{Hz}$ noise performance, a 3-order inside electrical modulator is necessary.

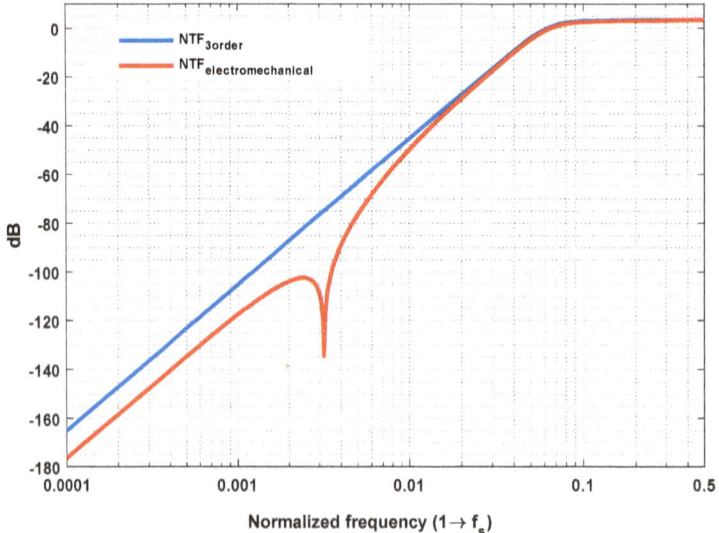

Figure 5. Spectrum of equivalent output quantization noise.

the magnitude and phase of vectors connecting its poles and zeros to point around the unit circle. As evident, the magnitude is minimum at dc, and starts increasing until the frequency reaches $0.5f_S$, resulting a limited variation from 1 α to $1 + \alpha$. Also, there is a positive phase angle φ varying with frequency, which starts from zero at dc, and reaches its maximum when vector $\vec{n}$ is perpendicular to the real axis, and falls to zero again when the frequency reaches $0.5f_S$. Thus, the name phase compensator [9] is derived from the fact that the gain variation is limited and concentrated around dc, but there is a positive phase lead which will exert its influence at crucial frequencies. The bode plot of phase compensator is shown in Figure 7b.

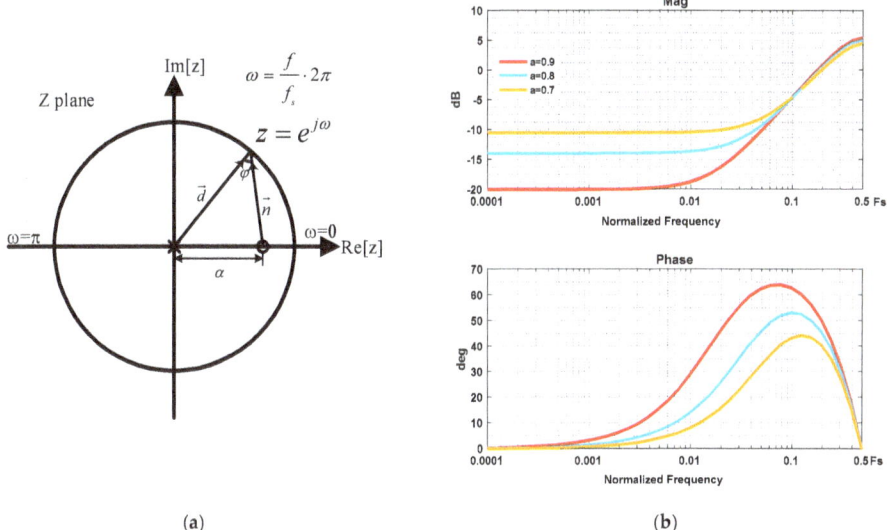

(a)   (b)

Figure 7. The frequency response of phase compensator: (**a**) the poles and zeros distribution; (**b**) the magnitude and phase response.

It can be found that a larger compensation factor will give a better phase compensation result, but at the expense of a larger gain loss in low frequency. An insufficient in-band loop gain means a reduction in quantization noise suppression and a larger residue displacement. Apart from the compensation factor, the sampling frequency is another factor influencing the result. A proper sampling frequency is preferred at which the phase lead is maximum, but it means that the sampling frequency cannot be chosen too high, in which case, the phase lead is not exerting its influence yet, which is also contradictory to performance consideration. Thus, the use of phase compensator is a compromise of performance and stability, the parameter of which should be carefully identified through a number of simulations.

The power spectrum density (PSD) plot of the compensated closed-loop is compared to that of open-loop configuration in Figure 8. As shown, the use of closed-loop configuration has extended the bandwidth, and the resonating peak is flattened by the use of phase compensator.

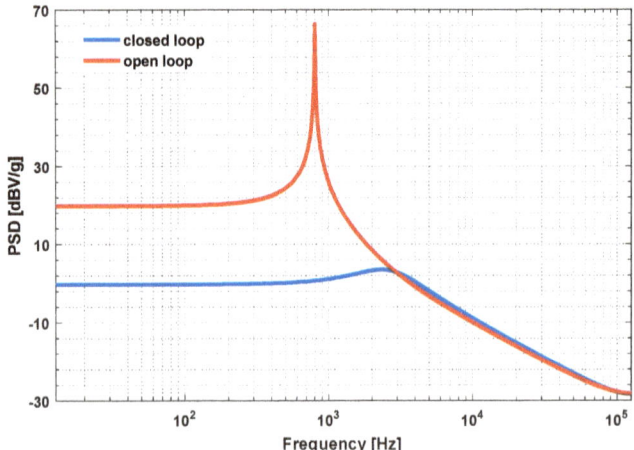

Figure 8. Compensated closed-loop response vs open-loop response.

3. BIST Function

In order to provide a cost-effective way for in situ self-test of harmonic distortion, a purely digital BIST circuitry is embedded in the system. There are two major challenges of designing a BIST circuitry for an accelerometer.

First is selecting the entrance of BIST stimulus. This is due to the fact that the sensing element is part of $\Sigma\Delta$ closed-loop, which could not be disconnected to impose electrostatic force, and extra driving electrode is not available in the general case. Thus, we should find an inner node in the feedback loop to apply the electrostatic force indirectly. The principle of entrance selection is that the applying point and observing point should cross over the sensing element, if not, the observed response will not reflect the characteristic of the sensing element. For example, if the BIST stimulus is directly added into the output node, due to the control of closed-loop, the output voltage will always equal to the input stimulus as long as the loop gain is sufficiently high, irrespective of any change in sensing element. In the proposed method, we choose the feedback point of the 3-order electrical filter as the entrance of BIST stimulus. The signal at this point is 1-bit quantized, thus, the use of digital excitation is convenient.

Consider that the BIST stimulus is a 1-bit $\Sigma\Delta$ modulated signal V_t. Then, by calculating the linear module shown in Figure 4, the transfer function from V_t to V_{out} can be expressed as

$$\frac{V_{out}}{V_t} = \frac{L_1(z)}{1 - L_1(z) - L_0(z)G(z)}, \tag{18}$$

which can be rearranged as

$$\frac{V_{out}}{V_t} = \frac{\left(\frac{L_1(z)}{1-L_1(z)}\right)}{1 - \left(\frac{L_0(z)}{1-L_1(z)}\right)G(z)}. \tag{19}$$

For in-band test signal V_t, the condition $L_1(z) \gg 1$ can hold, and by using Equation (12), the Equation (19) can be rewritten as

$$\frac{V_{out}}{V_t} \approx \frac{1}{1 - G(z)STF_{3-order}(z)} = \frac{1}{1 - G_x(z)} \approx -\frac{1}{G_x(z)}. \tag{20}$$

It can be found that the transfer function from V_t to V_{out} is the reverse of loop gain seen by residue displacement, which is the major determinant of the harmonic distortion of the whole loop. Thus, the value of loop gain and the harmonic distortion level can be determined by performing a single tone BIST test and observing the output response.

The other challenge that should be dealt with is to implement an embedded high-precision digital excitation source cost-effectively. There are multiple implementation methods of the digital excitation source, like direct digital synthesizing (DDS) or fixed-length recording technique [48]. However, the first method requires a large amount of hardware resources, and the second is lack of flexibility in signal control, thus, both of them are not suitable for implementation of on-chip BIST circuitry.

The BIST excitation source in our design is implemented using the $\Sigma\Delta$ resonating circuitry proposed in [49], and its block diagram is shown in Figure 9. The resonator incorporates a $\Sigma\Delta$ modulator in the feedback loop, resulting in concise architecture with an inherent 1-bit $\Sigma\Delta$ modulated output and no need for an area-consuming multibit multiplier. In order to enhance the signal to noise & distortion ratio (SNDR) performance to meet the BIST requirement, the $\Sigma\Delta$ modulator is implemented using 3-order cascode of integrators feedback (CIFB) architecture. The STF of it is set to 1 in order to minimize the introduced phase delay and simplify the selection of loop coefficient a_{12} and a_{21}.

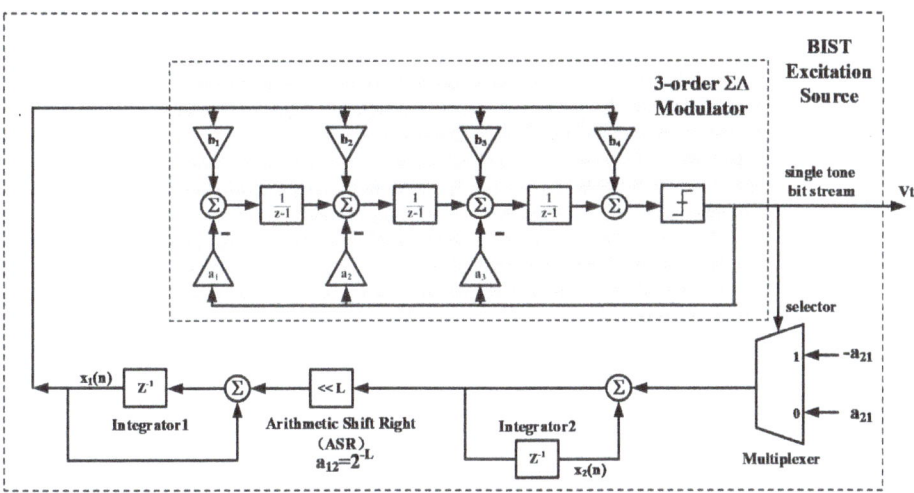

Figure 9. Block diagram of BIST excitation source.

For in-band signal, the loop gain of the resonator can be expressed as

$$G_R(z) = \frac{-a_{12}a_{21}z^{-1}STF_{3-order}}{(1 - z^{-1})^2}. \tag{21}$$

In order to keep a stable oscillation, according to the Barkhausen criterion, $G_R(z)$ should be equal to 1. The characteristic equation of the BIST stimulus generator becomes

$$z^{-2} + (a_{12}a_{21}STF_{3-order} - 2)z^{-1} + 1 = 0. \tag{22}$$

For an in-band signal, $STF_{4-order}$ can be considered as a constant equal to 1. To simplify the control mechanism, we choose $0 < a_{12}a_{21} < 2$, and the resonating frequency can be deduced:

$$\omega_0 = f_S \cos^{-1}(1 - \frac{a_{12}a_{21}}{2}), \tag{23}$$

where $f_S = 1/T_s$ is the sampling frequency of the system. It can be found that the resonating frequency is determined by the product of a_{12} and a_{21}. So far, only the in-band signal is concerned when performing above analysis. However, there is a quantization noise part which will inject additional energy into the loop and make the oscillation unstable. Thus, the value of a_{12} and a_{21} are chosen to be much less than 1, to alleviate this injection. This will limit the resonating frequency to a relatively low value, but is just suitable for a harmonic test.

Besides frequency, the amplitude of oscillation can be controlled by choosing the initial condition of the loop integrator, denoting the values of registers in integer1 and integer2 to be $x_1(n)$ and $x_2(n)$. For simplicity, the initial value $x_2(0)$ is chosen to be zero. The value of $x_1(n)$ at time node n = 0 and n = 1 can be expressed as

$$x_1(0) = A\sin(\phi), \tag{24}$$

$$x_1(1) = A\sin(\omega_0 T_s + \phi). \tag{25}$$

Using one iteration calculated from the model shown in Figure 9, note $x_2(0) = 0$, the relationship of $x_1(0)$ and $x_1(1)$ can be obtained:

$$\frac{x_1(1)}{x_1(0)} = \frac{A\sin(\omega_0 T_s + \phi)}{A\sin(\phi)} = 1 - a_{12}a_{21} \tag{26}$$

Expanding Equation (26), we can obtain

$$\cot(\phi) = \frac{1 - a_{12}a_{21} - \cos\omega_0 T_s}{\sin\omega_0 T_s}. \tag{27}$$

Using the relationship shown in Equation (23), and for oversampling systems, the condition $\omega_0 T_s \ll 1$ can hold, then we can obtain

$$\cot(\phi) = \frac{\cos\omega_0 T_s - 1}{\sin\omega_0 T_s}\bigg|_{\omega_0 T_s \to 0} = 0. \tag{28}$$

Then, the amplitude A and initial phase ϕ of the oscillation can be expressed:

$$\begin{cases} A = x_1(0) \\ \phi = \frac{\pi}{2} \end{cases} \tag{29}$$

In conclusion, the characteristic of BIST stimulus can be controlled by tuning the loop parameters, (e.g., tuning a_{12} and a_{21} for frequency control and tuning $x_1(0)$ for amplitude control). In order to avoid the using of multibit multiplier, a_{12} is set to 2^{-L} and realized by an arithmetic shifter to the right (ASR), resulting in a coarse frequency tuning. The value a_{21} and $x_1(0)$ is restored in on-chip registers that can be selected through a reserved digital interface for realizing a flexible test strategy.

4. Circuit Implementation Details

The schematic of proposed EM-$\Sigma\Delta$ accelerometer is shown in Figure 10. The system model shown in Figure 4 is implemented with switch capacitor (SC) circuit.

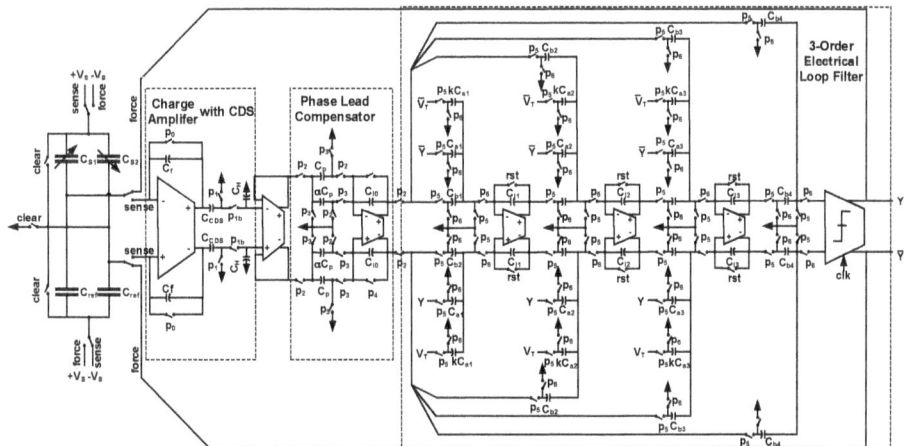

Figure 10. The schematic of proposed EM-ΣΔ accelerometer.

In the first stage, the capacitance change is converted to voltage by a charge amplifier. It has adopted an output correlated double sampling (CDS) technique, thus, low frequency in-band noise and dc offset from front-end amplifier are greatly reduced. The phase lead compensator is implemented using a summing amplifier. By summing the charge corresponding to no-delay in-phase signal, and delayed out-of-phase signal together, the transfer function shown in Equation (17) can be realized. The 3-order electrical loop filter is implemented by also using an SC circuit. In order to prevent the system from locking into saturation state in practical condition, a reset signal is added to clear the integrating capacitors, if needed. The BIST circuitry is implemented with verilog code, and synthesized to layout with electronics design automation (EDA) tools, and thus, is not shown in this schematic. The stimulus V_T generated by BIST circuitry is added into the loop through a replicated feedback path, and thus, the addition with system output is realized indirectly.

In order to remove the low frequency noise and offset in the front-end amplifier, the CDS technique is used. The circuit implementation and clock diagram are shown in Figure 11. At reset phase, each electrode of the sensing element is connected to the ground, in order to diminish the residue effect of previous feedback force. The charge on the feedback capacitor C_f is discharged too. Next, at phase A, the sensing element is charged by a pair of supply voltages, and the differential capacitance change is calculated by the SC circuit and stored on the capacitor C_{CDS}. After that, in phase B, the polarity of the charge voltage of the sensing element is inverted, inducing a negative voltage change at the output of charge amplifier. Before that, C_{CDS} is disconnected from the GND, leaving the node floating. Thus, after phase B, the output results perform a subtraction operation on the capacitor. Note that, the relative polarity of the low frequency noise and offset does not change, thus, they canceled each other out, due to the CDS operation. After that, the voltage at the floating node of C_{CDS} is only related to the measurand, and is sampled on the hold capacitor C_H at the next sampling phase.

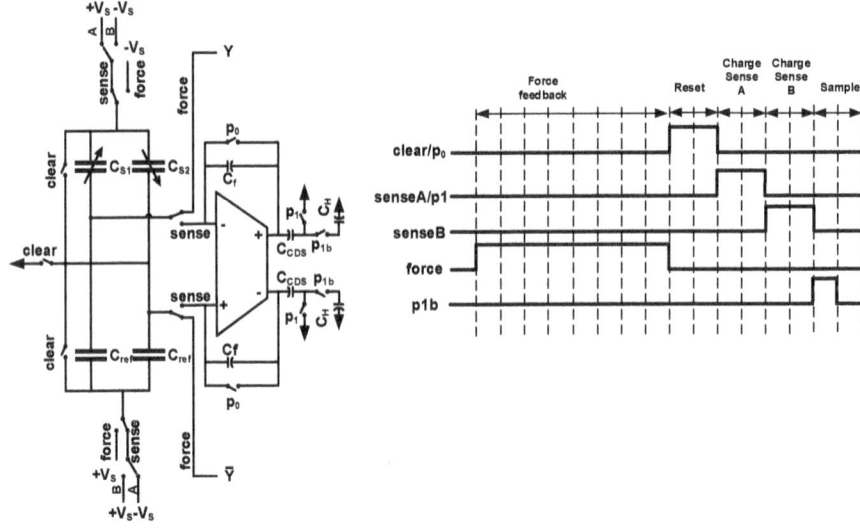

Figure 11. The schematic of front-end charge amplifier with CDS function.

As mentioned, in order to reduce the Brownian noise, the sensing element is sealed in a vacuum package, resulting in a highly underdamped frequency response. When the sensing element is incorporated into a closed-loop, a phase compensator with a transfer function $1 - \alpha z^{-1}$ should be added to provide some phase lead. The transfer function is realized using the SC circuit shown in Figure 12. In each cycle, the charge of previous cycle restored on C_{I0} is first cleared by p_4. In the next phase, p3, the inverse charge of previous cycle restored on αC_p is pushed onto the integrating capacitor C_{I0}, and the charge on C_p is cleared. Then, at phase p2, the input value of current cycle is sampled on C_p and transferred on to C_{I0} at the same time. Thus, at that time, the output value is the sum of the charge restored at previous cycle and the charge sampled at current cycle.

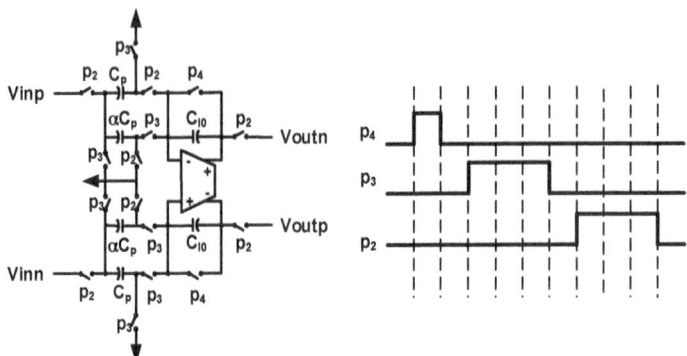

Figure 12. The schematic and timing diagram of phase compensator.

Then, the transfer function of this section can be written as

$$H_c(z) = \frac{C_p}{C_{I0}} - \alpha \frac{C_p}{C_{I0}} z^{-1}. \tag{30}$$

Normally, we choose $C_p = C_{I0}$, then the above function is the same as Equation (17). The ratio, α, of the sampling capacitor will set the compensation depth.

The schematic of the OTA used in proposed system is shown in Figure 13. In order to reduce the harmonic distortion, second stage with class AB output architecture has been used. A regulated cascade current source (M1~M5) is implemented as the tail current of input stage, in order to enhance impedance of it, and hence, the common mode and supply rejection performance. The common mode feedback circuit is using a parallel RC detector with an auxiliary amplifier, in order to enhance the working range and the common mode loop gain. It should be noted here that the parasitic capacitance of the sensing element will be charged and discharged at the same time, which will require additional current output ability and bandwidth margin. Thus, the front-end charge amplifier should be especially powerful and high-speed. Due to the use of two stage class-ab architecture, a 100 mA peak output current and 100 MHz gain bandwidth production (GBW) is easily obtained. The remaining amplifiers in Figure 10 are a scaled version of this architecture.

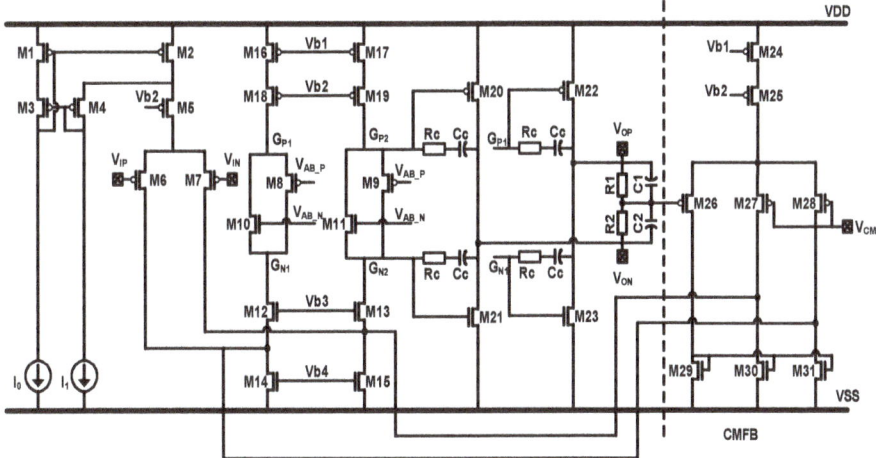

Figure 13. The schematic of proposed EM-$\Sigma\Delta$ accelerometer.

5. Results and Discussion

The complete electromechanical $\Sigma\Delta$ closed-loop interface circuit with BIST function is implemented using 0.35 µm CMOS BCD process. The interface chip contains switch-capacitor circuits as the analog part, and on-chip timing sequence and BIST circuit as the digital part. The reference clock is 2 MHz, which is generated by an off-chip crystal oscillator to obtain a stable timing reference. After on-chip phase adjustment, a sampling clock at 250 kHz is used as the main clock. The total area of the ASIC is 11.2 mm². The area of BIST circuit is 0.86 mm², and only occupies 1/13 of total area. And the power dissipation is 32 mW with a ±2.5 V supply voltage.

The prototype test board is shown in Figure 14. It is composed of a motherboard with supply regulator and control logic, and a daughter board with ASIC and sensing element. The sensing element is sealed in a vacuum ceramic package and connected to the ASIC on a print-circuit-board (PCB) board. On the daughter board, a pair of matching capacitors is needed to build the balanced capacitive bridge with the sensing element, and provide some calibration ability of the imbalance between the two sensing capacitors. The amplitude and frequency of on-chip BIST stimulus can be controlled by the logic signal generated on the motherboard, which will select one of the pre-load voltages stored in the registers, whereby the connector on the PCB and the reserved interface in the ASIC.

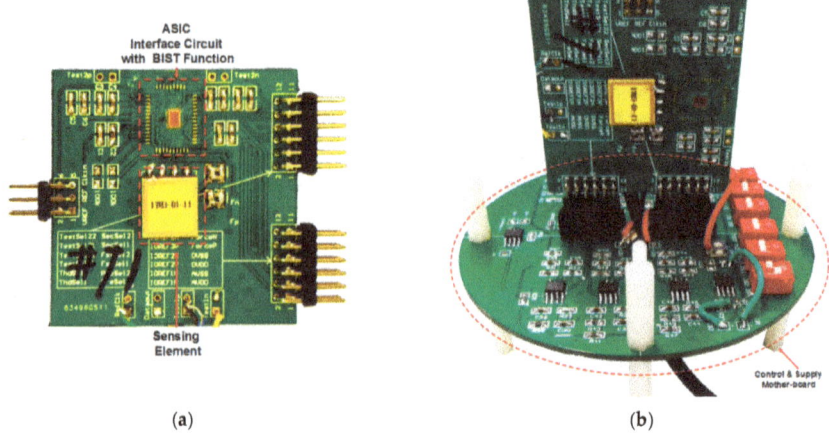

Figure 14. The photograph of prototype test board. (**a**) The daughter board with ASIC and sensing element; (**b**) The motherboard with the daughter board mounted.

The output bit stream of the system and BIST stimulus are brought out through shielded wires and captured by a logic analyzer. The calculated PSD of the output noise spectrum at static 0 g condition is shown in Figure 15. The PSD is normalized to a full-scale voltage of 2.5 V. The noise floor of the system is about -125 dB, which is equivalent to an input acceleration noise of 1.4 µg/$\sqrt{\text{Hz}}$ (with a measured sensitivity of 1.04 V/g). From Figure 13, an evident resonating peak due to the underdamping characteristic of the sensing element can be found. Before that frequency, the spectrum shows a fifth-order noise-shaping characteristic, and after that frequency, the noise-shaping characteristic degrades to a three-order one, which is consistent with the aforementioned analysis.

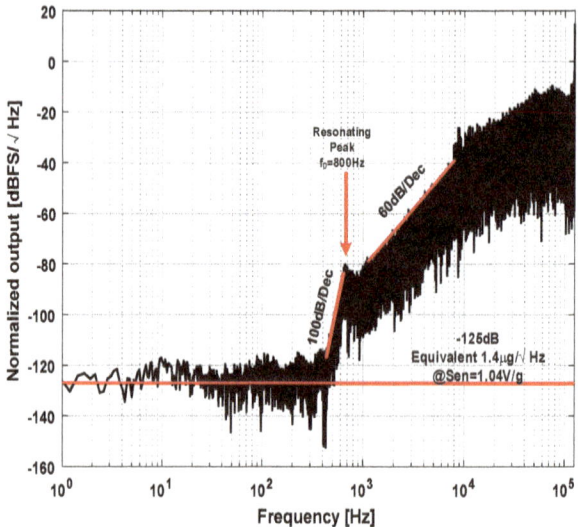

Figure 15. Normalized output noise spectrum at 0 g condition.

The static sensitivity and linearity of the system is identified by a series of static tests at different orientation in gravitational field. The prototype board is perpendicularly mounted on a dividing head, and the input gravitational acceleration can be calculated by the rotation angle readings. The static sensitivity and linearity test result is shown in Figure 16. As shown, the sensitivity of the closed-loop system is 1.04 V/g, and the static linearity is 0.708%.

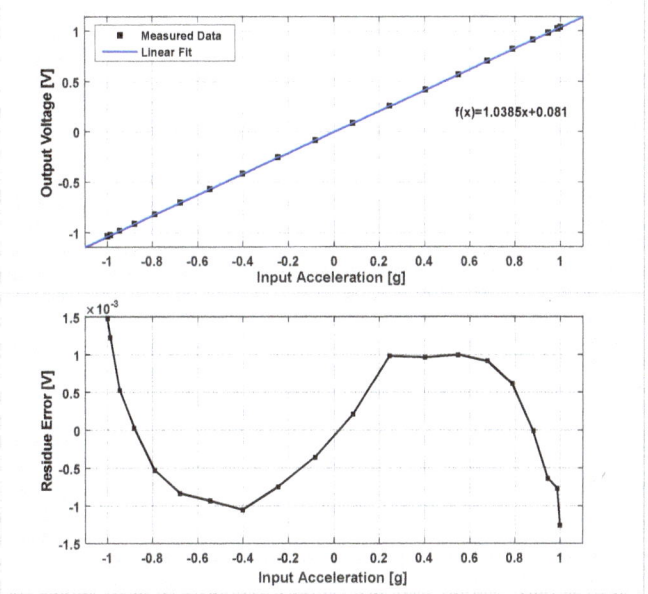

Figure 16. Static sensitivity and linearity results.

The harmonic distortion test is fulfilled by the on-chip BIST function. The sensing element is oriented at 0 g condition. A 1-bit ΣΔ modulated sinusoidal wave is generated by on-chip BIST resonator and injected into the loop to excite the proof mass to a constant oscillation. The amplitude and frequency of the BIST stimulus is set by the logic control signal generated on the mother board. Both the BIST stimulus and induced output bitstream is captured by a logic analyzer. Their calculated PSD is shown in Figure 17. It can be found that the noise and harmonic distortion of the BIST stimulus is well controlled by digital ΣΔ modulation technique. Furthermore, the noise floor of the output response in BIST test is slightly higher that of static test. This is due to the fact that the in-band quantization noise in the BIST stimulus is still injected into the loop, although most of it is suppressed by ΣΔ modulation. The third order harmonic of the BIST stimulus is -127 dB, while the third order harmonic measured in the output is -71 dB. This means that the loop gain is not sufficiently high, and the residue displacement does not get sufficient suppression, thus resulting in an obvious 3-order harmonic distortion. As shown in Equation (19), the loop gain of the closed feedback loop can be read by the ratio of BIST amplitude and output amplitude. By calculating the power of the signal bins, the amplitude ratio of BIST and output signal can be obtained, then, the loop gain is calculated as 13.95. As shown in Section 2.3, the feedback voltage of the 3-order electrical filter could give some latitude for gain enhancement.

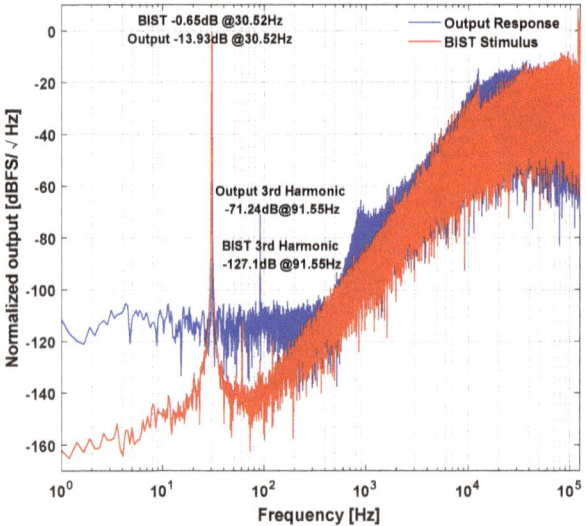

Figure 17. The power spectrum density (PSD) of the BIST stimulus and the output response induced by it.

The feedback voltage of the 3-order electrical filter is preserved for off-chip application. It is generated by the voltage regulators on the mother board. A series of BIST tests are carried out under different loop gain by adjusting the feedback voltage. The relationship between the measured loop gain, the harmonic distortion, and the noise floor is shown in Figure 18. It can be seen that as the loop gain increases, the harmonic distortion is effectively suppressed. However, the noise floor of the output will have a tendency to rise up. This is due to the fact that the scale of the feedback voltage will introduce an energy increase in front of the quantizer, and hence, equivalently decrease the loop gain, thus resulting in a relatively low noise suppressing ability. Both of the abovementioned two tendencies determine the minimum detectable harmonic distortion, since, when the noise floor rises close to the harmonic distortion level, the readings will not be accurate representations. As shown, the minimum securely detectable harmonic distortion is about −110 dB.

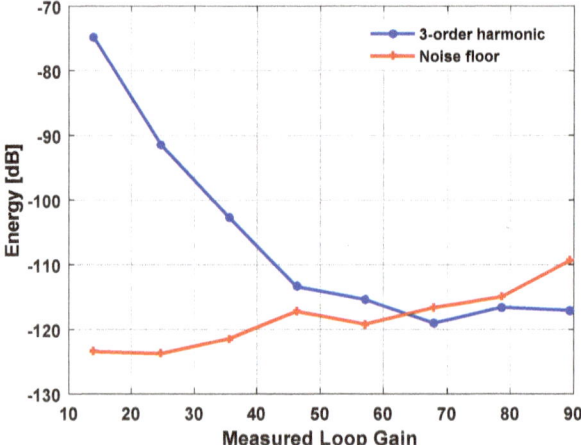

Figure 18. The relationship between the measured loop gain vs 3-order harmonic and output noise floor.

6. Conclusions

This paper has presented the design of a high-order electromechanical $\Sigma\Delta$ interface chip for a high-Q capacitive MEMS accelerometer. Most of our attention has been put on the loop nonlinearity. The source of nonlinearity is analyzed in detail, and we point out that after the linearizing effect provided by 1-bit $\Sigma\Delta$ modulation, the main source of nonlinearity comes from the residue displacement modulation effect. As we analyzed, this problem comes from the fact that the loop gain seen by the residue displacement is different with that seen by the quantization noise, and it is far less than the latter. We point out that enhancing of the loop gain is the key to solving this problem, but this compromises loop stability. Furthermore, the chip has integrated a digital BIST function aimed for on-chip dynamic non-linearity analysis. The proposed BIST method utilizes the $\Sigma\Delta$ nature of the interface. An electrical-only single bit signal is used as the BIST stimulus, which is $\Sigma\Delta$ modulated too, thus, the noise and linearity performance of the stimulus is easily assured. The use of 1-bit signal has also alleviated the need for a digital multiplier, resulting in area-efficient implementation, satisfying the requirement of an on-chip harmonic test. The BIST results show that the harmonic distortion can be effectively suppressed by enhancing the loop gain when scaling of the feedback voltage, and the minimum detectable 3-order distortion can be as low as -110 dB.

Although we have provided a glimpse of digital BIST method for EM-SD accelerometer, the on-chip test of the critical parameters (such as scale factor and bias) are left unsolved. This is due to the fact that those parameters are absolute values, but the practical relationship between the electrical stimulus and physical correspondent is hard to get, and will change with the environment variation. Whereas the harmonic distortion test is a relative test, minor variation in amplitude will not affect the test results. Thus, at present, only harmonic distortion is tested by the proposed BIST method. However, due to the flexibility of the proposed BIST method, different test mechanisms could be exploited. As the test results from various angles are collected, the other parameters may be predicted by a comprehensive analysis. This is a future object of this work.

Author Contributions: X.L. and L.Y. proposed the idea of the design. D.C., Y.W., Z.S., G.Z. fulfilled the detailed design, simulation, layout, and test. D.C. collected the results and wrote the paper.

Funding: This research received no external funding.

Acknowledgments: The authors would like to thank State Key Laboratory of Urban Water Resource and Environment (Harbin Institute of Technology) (No. 2016TS 06) for financial support of this research.

Conflicts of Interest: The authors declare no conflict of interest.

References

1. Zwahlen, P.; Dong, Y.; Nguyen, A.M.; Rudolf, F.; Stauffer, J.M.; Ullah, P.; Ragot, V. Breakthrough in high performance inertial navigation grade sigma-delta MEMS accelerometer. In Proceedings of the 2012 IEEE/ION Position, Location and Navigation Symposium, Myrtle Beach, SC, USA, 23–26 April 2012; pp. 15–19.

2. Petkov, V.P.; Boser, B.E. High-order electromechanical $\Sigma\Delta$ modulation in micromachined inertial sensors. *IEEE Trans. Circuits Syst. I Regul. Pap.* **2006**, *53*, 1016–1022. [CrossRef]

3. Jiangfeng, W.; Carley, L.R. Electromechanical $\Delta\Sigma$ modulation with high-Q micromechanical accelerometers and pulse density modulated force feedback. *IEEE Trans. Circuits Syst. I Regul. Pap.* **2006**, *53*, 274–287. [CrossRef]

4. Kulah, H.; Chae, J.; Yazdi, N.; Najafi, K. Noise analysis and characterization of a sigma-delta capacitive microaccelerometer. *IEEE J. Solid State Circuits* **2006**, *41*, 352–361. [CrossRef]

5. Chen, F.; Li, X.; Kraft, M. Electromechanical sigma delta modulators force feedback interfaces for capacitive MEMS inertial sensors: A review. *IEEE Sens. J.* **2016**, *16*, 6476–6495. [CrossRef]

6. Zhao, M.; Chen, Z.; Lu, W.; Zhang, Y.; Niu, Y.; Chen, G. A high-voltage closed-loop SC interface for a ±50 g capacitive micro-accelerometer with 112.4 dB dynamic range. *IEEE Trans. Circuits Syst. I Regul. Pap.* **2017**, *64*, 1328–1341.

7. Chen, F.; Chang, H.; Yuan, W.; Wilcock, R.; Kraft, M. Parameter optimization for a high-order band-pass continuous-time sigma-delta modulator MEMS gyroscope using a genetic algorithm approach. *J. Micromech. Microeng.* **2012**, *22*, 105006. [CrossRef]

8. Raman, J.; Rombouts, P.; Weyten, L. An unconstrained architecture for systematic design of higher order ΣΔ force-feedback loops. *IEEE Trans. Circuits Syst. I Regul. Pap.* **2008**, *55*, 1601–1614. [CrossRef]

9. Xu, H.; Liu, X.; Yin, L. A closed-loop ΣΔ interface for a high-Q micromechanical capacitive accelerometer with 200 Ng/$\sqrt{Hz}$ input noise density. *IEEE J. Solid State Circuits* **2015**, *50*, 2101–2112. [CrossRef]

10. Stauffer, J.M.; Dietrich, O.; Dutoit, B. RS9000, a novel MEMS accelerometer family for mil/aerospace and safety critical applications. In Proceedings of the IEEE/ION Position, Location and Navigation Symposium, Indian Wells, CA, USA, 4–6 May 2010; pp. 1–5.

11. Shkel, A.M. Precision navigation and timing enabled by microtechnology: Are we there yet? In Proceedings of the 2010 IEEE Sensors, Kona, HI, USA, 1–4 November 2010; pp. 5–9.

12. Yazdi, N.; Ayazi, F.; Najafi, K. Micromachined inertial sensors. *Proc. IEEE* **1998**, *86*, 1640–1659. [CrossRef]

13. Lutwak, R. Micro-technology for positioning, navigation, and timing towards PNT everywhere and always. In Proceedings of the 2014 International Symposium on Inertial Sensors and Systems (ISISS), Laguna Beach, CA, USA, 25–26 February 2014; pp. 1–4.

14. Hung, S.F.; Hong, H.C. A fully integrated BIST ΔΣ ADC using the in-phase and quadrature waves fitting procedure. *IEEE Trans. Instrum. Meas.* **2014**, *63*, 2750–2760. [CrossRef]

15. Hong, H.C.; Su, F.Y.; Hung, S.F. A fully integrated built-in self-test Σ-Δ ADC based on the modified controlled sine-wave fitting procedure. *IEEE Trans. Instrum. Meas.* **2010**, *59*, 2334–2344. [CrossRef]

16. Dianat, A.; Attaran, A.; Rashidzadeh, R.; Muscedere, R. Resonant-based test method for MEMS devices. In Proceedings of the 2014 21st IEEE International Conference on Electronics, Circuits and Systems (ICECS), Marseille, France, 7–10 December 2014; pp. 423–426.

17. Dianat, A.; Attaran, A.; Rashidzadeh, R. Test method for capacitive MEMS devices utilizing pierce oscillator. In Proceedings of the 2015 IEEE International Symposium on Circuits and Systems (ISCAS), Lisbon, Portugal, 24–27 May 2015; pp. 633–636.

18. Basith, I.I.; Kandalaft, N.; Rashidzadeh, R.; Ahmadi, M. Charge-controlled readout and BIST circuit for MEMS sensors. *IEEE Trans. Comput. Aided Des. Integr. Circuits Syst.* **2013**, *32*, 433–441. [CrossRef]

19. Deb, N.; Blanton, R.D. Built-in self-test of MEMS accelerometers. *J. Microelectromech. Syst.* **2006**, *15*, 52–68. [CrossRef]

20. Variyam, P.N.; Cherubal, S.; Chatterjee, A. Prediction of analog performance parameters using fast transient testing. *IEEE Trans. Comput. Aided Des. Integr. Circuits Syst.* **2002**, *21*, 349–361. [CrossRef]

21. Natarajan, V.; Bhattacharya, S.; Chatterjee, A. Alternate Electrical Tests for Extracting Mechanical Parameters of Mems Accelerometer Sensors. In Proceedings of the 24th IEEE VLSI Test Symposium, Berkeley, CA, USA, 30 April–4 May 2006; pp. 6–12.

22. Dumas, N.; Azais, F.; Mailly, F.; Nouet, P. A method for electrical calibration of MEMS accelerometers through multivariate regression. In Proceedings of the 2009 IEEE 15th International Mixed-Signals, Sensors, and Systems Test Workshop, Scottsdale, AZ, USA, 10–12 June 2009; pp. 1–6.

23. Ozel, M.K.; Cheperak, M.; Dar, T.; Kiaei, S.; Bakkaloglu, B.; Ozev, S. An electrical-stimulus-only BIST IC for capacitive MEMS accelerometer sensitivity characterization. *IEEE Sens. J.* **2017**, *17*, 695–708. [CrossRef]

24. Glueck, M.; Oshinubi, D.; Schopp, P.; Manoli, Y. Real-time autocalibration of MEMS accelerometers. *IEEE Trans. Instrum. Meas.* **2014**, *63*, 96–105. [CrossRef]

25. Frosio, I.; Pedersini, F.; Borghese, N.A. Autocalibration of triaxial MEMS accelerometers with automatic sensor model selection. *IEEE Sens. J.* **2012**, *12*, 2100–2108. [CrossRef]

26. Glueck, M.; Buhmann, A.; Manoli, Y. Autocalibration of MEMS accelerometers. In Proceedings of the 2012 IEEE International Instrumentation and Measurement Technology Conference Proceedings, Graz, Austria, 13–16 May 2012; pp. 1788–1793.

27. Rohac, J.; Sipos, M.; Simanek, J. Calibration of low-cost triaxial inertial sensors. *IEEE Instrum. Meas. Mag.* **2015**, *18*, 32–38. [CrossRef]

28. Ye, L.; Guo, Y.; Su, S.W. An efficient autocalibration method for triaxial accelerometer. *IEEE Trans. Instrum. Meas.* **2017**, *66*, 2380–2390. [CrossRef]

29. Zwahlen, P.; Balmain, D.; Habibi, S.; Etter, P.; Rudolf, F.; Brisson, R.; Ullah, P.; Ragot, V. In Open-loop and closed-loop high-end accelerometer platforms for high demanding applications. In Proceedings of the 2016 IEEE/ION Position, Location and Navigation Symposium (PLANS), Savannah, GA, USA, 11–14 April 2016; pp. 932–937.

30. Institute of Electrical and Electronics Engineers (IEEE). *IEEE Recommended Practice for Inertial Sensor Test Equipment, Instrumentation, Data Acquisition, and Analysis*; IEEE: Piscataway, NJ, USA, 2013; pp. 1–145.

31. Nessler, S.; Marx, M.; Manoli, Y. A self-test on wafer level for a MEMS gyroscope readout based on $\Delta\Sigma$ modulation. *IEEE Trans. Circuits Syst. I Regul. Pap.* **2018**, *65*, 870–880. [CrossRef]

32. Ezekwe, C.D.; Boser, B.E. Robust compensation of a force-balanced high-Q gyroscope. In Proceedings of the 2008 IEEE Sensors, Lecce, Italy, 26–29 October 2008; pp. 795–798.

33. Kraft, M.; Lewis, C.; Hesketh, T.; Szymkowiak, S. A novel micromachined accelerometer capacitive interface. *Sens. Actuators A Phys.* **1998**, *68*, 466–473. [CrossRef]

34. Balachandran, G.K.; Petkov, V.P.; Mayer, T.; Balslink, T. A 3-Axis gyroscope for electronic stability control with continuous self-test. *IEEE J. Solid State Circuits* **2016**, *51*, 177–186.

35. Dong, X.; Yang, S.; Zhu, J.; En, Y.; Huang, Q. Method of measuring the mismatch of parasitic capacitance in MEMS accelerometer based on regulating electrostatic stiffness. *Micromachines* **2018**, *9*, 128. [CrossRef]

36. Fan, D.; Liu, Y.; Han, F.; Dong, J. Identification and adjustment of the position and attitude for the electrostatic accelerometer's proof mass. *Sens. Actuators A Phys.* **2012**, *187*, 190–193. [CrossRef]

37. Zhou, W.; Yu, H.; Zeng, J.; Peng, B.; Zeng, Z.; He, X.; Liu, Y. Improving the dynamic performance of capacitive micro-accelerometer through electrical damping. *Microsyst. Technol.* **2016**, *22*, 2961–2969. [CrossRef]

38. Seeger, J.I.; Crary, S.B. In Stabilization of electrostatically actuated mechanical devices. In Proceedings of the International Solid State Sensors and Actuators Conference (Transducers '97), Chicago, IL, USA, 19 June 1997; pp. 1133–1136.

39. Bechtold, T.; Feng, L.; Schrag, G. *System-Level Modeling of MEMS*, 1st ed.; John Wiley & Sons: Weinheim, Germany, 2013; pp. 154–165.

40. Seeger, J.I.; Boser, B.E. Charge control of parallel-plate, electrostatic actuators and the tip-in instability. *J. Microelectromech. Syst.* **2003**, *12*, 656–671. [CrossRef]

41. Fargas-Marques, A.; Casals-Terre, J.; Shkel, A.M. Resonant pull-in condition in parallel-plate electrostatic actuators. *J. Microelectromech. Syst.* **2007**, *16*, 1044–1053. [CrossRef]

42. Rocha, L.A.; Cretu, E.; Wolffenbuttel, R.F. Using dynamic voltage drive in a parallel-plate electrostatic actuator for full-gap travel range and positioning. *J. Microelectromech. Syst.* **2006**, *15*, 69–83. [CrossRef]

43. Nielson, G.N.; Barbastathis, G. Dynamic pull-in of parallel-plate and torsional electrostatic MEMS actuators. *J. Microelectromech. Syst.* **2006**, *15*, 811–821. [CrossRef]

44. Wolfram, H.; Dotzel, W. Stability analysis of a MEMS acceleration sensor. In Proceedings of the 2006 International Conference on Applied Electronics, Pilsen, Czech Republic, 6–7 September 2006.

45. Veijola, T. Equivalent Circuit Models for Micromechanical Inertial Sensors. Available online: https://pdfs. semanticscholar.org/0cf6/7a4357cd4c1635dbdf6bfc116e936474177f.pdf (accessed on 30 August 2018).

46. Xuesong, J. Capacitive Position-Sensing Interface for Micromachined Inertial Sensors. Ph.D. Thesis, University of California, Berkeley, CA, USA, 2003.

47. Veillette, B.R.; Roberts, G.W. On-chip measurement of the jitter transfer function of charge-pump phase-locked loops. *IEEE J. Solid State Circuits* **1998**, *33*, 483–491. [CrossRef]

48. Lu, A.K.; Roberts, G.W.; Johns, D.A. A high-quality analog oscillator using oversampling D/A conversion techniques. *IEEE Trans. Circuits Syst. II Analog. Digit. Signal Process.* **1994**, *41*, 437–444. [CrossRef]

49. Richard Schreier, G.C.T. *Understanding Delta-Sigma Data Converters*, 1st ed.; John Wiley & Sons, Inc.: Hoboken, NJ, USA, 2005; pp. 115–117.

Article

MEMS Inertial Sensors Based Gait Analysis for Rehabilitation Assessment via Multi-Sensor Fusion

Sen Qiu [1,*], Long Liu [1,2], Hongyu Zhao [1], Zhelong Wang [1] and Yongmei Jiang [3]

[1]   School of Control Science and Engineering, Dalian University of Technology, Dalian 116024, China;
     liulong@neusoft.edu.cn (L.L.); zhaohy@dlut.edu.cn (H.Z.); wangzl@dlut.edu.cn (Z.W.)
[2]   Dalian Neusoft University of Information, Dalian 116023, China
[3]   Dalian Medical University, Dalian 116027, China; natalie@mail.dlut.edu.cn
*   Correspondence: qiu@dlut.edu.cn

Received: 31 July 2018; Accepted: 30 August 2018; Published: 3 September 2018

Abstract: Gait and posture are regular activities which are fully controlled by the sensorimotor cortex. In this study, fluctuations of joint angle and asymmetry of foot elevation in human walking stride records are analyzed to assess gait in healthy adults and patients affected with gait disorders. This paper aims to build a low-cost, intelligent and lightweight wearable gait analysis platform based on the emerging body sensor networks, which can be used for rehabilitation assessment of patients with gait impairments. A calibration method for accelerometer and magnetometer was proposed to deal with ubiquitous orthoronal error and magnetic disturbance. Proportional integral controller based complementary filter and error correction of gait parameters have been defined with a multi-sensor data fusion algorithm. The purpose of the current work is to investigate the effectiveness of obtained gait data in differentiating healthy subjects and patients with gait impairments. Preliminary clinical gait experiments results showed that the proposed system can be effective in auxiliary diagnosis and rehabilitation plan formulation compared to existing methods, which indicated that the proposed method has great potential as an auxiliary for medical rehabilitation assessment.

Keywords: MEMS sensors; gait analysis; rehabilitation assessment; body sensor network

---

1. Introduction

Human walking contains important physiology, kinematic and dynamic information. There are many application prospects of human gait analysis in real life, such as monitoring the patient's recovery progress in clinical practice, the control strategy of bionic robots, etc. For instance, stroke is a common neurodegenerative condition with a principal symptom of progressive limbs movement disorder. The prevalence of this neurological disease has been estimated at 120 affected individuals per 100,000 and 75% of the victims suffer from sequelae (mostly involve gait dysfunction) afterwards according to epidemiological statistics [1–3]. An insightful example of gait disorder can be observed in stroke—note that this study is not merely focused on stroke. However, a great majority of subjects who took part in the gait analysis experiment suffer from stroke. Furthermore, stroke patients share most gait symptoms with common neurological disorders in clinical practice. Specifically, stroke is a chronic neurological disorder associated with hemiplegia, lack of balance and abnormal gait. Therefore, objective and accurate inspection of gait parameters (as shown in Table 1) is of great help to a neurologist for appropriate assessment and diagnosis of stroke patients. To this end, the findings of this study could be helpful for revealing the pathogenesis of gait disorders. Stroke normally results in gait asymmetry, which can be reflected in the differences of gait phase partition between two feet due to paresis and imbalance. To sum up, it is crucial to discover the main components of gait disorders. In view of this situation, it is critical to develop an objective and quantitative approach to assess the patients' physical condition. This paper established a wearable gait analysis platform based on a MEMSsensor and body sensor network. The platform can be used to collect acceleration, angular velocity and the geomagnetic signals in the process of walking movement. Accurate gait parameters can be calculated through a sensor data fusion algorithm and error correction process.

Table 1. Typical spatio-temporal gait parameters.

Gait Parameter	Description
Stride length (m)	Distance between two consecutive footprint of the same foot.
Stride speed (m/s)	Stride length divided by walking cycle.
Stride frequency	Number of steps taken per minute during walking.
Walking cycle (s)	Duration of a single stride, inversely proportional to cadence.
Stance time (s)	Duration of stance phase when feet contact with the ground, starting with initial-contact (IC) and ending with foot-off (FO) of the same foot.
Swing time (s)	Duration of swing phase when feet swing above the ground, starting with FO and ending with IC.
Clearance (m)	Foot elevation in swing phase, which reflects the muscular strength of lower limbs and can be diversified as maximum and minimum foot elevation.
Plantar & dorsiflex (degrees)	The angle between the dorsum of the foot and the back of the leg.
Knee ROM (degrees)	Range of knee flexion during a single stride.

With the rapid development of modern medical technology, the concepts of medical treatment have been gradually changed to "Prevention first". In this case, it is quite necessary to conduct the acquisition and processing of health information in advance, so that early medical diagnosis and intervention are feasible. Meanwhile, in order to implement ambulatory monitoring without affecting the subjects' normal physiological activities, the traditional wire communication mode gradually shifts to a wireless mode, which is likely to be more persuasive. Moreover, innovative sensing technology is indispensable for continuous monitoring since the medical monitoring equipment has been developed towards microscale and a long-term span. To this end, the emerging Body Sensor Network (BSN) might serve as a peripheral node of the Internet of Medical Things or even the ubiquitous network.

2. Related Works

Quantitative gait analysis system mainly including a camera system [4–6], electromyography measuring system [7] and force platform [8,9]. The camera system consists of multiple high resolution cameras located in an indoor space, the orientation and position information of the target subject can be calculated using attached highlighting reflective spots. The electromyography measuring system detects human lower limb muscle signals by surface electromyography in the waking process; force platform reflects the change of plantar pressure during walking. However, the applications of above gait analysis system are limited in clinical practice, and the main reasons lie in three aspects. Firstly, the systems are expensive, which might be a barrier to routine use. Secondly, the usage of the systems are complex and require special operation, and it usually takes hours to complete the whole gait measurement process. Finally, specific space is normally needed to perform gait analysis using the above systems. In particular, the camera system may need more than 100 hundred square meters [10–13]. Table 2 lists a brief comparison of mainstream gait analysis methods.

Considerable research has been conducted into the progression of gait dysfunction through the various stages of stroke. Specifically, stroke subjects experience decreased stride length, cadence and walking speed, significant variability in stride length and gait cycle, and Walking imbalances [10,14–16]. Chang et al. [17] employed a specialized wearable system and found that stroke subjects demonstrated decreased gait velocity, stride length and prolonged double support phase. They further identified a high correlation between these gait parameters and age of onset. Further investigation into gait impairments in subjects with neurological diseases has also indicated that the degree of gait abnormality and the disease progression. Previous research has highlighted the advantages of quantitative gait analysis in gait diagnosis; however, laboratory-based systems such as optical tracking and plantar pressure measurement are typically expensive and are not available in ordinary clinical settings [18–20]. Therefore, significant interests have increased rapidly in the development of alternative gait analysis tools.

Table 2. Comparison of mainstream gait analysis method.

Items	Observation Method	Optical System	Inertial Body Sensor Network (BSN)
Objectivity	subjective	objective	objective
Robustness	poor	sensitive to occlusion	very stable
Repeatability	poor	high	high
Efficiency	medium	low	high
Set-up time	several minutes	half-hour	several minutes
Usability	high	low	high
Visual text	no	partial	fully

With the maturity of microelectromechanical systems and the development of information fusion technologies, the application of inertial motion analysis technology is becoming more and more extensive [8,21–25]. Due to the noticeable advantages of small size and low cost, wearable sensors can be mounted directly on the body segment with no need for specified test environment [24,26–28]. Such system may also serve as a good supplement of the gold standard including optical system and plantar pressure monitoring system. In previous studies, we have adopted a wearable inertial sensor in a walking distance calculation and walking pattern classification [29–31]. Ambulatory measurement of the participant's trunk inclination using inertial measurement unit (IMU) was carried out by Farris et al. [3]. Bao et al. [32] developed a smart shoe for gait analysis using force sensitive resistors and IMU sensors. Luinge et al. [33] proposed the estimation of arm orientation by wearable inertial sensors. Dejnabadi et al. [34] introduced an approach to accurate measurement of joint angles based on IMU. However, due to their inability to detect heading reference, inertial based systems generally fail to measure differential orientation, a prerequisite for computing the 3D knee flexion angle recommended by the Internal Society of Biomechanics [35]. Roetenberg et al. [36] developed an ambulatory position and orientation tracking method fusing magnetic and inertial sensing. Since magnetometers measure the strength and direction of the local magnetic field, the geographical north direction can be found. In this case, the initial heading orientation can be obtained with the supplement of a magnetometer. Moreover, the system remains self-contained, which means it does not rely on any external infrastructure [37]. In addition, there are already wireless IMU BSN commercial products such as Trigno Avanti (Delsys Inc., Natick, MA, USA), Mvn Suit (XSens Inc., Enschede, The Netherlands), Perception Neuron (Noitom Inc., Beijing, China) and iSen (STT Systems Inc., San Sebastian, Spain). The current limitations of the state-of-the-art mentioned in the literature are the sensor alignment and integral error. Cost-effectiveness and stability are two other concerns. Moreover, little research about the follow-up monitoring of patients' lower limbs has been carried out. Therefore, the contributions of this paper include the sensor alignment method and the availability of follow-up monitoring of patients' key gait parameters.

The rest of this paper is organized as follows: Section 3 describes the structure of the proposed gait analysis system and the methodology used to estimate the gait parameters during walking; experimental results are given in Section 4; and the potential applications of gait analysis are discussed in Section 5, which concludes the paper as well.

3. Materials and Methods

3.1. System Setup

An ambulatory gait analysis system has been developed based on IMU. We have assembled the IMU from accelerometers/gyroscopes chipsets including ST Microelectronics (Geneva, Switzerland) and Analog Devices (Boston, MA, USA). A multi-sensor fusion algorithm is used to estimate gait parameters. The inertial measurement unit (IMU) can be strapped on both feet, shank and thigh via adjustable elastic straps with hook & loop. We chose straps over housing shells due to their flexible structure, strong adaptability, lower cost and good maneuverability. The installation is quite simple, which can be finished in several minutes. We have designed a fastener with which the sensor nodes can be fixed on the specified location firmly, ensuring the estimation accuracy. No special laboratory is needed, and the gait measurement can be performed just in the corridor or in the ward. The patients can even keep follow-up monitoring in the community after they discharged from the hospital. The performance-to-price ratio is relatively high and it is convenient to automatically generate a gait diagnostic report. Moreover, useful contrastive analysis can be made with repeated gait analysis. The principle and structure of the proposed gait analysis system is shown in Figure 1.

The sensor array includes triaxial accelerometers, gyroscope and magnetometer. Sensor performance specification is shown in Table 3. Raw sensor data is logged at 100 Hz and then forwarded to a receiver via a 2.4 GHz wireless network. The actual linear motion acceleration is used to calculate position. Note that the gravity component ($\mathbf{g} = [0, 0, 9.81]^T \text{m/s}^2$) is normally eliminated from the resultant acceleration signal before estimating the position by integral operation of actual motion acceleration two times.

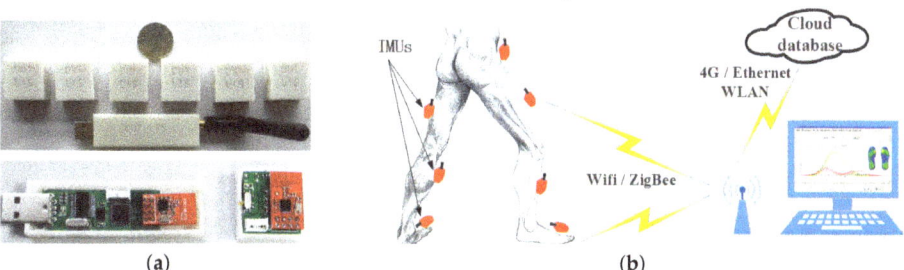

(a)   (b)

Figure 1. The principle and structure of the proposed gait analysis system (**a**) self-made motion tracking sensor nodes; (**b**) gait analysis scenario.

Table 3. Sensor performance specification.

Unit	Accelerometer	Gyroscope	Magnetometer
Dimensions	3 axes	3 axes	3 axes
Dynamic Range	$\pm 50 \text{ m/s}^2$	$\pm 1200 \text{ deg/s}$	$\pm 750 \text{ mGauss}$
Bandwidth (Hz)	30	40	10
Linearity (% of FS)	0.2	0.1	0.2
Bias stability (unit 1σ)	0.02	1	0.1
Alignment Error (deg)	0.1	0.1	0.1

3.2. Accelerometer Non-Orthogonal Error Estimation

In general, the accelerometer orthogonal error is less than 1°, which is normally indicated in the product technical manual. Orthogonal errors exist in three-axis, as shown in Figure 2, and the non-orthogonal error can be expressed as follows:

$$E = \begin{bmatrix} \cos\alpha & 0 & \sin\alpha \\ \sin\beta\cos\gamma & \cos\beta\cos\gamma & \sin\gamma \\ 0 & 0 & 1 \end{bmatrix}. \tag{1}$$

Formula (1) can be simplified using the approximation of the trigonometric function value and then we have the following equation:

$$E \approx \begin{bmatrix} 1 & 0 & \alpha \\ \beta & 1 & \gamma \\ 0 & 0 & 1 \end{bmatrix}. \tag{2}$$

The accelerometer merely senses gravity under stationary state. A relationship exists between accelerometer observations g_s and the true gravitational acceleration $\pm bg$ as follows:

$$\pm bg = E^{-1}g_s = \begin{bmatrix} 1 & 0 & -\alpha \\ -\beta & 1 & \alpha\beta - \gamma \\ 0 & 0 & 1 \end{bmatrix}. \tag{3}$$

Meanwhile,

$$\frac{g_s^T E^T E g_s}{\|g_s\|^2} = 1. \tag{4}$$

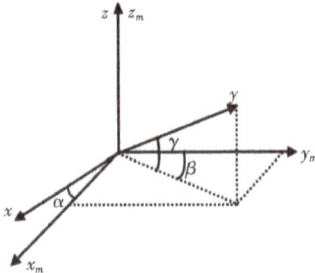

Figure 2. Non-orthogonal angle error of tri-axis accelerometer.

After establishing the non-orthogonal error correction model, the ellipsoid fitting method is introduced to calculate the non-orthogonal error angle. We can get three components of gravity vector $\pm bg$ with respect to three sensitive axes as follows:

$$G_g = \|\pm bg\| [-\sin\theta, \sin\phi\cos\theta, \cos\phi\cos\theta]^T, \tag{5}$$

while the accelerometer observations $G_s = [a_x, a_y, a_z]T$ should satisfy the following equation theoretically:

$$G_s^T G_s = \|\pm bg\|^2, \tag{6}$$

which means the distribution of measured values is a spherical with the radius of $\|g\|^2$; however, actually, the accelerometer measurement distribution is an ellipsoid due to the existence of nonorthogonal errors. Set $O = [a, b, c, f, g, h, p, q, r, d]^T$ and quadric equation can be written as follows:

$$ax^2 + by^2 + cz^2 + 2fyz + 2gxz + 2px + 2qy + 2rz + d = 0. \tag{7}$$

Then, we can get the coefficient matrix from Equation (4) after ellipsoid fitting,

$$\begin{bmatrix} a & d & c \\ d & b & f \\ e & f & c \end{bmatrix} = \frac{E^T E}{\|g_s\|^2}. \tag{8}$$

Define $A = a + b + c$, $B = ab + bc + ac - f^2 - g^2 - h^2$, if $4B - A^2 > 0$. Then, Equation (7) is the description of ellipsoid surface. Select n group of accelerometer measurements, $[x_i, y_i, z_i]$, $i = 1, \cdots, n$, denote $C = [X_1, X_2, \cdots, X_n]$ and $X_i = [x_i^2, y_i^2, z_i^2, 2y_i z_i, 2x_i z_i, 2y_i x_i, 2x_i, 2y_i, 2z_i, 1]^T$, and then the ellipsoid fitting is converted into the following constraints:

$$\begin{cases} 4B - A^2 = 1, \\ min(O^T C^T CO). \end{cases} \tag{9}$$

Define the coefficient matrix as follows:

$$M_0 = \begin{bmatrix} -1 & 1 & 1 & 0 & 0 & 0 \\ 1 & -1 & 1 & 0 & 0 & 0 \\ 1 & 1 & -1 & 0 & 0 & 0 \\ 0 & 0 & 0 & -4 & 0 & 0 \\ 0 & 0 & 0 & 0 & -4 & 0 \\ 0 & 0 & 0 & 0 & 0 & -4 \end{bmatrix}, \tag{10}$$

$$M = \begin{bmatrix} M_0 & 0_{6\times4} \\ \pm b0_{4\times6} & \pm b0_{4\times4} \end{bmatrix}. \tag{11}$$

Lagrangian function can be introduced to convert the constraint problem to the following equation:

$$\begin{cases} O^T MO = 1, \\ C^T CO = \lambda MO. \end{cases} \tag{12}$$

It has been proved that this constraint problem has a unique solution in the field of mathematics [38]. Therefore, the ellipsoid coefficient vector can be determined, and then the ellipsoid radius and the nonorthogonal error angle can be calculated based on the ellipsoid coefficient. Define the symmetrical coefficient matrix and the translation vector S $T = [2p, 2q, 2r]^T$,

$$S = \begin{bmatrix} a & h & g \\ h & b & f \\ g & f & c \end{bmatrix}. \tag{13}$$

Then, the transformation $H' = N^T H + R$ can convert the quadric equation into a standard ellipsoid equation:

$$\frac{x^2}{a'^2} + \frac{y^2}{b'^2} + \frac{z^2}{c'^2} = 1, \tag{14}$$

where N is the eigenvector of matrix S, denote D as the main diagonal matrix of S, and then we have $S = NDN^T$ and $R = -(2D)^{-1}NT$. Ellipsoid radius can be obtained by the following formula:

$$\begin{cases} a' = \sqrt{\frac{-(NF)^T R}{(2-d)D(1,1)}}, \\ b' = \sqrt{\frac{-(NF)^T R}{(2-d)D(2,2)}}, \\ c' = \sqrt{\frac{-(NF)^T R}{(2-d)D(3,3)}}. \end{cases} \tag{15}$$

After calculating the ellipsoid coefficient, we can calculate the nonorthogonal error angle by the analytic method. Equation (3) can be used to compensate the non-orthogonal error of the accelerometer.

3.3. Stance Phase Detection by Decision Level Data Fusion

Sensor drift is an inherent property that results in linear growing integration errors in attitude and position estimation. In particular, position errors grow proportional to the square of the acceleration error. To this end, the widely used Zero Velocity Updating (ZVU) method is adopted in this paper. Though the valid interval of ZVU algorithm is illustrated in literature [31,39–41], the effectiveness of the ZVU technique largely relies on the stance phase detection as shown in Figure 3. This paper takes two criteria to determine stance phase in each gait cycle. These key periods are determined by calculating the squared Euclidean norm of acceleration values, as shown in formula (16):

$$V = \sqrt{(a_x/\|\mathbf{g}\|)^2 + (a_y/\|\mathbf{g}\|)^2 + (a_z/\|\mathbf{g}\|)^2}, \tag{16}$$

where a_x, a_y and a_z represent the triaxial acceleration measurements of foot sensors:

$$M = \frac{1}{N} \sum_{i=j-N}^{i=j} ((S_i - \bar{S}_N)^2, \tag{17}$$

where $\bar{S}_N$ is the mean of S_i over N samples.

Meanwhile, angular rate energy E_{gyro} is adopted as the other criterion (18). The second moment is used to detect stance phase, which is defined in the following formula [39]:

$$E = \frac{1}{\sigma_\omega^2 W} \sum_{i=j}^{j+W-1} \|\omega_i\|^2, \tag{18}$$

$$\hat{R} = \begin{cases} 1, V < \lambda_1 \cap E < \lambda_2, \\ 0, \text{ other value,} \end{cases} \tag{19}$$

where W is the window size selected according to the sensors sampling rate; $\omega_i = [\omega_{x,i}, \omega_{y,i}, \omega_{z,i}]^T$ is the triaxial angular velocity vector; and σ_ω^2 is the gyroscope noise variance. λ_1 and λ_2 are empirically predefined thresholds. The detection results ($\hat{R}$) are sequences consisting of "zero" and "one". The algorithm continually finds the interval when ZVU is valid and updates the corresponding $\mathbf{v}_G(t)$ as $[0,0,0]^T$ based on the two indicators above.

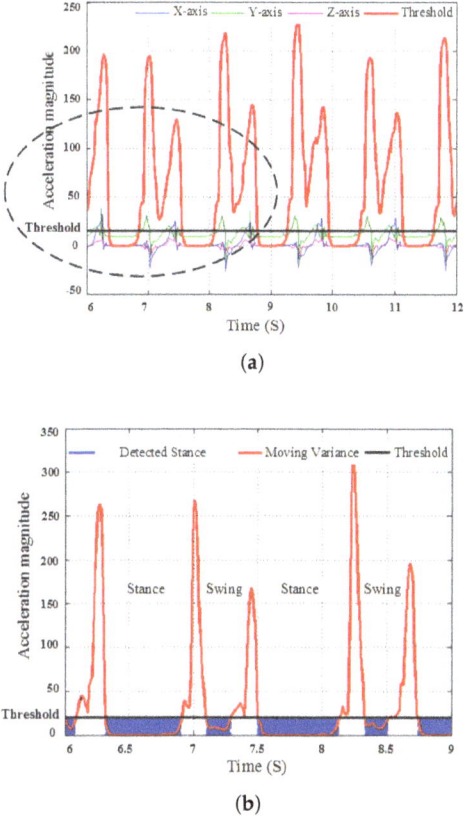

Figure 3. Stance phase detection (**a**) stance phase detection by raw data; (**b**) close-up view.

3.4. Knee Angle Estimation

Knee flexion angle can be calculated by fusing the multi-sensor data. Figure 4a demonstrates the calculation principle and Figure 4b,c show the calculated swing angles and knee flexion based on the gyroscope observations. In addition, the calculation method of knee flexion angle is as follows:

$$\theta_{knee} = \theta_{thigh} - \theta_{shank} + \theta^{ini} = \int (\omega_{thigh} - \omega_{shank}) dt + \theta^{0}_{thigh} - \theta^{0}_{shank},\tag{20}$$

where θ^{0}_{thigh} and θ^{0}_{shank} are initial angles between lower limbs (thigh and shank, respectively) and the gravity direction when the subjects are standing still at ease, which can be calculated by the measurements of the accelerometer as proposed in the previous study [29,30]. ω_{thigh} and ω_{shank} are angular velocity values of thigh fixed and shank fixed sensors.

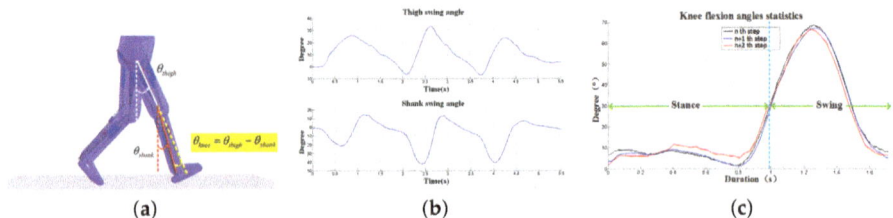

(a)   (b)   (c)

Figure 4. Knee angle estimation results (a) calculation principle of knee; (b) swing angle of thigh and shank; (c) knee flexion flexion statistics.

3.5. Attitude Estimation and Quaternion Correction in IMUs via Sensor Fusion

IMUs refer to sensor modules consisting of three-dimensional accelerometers, gyroscopes and magnetometers (in some cases magnetometers are not included). According to physical and dynamical theory, acceleration measurements can be integrated once to acquire linear velocity and twice to obtain relative position change based on the previous observations. Likewise, angular velocity observations from gyroscopes can be integrated once to estimate the attitude change between two consecutive measurements. In practice, inertial sensors are prone to be disturbed by system noises, drift and measurement errors. All these factors would cause significant integration errors when the raw sensor data are used for integration. In some cases, when the subject moves slowly or stays static, one can merely adopt accelerometers or inclinometers to directly determine the 3D attitude with acceptable results, which avoids the integration operation. However, it is inevitable that external acceleration applied to the accelerometers could ruin the attitude detection based on the calculation of gravity vector components in these orthogonal planes. In most cases, multiple sensor data fusion is necessary to determine 3D attitude. Since the collected data are discrete, we need to perform interpolation between q_m and q_n and the interpolation principle is shown in Figure 5.

$$q_t = q_m + (q_n - q_m) \times t, t \in [0,1]. \tag{21}$$

Simple linear interpolation is valid in some cases. However, it can not effectively describe the curve between q_m and q_n. To ensure that the angle θ between q_m and q_t is linear, i.e., $\theta(t) = (1-t)\theta + t\theta$, we then choose slerp function $slerp(q_m, q_n, t)$ to conduct smooth interpolation of quaternions as follows:

$$slerp(q_m, q_n, t) = \frac{\sin(1-t)\theta}{\sin\theta} q_m + \frac{\sin t\theta}{\sin\theta} q_n. \tag{22}$$

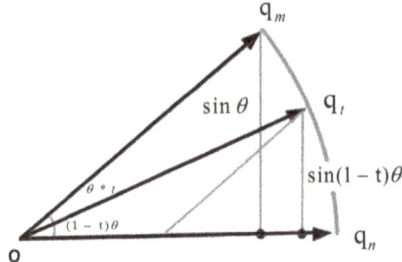

Figure 5. Quaternion interpolation.

Normally standardized operation is necessary:

$$q_t - \frac{q_m + (q_n - q_m) \times t}{\|q_m + t(q_n - q_m)\|}. \tag{23}$$

Moreover, the quaternion number can also avoid the singular point problem of Euler angle representation [27,42], when the pitch angle reaches $\pm 90°$. The three-dimensional attitude represented by quaternions are:

$$\phi = \arctan\left(\frac{2\,(q_2 q_3 + q_0 q_1)}{q_0^2 - q_1^2 - q_2^2 + q_3^2}\right), \tag{24}$$

$$\theta = \arcsin\left(-2\,(q_1 q_3 - q_0 q_2)\right), \tag{25}$$

$$\psi = \arctan\left(\frac{2\,(q_1 q_2 + q_0 q_3)}{q_0^2 + q_1^2 - q_2^2 - q_3^2}\right). \tag{26}$$

The attitude of the updated sensor can be obtained by solving the differential equation of quaternions. The differential equation of quaternions can be expressed as:

$$\begin{bmatrix} \dot{q}_0 \\ \dot{q}_1 \\ \dot{q}_2 \\ \dot{q}_3 \end{bmatrix} = \frac{1}{2} \begin{bmatrix} 0 & -\omega_x & -\omega_y & -\omega_z \\ \omega_x & 0 & \omega_z & -\omega_y \\ \omega_y & -\omega_z & 0 & \omega_x \\ \omega_z & \omega_y & -\omega_x & 0 \end{bmatrix} \begin{bmatrix} q_0 \\ q_1 \\ q_2 \\ q_3 \end{bmatrix}. \tag{27}$$

On the basis of the widely used sensor coordinate transformation, we can get the updated attitude quaternion with inevitable errors. As for a certain vector, its magnitude should be the same though it is expressed in different coordinate frames. The magnitude deviation caused by coordinate transformation can be adopted to adjust the rotation matrix. This paper uses two reference vectors (gravity vector and magnetic vector) to modify the quaternion. In the static state (without linear acceleration), gravity vector $[0,0,1]^T$ is converted into $[c_x, c_y, c_z]^T$ after coordinate transformation, while accelerometer observations are $[a_x, a_y, a_z]^T$. Then, $[c_x, c_y, c_z]^T$ and $[a_x, a_y, a_z]^T$ both represent the gravity vector in the sensor frame. Then, we can obtain error matrix e_g^s by multiplying both vectors:

$$e_g^s = \begin{bmatrix} e_x \\ e_y \\ e_z \end{bmatrix} = \begin{bmatrix} a_x \\ a_y \\ a_z \end{bmatrix} \times \begin{bmatrix} c_x \\ c_y \\ c_z \end{bmatrix} = \begin{bmatrix} a_y * c_z - a_z * c_y \\ a_z * c_z - a_x * c_z \\ a_x * c_y - a_y * c_x \end{bmatrix}. \tag{28}$$

The error matrix can be used to correct the attitude quaternion. Proportional-Integral (PI) feedback adjustment is introduced hereby,

$$\omega_t = \omega_{t-1} + k_p * e_g^s + k_i * e_g^s * \tau, \tag{29}$$

where ω_t is a three-axis angular velocity component, which can be used to correct quaternions combined with skew symmetric matrix. k_p and k_i are proportion coefficient and integral coefficient, respectively. Both parameters are ascertained after repeated experiments on different specific groups.

Moreover, in practice, the measurement error of MEMS magnetometer is non-negligible, the magnetometer measurement error mainly includes environmental interference and inherent error [12,37]. Researchers have conducted various calibration methods using high precision instruments and equipments [38,43,44], however, these methods are time-consuming and the calibration effect largely relies on the precision of equipment. This work uses a calibration method which allows non-experts to easily implement the calibration procedure. To calibrate the magnetometer, sensor nodes were rotated along "Figure of eight knot" trajectory for several times before they were mounted on human lower limbs. Then the outputs were wirelessly collected and be used to perform the magnetometer correction. The correction of magnetometer should be performed near the body segment where the sensor was mounted in an indoor experimental scene [45]. In conclusion, the flowchart of the proposed gait parameters estimation approach can be summarized in Figure 6.

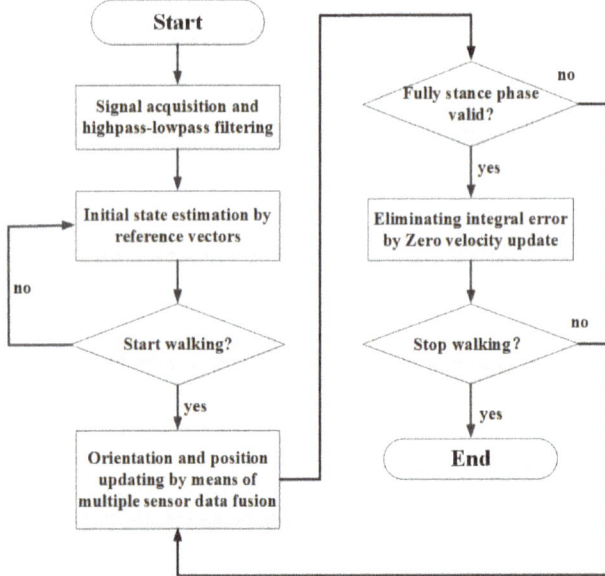

Figure 6. Flowchart of the proposed approach.

4. Experimental Results

The proposed gait analysis system has been adopted to carry out preliminary clinical trials at the Second Affiliated Hospital of Dalian Medical University. Gait data used in this study consist of walking stride time series from 30 healthy adults (22–45 years old), 20 patients with stroke (46–77 years old), and 20 patients with joint disease (30–58 years old). All subjects were instructed to walk continuously on level ground along an obstacle-free corridor for more than 15 m. Note that we have conducted walking trials including U-turns and stair climbing into the evaluation and the proposed method works well. Straight walking is a simplified case suggested by the clinician, and straight walking is mostly adopted in the observational method in clinical practice.

According to the obtained gait parameters in Table 4, the stride lengths, stride speed and feet clearance are relatively low in both neurological and arthropathy patients, which are also consistent with clinical observation. Similarly, mean values of stance time associated with patients are significantly higher than those of healthy subjects. The results from statistical results presented in the table indicate strong evidence of the capability of the typical gait parameters in characterizing the walk of healthy subjects and patients. In this regard, these gait-related symptoms can be explained by clinicians for diagnostic and treatment purposes because the disease progression can be quantificationally monitored via these observations.

Table 4. Gait parameters comparison for healthy subjects and patients. Results are presented as mean (±SD).

Parameter	Healthy	Neurological	Arthropathy
Stride length (m)	1.21 ± 0.13	0.68 ± 0.35	0.73 ± 0.29
Stride speed (m/s)	0.94 ± 0.15	0.71 ± 0.26	0.86 ± 0.21
Stride frequency	92 ± 9	64 ± 22	72 ± 17
Walking cycle (s)	1.32 ± 0.08	1.68 ± 0.13	1.48 ± 0.07
Stance time (s)	0.86 ± 0.05	1.14 ± 0.11	0.99 ± 0.08
Swing time (s)	0.46 ± 0.03	0.54 ± 0.06	0.49 ± 0.04
Clearance (m)	0.22 ± 0.04	0.08 ± 0.07	0.14 ± 0.06
Knee ROM (degrees)	65 ± 9	39 ± 17	46 ± 22

4.1. Knee Flexion Monitoring

Table 5 presents the maximum of joint angle during walking. The knee flexion angle is constantly positive; the positive ankle joint angle represents dorsiflexion and a negative value signifies plantarflexion. In the course of stance phase, the knee flexion angle increases; meanwhile, the ankle dorsiflexion turns to plantarflexion; note that the maximum ankle plantarflexion appears in the final stage of stance phase; in the course of swing phase, knee flexion angles reach the maximum while ankle plantarflexion turns to dorsiflexion, followed by the next stance phase.

Table 5. Joint angle calculation results of a healthy subject and a typical stroke patient. Results are presented as mean (±SD).

Joint Angle °	Heel Strike	Foot Flat	Heel Off	Swing
Knee joint (Healthy subject)	7.2 ± 3.6	19.9 ± 5.1	39.4 ± 6.1	66.2 ± 5.4
Knee joint (Stroke Patient)	9.8 ± 4.5	14 ± 4.2	22.3 ± 5.8	38.1 ± 7.1
Ankle joint (Healthy subject)	8.7 ± 4.4	15.3 ± 4.3	−19.5 ± 6.8	−8.3 ± 4.2
Ankle joint (Stroke Patient)	9.3 ± 4.5	11.4 ± 5.1	−16.9 ± 5.6	−7.2 ± 3.9

Knee flexion range of motion (ROM) is often evaluated using a goniometer in rehabilitation clinics or in hospital wards. The more knee ROM regained during the therapeutic process, the better knee recovery would be affirmed and the sooner early discharge could be guaranteed. In this study, we conducted data collection and knee ROM analysis on 30 healthy subjects, 20 neurological patients (mainly stroke patients) and 20 arthropathy patients, respectively. The first data collection is performed before medical treatments, while the remaining data collections occurred at two weeks and six weeks after treatments, respectively. Figure 7 illustrates the knee ROM recovery status of one typical arthropathy patient who received minimally invasive surgery and one typical stroke patient undergoing rehabilitation training, respectively. The results show that both patients recovered significantly in terms of knee ROM after receiving six weeks of medical treatments. By six weeks after minimally invasive surgery, the knee ROM of arthropathy patient almost returns to the normal range and the gait symmetry is much better when the pains were alleviated.

The one-way analysis of variance (ANOVA) results of knee ROM between bilateral lower limbs are shown in Tables 6 and 7. We can conclude that patients showed large standard deviations in knee ROM, which is a significant feature different from healthy subjects. In this paper, the hypothesis is the symmetrical (balanced) bilateral knee angle ROM, p-value < 0.05 should be interpreted as the hypothesis is true, and the hypothesis is invalid for subjects with p-value > 0.05. Results showed that, as for arthropathy patients, no significant knee ROM difference exists on bilateral lower limbs after six weeks' treatment based on the ANOVA analysis results (p-value = 0.0046); however, the stroke patient still has significant knee ROM different on bilateral lower limbs after six weeks of treatment (p-value = 0.8637). In fact, many stroke patients still have obvious asymmetry between bilateral knee ROM even after several months, though the symptoms may be relieved to a great extent.

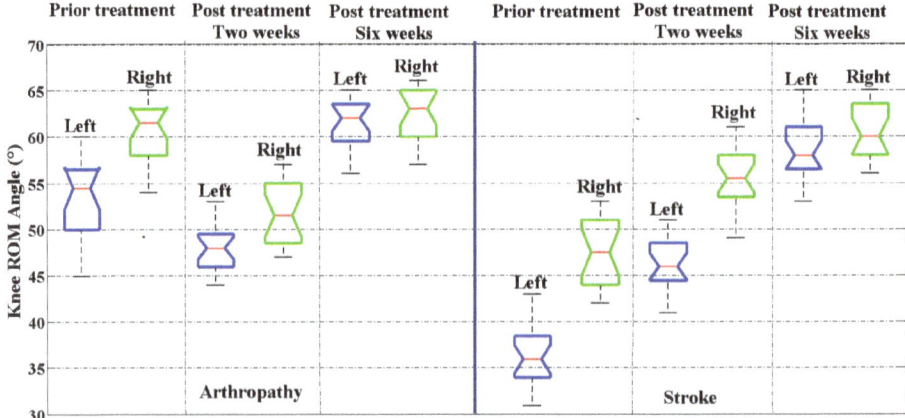

Figure 7. Knee range of motion (ROM) recovery history before and after medical treatments for an arthropathy patient and a stroke patient, respectively.

Table 6. Analysis of variance (ANOVA) table of bilateral knee range of motion (ROM) for an arthropathy patient (SS: Sum of squares of variance; df: Degree of freedom of variance; MS = SS/df; F: F test statistic).

Item	Source	SS	df	MS	F	*p*-Value
	Columns	285.1	19	15.0053	0.39	0.9787
Prior treatment	Error	778	20	38.9		
	Total	1063.1	39			
	Columns	260.275	19	13.6987	1.26	0.3061
Post treatment 2 weeks	Error	217.5	20	10.875		
	Total	477.775	39			
	Columns	227.275	19	11.9618	3.39	0.0046
Post treatment 6 weeks	Error	70.5	20	3.525		
	Total	297.775	39			

Table 7. Analysis of variance (ANOVA) table of bilateral knee range of motion (ROM) for a stroke patient.

Item	Source	SS	df	MS	F	*p*-Value
	Columns	304.28	19	16.0145	0.23	0.9988
Prior treatment	Error	1385.5	20	69.275		
	Total	1689.78	39			
	Columns	203.9	19	10.7316	0.21	0.9993
Post treatment 2 weeks	Error	1006	20	50.3		
	Total	1209.9	39			
	Columns	165.6	19	8.7158	0.6	0.8637
Post treatment 6 weeks	Error	290	20	14.5		
	Total	455.6	39			

4.2. Feet Clearance Monitoring

According to Figure 8a, it is observed that there exists foot elevation asymmetry (left foot: ~15 cm, right foot: ~5 cm) of a stroke patient with hemiplegia symptoms on the right foot, which is consistent with clinical observation, indicating that the patient loses the ability to keep balance. Figure 8b presents the foot elevation asymmetry of an arthropathy patient, which presents the gait walking disorder from the perspective of feet elevation statistics.

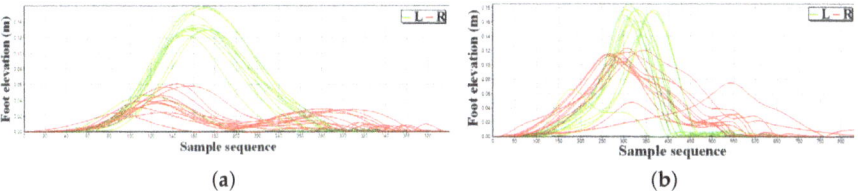

(a) (b)

Figure 8. Gait clearance comparison (**a**) a stroke patient; (**b**) an arthropathy patient.

5. Discussion and Conclusions

This paper aims to provide ambulatory and robust measurements of human gait, and we adopted body-worn sensors to estimate gait parameters. Digitalized and objective gait information can act as desirable guidance for making and adjusting rehabilitation plans, and the results of the preliminary clinical gait analysis experiments have also verified the accuracy of this method for human limbs motion capturing. With no pressure sensor for the stance phase detection and no optical device for integral error elimination, the pure ZVU-aided gait analysis system using body-worn IMU can achieve a good auxiliary diagnosis performance.

Experimental studies have been presented for an Magnetic Angular Rate and Gravity (MARG) unit with reference measurements obtained via a precision optical measurement system, i.e., the NDI Polaris Spectra System (Northern Digital Inc., Waterloo, ON, Canada). The proposed gait analysis system accuracy was validated and the three-dimensional position estimation error is less than 0.015 m, as shown in Figure 9. A comparison experiment with an optical system has demonstrated the accuracy and feasibility of the proposed principle of error correction, and the results of the preliminary clinical gait analysis experiments have also verified the accuracy of this method for human limbs motion capturing.

The current work provides new insights to better understand the biomechanics of walking due to neurological diseases. In addition, they appear to be valuable tools that can highlight differences in gait dynamics with respect to stroke patients. In this regard, measurement of gait may possibly afford pertinent clinical information on neuromotor conditions, characterization of some neurological disorders, and rehabilitation. A better understanding of gait differences based on etiology of amputation or fall history may provide useful information to help guide prosthetic prescription or rehabilitation interventions.

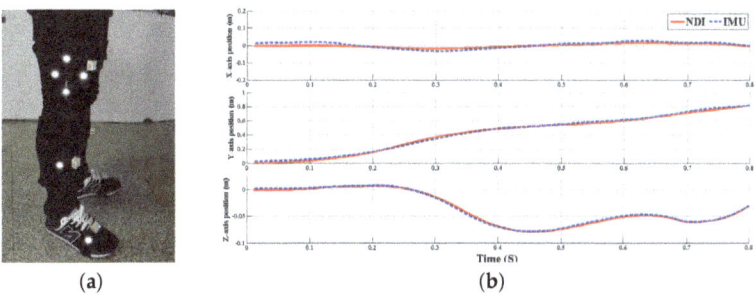

(a) (b)

Figure 9. System accuracy validation by an NDI Polaris Spectra System (**a**) sensor placement and reflection points of optical system; (**b**) error statistics of a three-dimensional thigh sensor position estimation for random lower limbs' movement.

Author Contributions: S.Q. drafted the manuscript and performed the experiment and analysis; L.L. was responsible for the design of the experiment, data analysis and interpretation of the results; H.Z. and Z.W. proposed the idea and were responsible for equipment configuration, data visualization, and manuscript revision; Y.J. organized the clinical trials.

Funding: This research was funded by the National Natural Science Foundation of China under Grants 61803072 and 61873044, and the China Postdoctoral Science Foundation under Grant 2017M621132, and the Fundamental Research Funds for the Central Universities under Grant DUT18RC(4)034, and the Liaoning Key R & D Guidance Project under Grant 2017225078.

Acknowledgments: The authors would like to express their thanks to the members of LIS (laboratory of Intelligent system, Dalian University of Technology).

Conflicts of Interest: The authors declare no conflict of interest.

Abbreviations

The following abbreviations are used in this manuscript:

MEMS	Micro-Electro-Mechanical Sensor
BSN	Body Sensor Network
IMU	Inertial Measurement Unit
MARG	Magnetic Angular Rate and Gravity
ZVU	Zero Velocity Updating

References

1. Albert, M.V.; Azeze, Y.; Courtois, M.; Jayaraman, A. In-lab versus at-home activity recognition in ambulatory subjects with incomplete spinal cord injury. *J. NeuroEng. Rehabil.* **2017**, *14*, 1–6. [CrossRef] [PubMed]
2. Baghdadi, A.; Cavuoto, L.A.; Crassidis, J.L. Hip and Trunk Kinematics Estimation in Gait through Kalman Filter using IMU Data at the Ankle. *IEEE Sens. J.* **2018**, *18*, 4253–4260. [CrossRef]
3. Farris, R.J.; Quintero, H.A.; Murray, S.A.; Member, S.; Ha, K.H.; Hartigan, C.; Goldfarb, M. A Preliminary Assessment of Legged Mobility Provided by a Lower Limb Exoskeleton for Persons With Paraplegia. *IEEE Trans. Neural Syst. Rehabil. Eng.* **2014**, *22*, 482–490. [CrossRef] [PubMed]
4. Zhou, H.; Hu, H. Reducing drifts in the inertial measurements of wrist and elbow positions. *IEEE Trans. Instrum. Meas.* **2010**, *59*, 575–585. [CrossRef]
5. Prakash, C.; Gupta, K.; Mittal, A.; Kumar, R.; Laxmi, V. Passive marker based optical system for gait kinematics for lower extremity. *Procedia Comput. Sci.* **2015**, *45*, 176–185. [CrossRef]
6. Tian, Z.; Fang, X.; Zhou, M.; Li, L. Smartphone-based indoor integrated WiFi/MEMS positioning algorithm in a multi-floor environment. *Micromachines* **2015**, *6*, 347–363. [CrossRef]
7. Park, S.Y.; Lee, S.Y.; Kang, H.C.; Kim, S.M. EMG analysis of lower limb muscle activation pattern during pedaling: Experiments and computer simulations. *Int. J. Precis. Eng. Manuf.* **2012**, *13*, 601–608. [CrossRef]
8. Qiu, S.; Wang, Z.; Zhao, H.; Liu, L.; Jiang, Y.; Li, J.; Fortino, G. Body Sensor Network based Robust Gait Analysis: Toward Clinical and at Home Use. *IEEE Sens. J.* **2018**, *18*, 1–9.
9. Yu, L.; Zheng, J.; Wang, Y.; Song, Z.; Zhan, E. Adaptive method for real-time gait phase detection based on ground contact forces. *Gait Posture* **2015**, *41*, 269–275. [CrossRef] [PubMed]
10. Qiu, S.; Wang, Z.; Zhao, H.; Hu, H. Using Distributed Wearable Sensors to Measure and Evaluate Human Lower Limb Motions. *IEEE Trans. Instrum. Meas.* **2016**, *65*, 939–950. [CrossRef]
11. Brzostowski, K. Toward the Unaided Estimation of Human Walking Speed Based on Sparse Modeling. *IEEE Trans. Instrum. Meas.* **2018**, *67*, 1389–1398. [CrossRef]
12. Qiu, S.; Wang, Z.; Zhao, H.; Liu, L.; Jiang, Y. Using Body-Worn Sensors for Preliminary Rehabilitation Assessment in Stroke Victims with Gait Impairment. *IEEE Access* **2018**, *6*, 31249–31258. [CrossRef]
13. Ren, M.; Guo, H.; Shi, J.; Meng, J. Indoor pedestrian navigation based on conditional random field algorithm. *Micromachines* **2017**, *8*, 320. [CrossRef]
14. Barth, J.; Oberndorfer, C.; Pasluosta, C.; Schülein, S.; Gassner, H.; Reinfelder, S.; Kugler, P.; Schuldhaus, D.; Winkler, J.; Klucken, J.; Eskofier, B.M. Stride segmentation during free walk movements using multi-dimensional subsequence dynamic time warping on inertial sensor data. *Sensors* **2015**, *15*, 6419–6440. [CrossRef] [PubMed]

15. Ahmadi, A.; Destelle, F.; Unzueta, L.; Monaghan, D.S.; Linaza, M.T.; Moran, K.; O'Connor, N.E. 3D human gait reconstruction and monitoring using body-worn inertial sensors and kinematic modeling. *IEEE Sens. J.* **2016**, *16*, 8823–8831. [CrossRef]

16. Anwary, A.R.; Yu, H.; Vassallo, M. Optimal Foot Location for Placing Wearable IMU Sensors and Automatic Feature Extraction for Gait Analysis. *IEEE Sens. J.* **2018**, *18*, 2555–2567. [CrossRef]

17. Chang, H.C.; Hsu, Y.L.; Yang, S.C.; Lin, J.C.; Wu, Z.H. A Wearable Inertial Measurement System With Complementary Filter for Gait Analysis of Patients With Stroke or Parkinson Disease. *IEEE Access* **2016**, *4*, 8442–8453. [CrossRef]

18. Leal-Junior, A.G.; Frizera, A.; Avellar, L.M.; Marques, C.; Pontes, M.J. Polymer Optical Fiber for In-Shoe Monitoring of Ground Reaction Forces during the Gait. *IEEE Sens. J.* **2018**, *18*, 2362–2368. [CrossRef]

19. Jiménez Ruiz, A.R.; Seco Granja, F.; Prieto Honorato, J.C.; Guevara Rosas, J.I. Accurate pedestrian indoor navigation by tightly coupling foot-mounted IMU and RFID measurements. *IEEE Trans. Instrum. Meas.* **2012**, *61*, 178–189. [CrossRef]

20. Lan, K.C.; Shih, W.Y. Using smart-phones and floor plans for indoor location tracking. *IEEE Trans. Hum. Mach. Syst.* **2014**, *44*, 211–221.

21. Fourati, H. Heterogeneous Data Fusion Algorithm for Pedestrian Navigation via Foot-Mounted Inertial Measurement Unit and Complementary Filter. *IEEE Trans. Instru. Meas.* **2015**, *64*, 221–229. [CrossRef]

22. Zhao, H.; Wang, Z.; Qiu, S.; Shen, Y.; Zhang, L.; Tang, K. Heading Drift Reduction for Foot-Mounted Inertial Navigation System via Multi-Sensor Fusion and Dual-Gait Analysis. *IEEE Sens. J.* **2018**, *18*, 1–8.

23. Wang, Z.L.; Zhao, H.Y.; Qiu, S.; Gao, Q. Stance phase detection for ZUPT-aided foot-mounted pedestrian navigation system. *IEEE/ASME Trans. Mechatron.* **2015**, *20*, 3170–3181. [CrossRef]

24. Wu, D.; Wang, Z.; Chen, Y.; Zhao, H. Mixed-kernel based weighted extreme learning machine for inertial sensor based human activity recognition with imbalanced dataset. *Neurocomputing* **2016**, *190*, 35–49. [CrossRef]

25. Fortino, G.; Di Fatta, G.; Pathan, M.; Vasilakos, A.V. Cloud-assisted body area networks: State-of-the-art and future challenges. *Wirel. Netw.* **2014**, *20*, 1925–1938. [CrossRef]

26. Wang, Z.; Zhao, C.; Qiu, S. A system of human vital signs monitoring and activity recognition based on body sensor network. *Sens. Rev.* **2014**, *34*, 42–50. [CrossRef]

27. Qiu, S.; Yang, Y.; Hou, J.; Ji, R. Ambulatory estimation of 3D walking trajectory and knee joint angle using MARG Sensors. In Proceedings of the Fourth International Conference on Innovative Computing Technology (INTECH), London, UK, 13–15 Augsut 2014; pp. 191–196.

28. Gravina, R.; Alinia, P.; Ghasemzadeh, H.; Fortino, G. Multi-Sensor Fusion in Body Sensor Networks: State-of-the-art and research challenges. *Inf. Fusion* **2016**, *35*, 68–80. [CrossRef]

29. Wang, Z.; Qiu, S.; Cao, Z.; Jiang, M. Quantitative assessment of dual gait analysis based on inertial sensors with body sensor network. *Sens. Rev.* **2013**, *33*, 48–56. [CrossRef]

30. Qiu, S.; Wang, Z.; Zhao, H. Heterogeneous data fusion for three-dimensional gait analysis using wearable MARG sensors. *Int. J. Comput. Sci. Eng.* **2017**, *14*, 222–233. [CrossRef]

31. Qiu, S.; Wang, Z.; Zhao, H.; Qin, K.; Li, Z.; Hu, H. Inertial/magnetic sensors based pedestrian dead reckoning by means of multi-sensor fusion. *Inf. Fusion* **2018**, *39*, 108–119. [CrossRef]

32. Bao, S.D.; Meng, X.L.; Xiao, W.; Zhang, Z.Q. Fusion of inertial/magnetic sensor measurements and map information for pedestrian tracking. *Sensors* **2017**, *17*, 340. [CrossRef] [PubMed]

33. Luinge, H.J.; Veltink, P.H.; Baten, C.T.M. Ambulatory measurement of arm orientation. *J. Biomech.* **2007**, *40*, 78–85. [CrossRef] [PubMed]

34. Dejnabadi, H.; Jolles, B.M.; Aminian, K. A new approach to accurate measurement of uniaxial joint angles based on a combination of accelerometers and gyroscopes. *IEEE Trans. Biomed. Eng.* **2005**, *52*, 1478–1484. [CrossRef] [PubMed]

35. Favre, J.; Aissaoui, R.; Jolles, B.M.; de Guise, J.A.; Aminian, K. Functional calibration procedure for 3D knee joint angle description using inertial sensors. *J. Biomech.* **2009**, *42*, 2330–1335. [CrossRef] [PubMed]

36. Roetenberg, D.; Luinge, H.; Slycke, P. *Xsens MVN : Full 6DOF Human Motion Tracking Using Miniature Inertial Sensors*; Xsens Technologies: Enschede, The Netherlands, 2013; pp. 1–9.

37. Wang, Z.; Li, J.; Wang, J.; Zhao, H.; Qiu, S.; Yang, N.; Shi, X. Inertial Sensor-Based Analysis of Equestrian Sports between Beginner and Professional Riders under. *IEEE Trans. Instrum. Meas.* **2018**, *14*, 1–13.

38. Gheorghe, M.V.; Member, S.; Bodea, M.C.; Member, L.S. Calibration Optimization Study for Tilt-Compensated Compasses. *IEEE Trans. Instrum. Meas.* **2018**, *67*, 1486–1494. [CrossRef]

39. Skog, I.; Händel, P.; Nilsson, J.O.; Rantakokko, J. Zero-velocity detection—An algorithm evaluation. *IEEE Trans. Bio-Med. Eng.* **2010**, *57*, 2657–2666. [CrossRef] [PubMed]

40. Bebek, O.; Suster, M.a.; Rajgopal, S.; Fu, M.J.; Huang, X.; Cavusoglu, M.C.; Young, D.J.; Mehregany, M.; Van Den Bogert, A.J.; Mastrangelo, C.H. Personal navigation via shoe mounted inertial measurement units. *IEEE Trans. Instrum. Meas.* **2010**, *59*, 3018–3027. [CrossRef]

41. Martinez-Hernandez, U.; Mahmood, I.; Dehghani-Sanij, A.A. Simultaneous Bayesian Recognition of Locomotion and Gait Phases with Wearable Sensors. *IEEE Sens. J.* **2018**, *18*, 1282–1290. [CrossRef]

42. Gouwanda, D.; Gopalai, A.A.; Khoo, B.H. A Low Cost Alternative to Monitor Human Gait Temporal Parameters-Wearable Wireless Gyroscope. *IEEE Sens. J.* **2016**, *16*, 9029–9035. [CrossRef]

43. Wang, C.; Qu, X.; Zhang, X.; Zhu, W.; Fang, G. A fast calibration method for magnetometer array and the application of ferromagnetic target localization. *IEEE Trans. Instrum. Meas.* **2017**, *66*, 1743–1750. [CrossRef]

44. Vargas-Valencia, L.; Elias, A.; Rocon, E.; Bastos-Filho, T.; Frizera, A. An IMU-to-Body Alignment Method Applied to Human Gait Analysis. *Sensors* **2016**, *16*, 2090. [CrossRef] [PubMed]

45. Choe, N.; Zhao, H.; Qiu, S.; So, Y. A Sensor-to-Segment Calibration Method for Motion Capture system based on low cost mimu. *Measurement* **2018**. [CrossRef]

 *micromachines*

Article

Design of Ensemble Stacked Auto-Encoder for Classification of Horse Gaits with MEMS Inertial Sensor Technology

Jae-Neung Lee, Yeong-Hyeon Byeon and Keun-Chang Kwak *

Department of Control and Instrumentation Engineering, Chosun University, 375 Seosuk-dong,
Gwangju 501-759, Korea; ljn1321@daum.net (J.-N.L.); qasdfghjt@daum.net (Y.-H.B.)
* Correspondence: kwak@chosun.ac.kr; Tel.: +82-62-230-6086

Received: 17 July 2018; Accepted: 12 August 2018; Published: 17 August 2018

Abstract: This paper discusses the classification of horse gaits for self-coaching using an ensemble stacked auto-encoder (ESAE) based on wavelet packets from the motion data of the horse rider. For this purpose, we built an ESAE and used probability values at the end of the softmax classifier. First, we initialized variables such as hidden nodes, weight, and max epoch using the options of the auto-encoder (AE). Second, the ESAE model is trained by feedforward, back propagation, and gradient calculation. Next, the parameters are updated by a gradient descent mechanism as new parameters. Finally, once the error value is satisfied, the algorithm terminates. The experiments were performed to classify horse gaits for self-coaching. We constructed the motion data of a horse rider. For the experiment, an expert horse rider of the national team wore a suit containing 16 inertial sensors based on a wireless network. To improve and quantify the performance of the classification, we used three methods (wavelet packet, statistical value, and ensemble model), as well as cross entropy with mean squared error. The experimental results revealed that the proposed method showed good performance when compared with conventional algorithms such as the support vector machine (SVM).

Keywords: motion analysis; auto-encoder; dance classification; deep learning; self-coaching; wavelet packet; classification of horse gaits

---

1. Introduction

Riding is an action that includes horse riding or modern equestrian dressage. There are various kinds of horse riding styles, such as show jumping, horse therapy, and so forth. Normally, horse riding requires the skills taught by the coach. However, with the development of technology, motion capture technology has developed and might replace the coach's role. Motion capture technology is largely divided into acoustical, mechanical, magnetic, and optical sensor. Speaking of their disadvantages, it is difficult for us to collect precise motion using an acoustical sensor, and their movement is restricted because the mechanical type has to wear heavy equipment. Afterwards, optical equipment requires expensive equipment and has a large influence on ambient lighting. Finally, sensors based on magnetic sensors are also sensitive to iron, but horse riding is not closely related to iron. For that reason, magnetic sensors were used.

Normally, horse riding is not taught one-on-one. That is to say, compared with other popularization sports (swimming, badminton, tennis, health), it is very costly. Therefore, horse-riding teaching is acknowledged as an aristocratic sport. For the sake of reducing the education cost of horse riding, a self-coaching system has been designed. In order to do self-coaching system research, the gaits of a horse are classified into different parts: walk, sitting trot, rising trot, and canter. Posture coaching is designed and adjusted according to the gait analysis. In this paper, we do self-coaching research by the means of analyzing horse gait classification. Then, we collect feedback information and show it in the form of voice.

In the past, various designs of the developer were required to extract data characteristics, but deep learning has a positive effect on the developer, because data features can be extracted by themselves. Also, deep learning technology is classified into five main fields of application including natural language processing, customer-centric management, image recognition, and speech recognition [1]. An AE is a type of deep learning method applied in many fields such as video, audio, and text mining. Moreover, an AE is effectively used to develop application services such as motion classification and action recognition. In addition, an AE can extract high-level features from data to train complex features in images [2].

There are application categories of an AE that are explained as follows. First, in the field of human motion on accelerometer with MEMS (Microelectromechanical systems), Fourati [3] studied a viable quaternion-based complementary observer (CO) that is designed for rigid body attitude estimation. In particular, the authors used XSENS to analyze human leg, head, and arm movements. Research on the joints of the human body is necessary when performing posture coaching of horse riding. It can constitute dynamic coaching for the movement of a person. Fourati [4] proposed a foot-mounted zero velocity update (ZVU) aided inertial measurement unit (IMU) filtering algorithm for pedestrian tracking in an indoor environment. In particular, the authors developed a coordinate and navigation system to visualize foot motion from 0.12 s to 0.72 s. Although the sensor is attached mainly to the top of the foot, it shows a waveform of a value similar to the data of the sensor attached to the waist of the rider. Since the inertial sensor requires attachment to the body, the user may feel inconvenience. The sensor requires attachment to the body. For that reason, there are a number of studies that have been applied to motion analysis using images through cameras. However, inertia sensors should be used to achieve precise posture analysis [5–11].

Second, in the field of human motion on deep learning with 2D data, Sarafianos [12] consider the automated recognition of human actions in surveillance videos. Convolution neural network (CNN) is mainly used in 2D, but the author suggested CNN, which is also applicable to 3D. We improved the accuracy of action classification by using frame information in actions. Hasan [13] proposed a continuous activity learning framework for streaming videos by intricately tying together deep hybrid feature models and active learning. Hassan allows us to automatically select the most suitable features and take the advantage of incoming unlabeled instances to improve the existing model incrementally. The authors extracted the characteristics of the encoder through two types of data, not just one. Hong [14] proposed a novel feature extractor with deep learning. It is based on denoising AE and improves traditional methods by adopting locality preserved restriction. The author used data that combines motion data with silhouette data. In previous studies, the AE has been actively studied to fuse two or more data [15–23].

In the sports coaching field, Saponara [24] presented a wearable biometric performance measurement system for combat sports, Dhruv [25] researched new integrated technologies that allow coaches, physicians, and trainers to better understand the physical demands of athletes in real time. Extant studies [26] can be a weakness in terms of reliability because of insufficient data. In this study, we constructed a big database, and it is possible to extract a reasonable performance. As data increased, adaptive neuro-fuzzy inference system (ANFIS) [26] faced limitations in performance and time. Therefore, we could solve problems by applying deep learning. Additionally, the ESAE is used for classifying horse gaits. AE can be converted into an ensemble form, which can have a synergistic effect on performance enhancement. In this paper, we propose an ESAE based on motion data of the horse rider. The reason for the classification of horse gaits is that the posture changes according to the horse gaits. The results showed that the angle of each joint was different about gaits of the horse. So, the standard of coaching can be different for each gait. In this paper, the goal is to construct the motion and position database (DB) of the expert rider and compare with the amateur to detect false motion for each horse gaits. This paper also aims to provide real-time horse-riding coaching by providing feedback to the user about posture. The rider's posture is different for each of the four types of horse gaits. For this reason, the motion of an expert can serve as an example for a beginner. This paper is organized as follows. In Section 2, we describe related research on AEs and SAEs. Algorithms are described in terms of mathematical theorems and concepts. Section 3 describes the proposed method. We describe a feature extraction method using a wavelet packet, and the methods applying five statistical values (maximum value, minimum value, average, variance, and standard deviation) are described. Section 4 describes the DB construction method and the outline of the experiment. In addition, it describes how to build a horse-riding DB. Experiments using the proposed method and comparison algorithm are shown. Section 5 summarizes the conclusions and future challenges.

2. Auto-Encoder and Its Variants

2.1. Auto-Encoder

AE belongs to one of the unsupervised learning algorithms. It is a neural network that aims to produce an output of X′ similar to the input data of X. In other words, AEs are composed of encoding (compression) and decoding (recovery), so data reconstruction is the purpose of AEs. Figure 1 shows the simple structure of the AE.

Figure 1. Simple structure of AE.

Generally, the encoding process is designed using fewer nodes than the number of nodes in the previous step. For example, if the number of hidden layers of the first AE is set to 40, the number of second AE is set to 30 less than 40. Through the decoding process, the number of hidden layers becomes 40, and the sizes of input data and output data become equal. Figure 2 shows the structure of AE with softmax classifier. Table 1 shows symbols used in defining the learning operation.

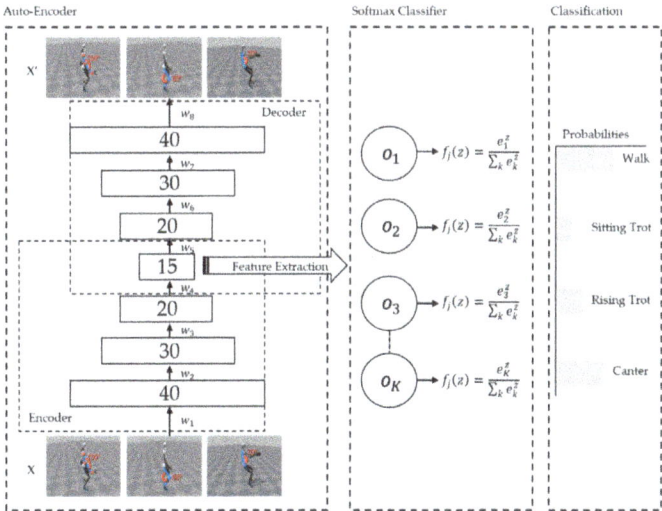

Figure 2. Structure of AE with softmax classifier.

Table 1. Symbols used in defining the learning operation.

Symbol	Definition	Symbol	Definition
I	Size of input layer (= output)	J	Size of hidden layer
v_{ij}	The j-th weight of the i-th output layer neuron	z_{nj}	The output value of the j-th hidden layer neuron for the n-th learning vector
η	Learning rate	x'_{ni}	The output value of the i-th output layer neuron for the n-th learning vector
x_{ni}	The i-th element of the n-th learning vector	θ_{ij}	Bias of the j-th hidden layer neuron
w_{ji}	The j-th weight of the j-th hidden layer neuron	b_i	Bias of i-th output layer neuron

AE activates the unit z of the hidden layer in the same way as the multi-layer perceptron (weighted sum). z is described as Equation (1).

$$z = f(x) = \sigma(Wx + b) \tag{1}$$

σ is an active function, mainly a sigmoid function and ReLu function. Next, the auto-encoder decoding step is a result of projecting the new weight value and summing the bias values with z obtained in Equation (1). X' is defined as Equation (2).

$$X' = g(y) = \sigma'(W'z + b') \tag{2}$$

The difference between the input data X and the output data X' is minimized through the objective function L.

$$L(X, X') = ||X - X'||^2 = ||X - \sigma'(W'(z) + b\prime||^2 \tag{3}$$

The objective function L is a cross entropy, and it is defined as shown in Equation (4).

$$L_H(X, X') = -X_k \log X'_x + (1 - X_k) \log(1 - X'_k] \tag{4}$$

Partial differentiation of L into weights is shown in Equation (5) using chain rules.

$$\frac{\partial L}{\partial y_i} \frac{\partial y_i}{\partial x_j} \frac{\partial x_i}{\partial w_{ij}} = -(X_i - X'i)x_j \tag{5}$$

Figure 3 shows process of training AE.

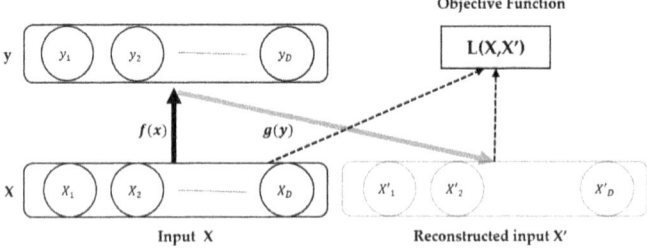

Figure 3. Process of training AE.

Equation (6) represents a weight of output layer of learning operation. The weight of output layer means the connection strength between the hidden layer and the output layer.

$$v_{ij}(t+1) = v_{ij}(t) + \eta(X_{ni} + x'_{ni})(1 - x'_{ni})x_{ni}z_{nj} \tag{6}$$

Equation (7) is a learning operation of the hidden layer weight, and the weight of hidden layer means the connection strength between the input and the hidden layer.

$$w_{ji}(t+1) = w_{ji}(t) + \eta x_{ji} z_{nj}(1 - z_{nj}) \sum_{i}^{l} \delta_{ni} v_{ij} \tag{7}$$

$$\delta_{ni} = (x_{ni} - x'_{ni})(1 - x'_{ni}) x'_{ni}. \tag{8}$$

The bias learning operation of the output layer neuron is shown in Equation (9).

$$b_i(t+1) = b_i(t) + \eta(x_{ni} - x'_{ni})(1 - x'_{ni}) x'_{ni} \tag{9}$$

Finally, the bias learning operation of the hidden layer neuron is shown in Equation (10).

$$\theta_j(t+1) = \theta_j(t) + \eta(1 - z_{nj}) z_{nj} \sum_{i}^{l} \delta_{ni} w_{ij} \tag{10}$$

Pseudocode of Auto-Encoder

Procedure Auto-Encoder

1. Reading:
   Signal ← horse data
   ← sampling factor
   Target signal ← signal samples by K
2. Initialization of variable:
   Hidden node ← select the number of layers of AE
   Max epochs ← select the number of layers of AE
   Weight ← select the number of layers of AE
   Sparsity regularization ← select the number of layers of AE
3. Training the AE network (minimum mean square error sense):
   3.1 Feedforward
       (a) Calculation of outputs at each layer by feeding signal as input to the network.
       (b) Calculation of error at final layer with reference to target signal
   3.2 Backpropagation of error
       (a) Calculation of error
       (b) Back propagating data (error) to previous layers
   3.3 Gradients calculation
       (a) Calculation of gradients for all weights, biases in all the layers
   3.4 Update
       (a) Update the parameters using gradient descent mechanism as
   New parameters ← old parameters—gradients
4. Testing
       (a) Test the network using signal against target signal

       (b) Calculate the error as error ← predicted output—target signal
5. Repeat until convergence
           if |error| ≥ tolerance level,
               Go to Step.3
   else
               Finish training

2.2. Stacked Auto-Encoder

The stacked auto-encoder (SAE) is a neural network consisting of multiple layers of AE, in which the outputs of each layer are connected to the inputs of the successive layer. SAE requires an enormous amount of computation and risks falling into the local minimum when learning the weights as the number of layers and nodes increases. There is also a problem of vanishing gradient (VG), in which weights are gradually reduced in the process of updating small values continuously. Therefore, a designer can build a network by stacking AEs according to performance.

The generated network can extract important features from the input data. The parameters of each layer node compare the output and the input data in the output layer through the hidden layer. It can be found that the parameters are determined according to the output data; also, the input data runs in the same way. Once the parameters of a hidden layer are determined, the output layer is removed, and the output of the trained hidden layer is used as input data to design another AE that has a hidden layer and an output layer. An SAE is not a deep generative model. The reason is that RBM depends on probability, and it anticipates test data with input data. However, an SAE trains the model in a deterministic manner. It trains $h = s(Wx + b)$, not $p(h = 0, 1) = s(Wx + b)$. The advantage of SAE is that the learning speed is fast, and the properties of the deep neural network can be adjusted. SAE stacks block structurally. Primarily, the hidden layer is activated by the input data, and then the active hidden layer reconstructs the input data.

The error between the input data and the reconstructed data is reduced by using the objective function. The parameters are updated according to activate the hidden layer. In the SAE network, the labels of the sample are also added into the softmax classifier, and the network parameters are tuned using the BP algorithm. Figure 4 shows the structure of an SAE. The target function for fine-tuning the network parameters is as follows:

$$J(W_k|_{k-1}^M, b_k|_{k-1}^M) = \text{argmin} \sum_{i=1}^N ||y_i - gn(fn(\ldots f_1(x_i)))||_2^2 \tag{11}$$

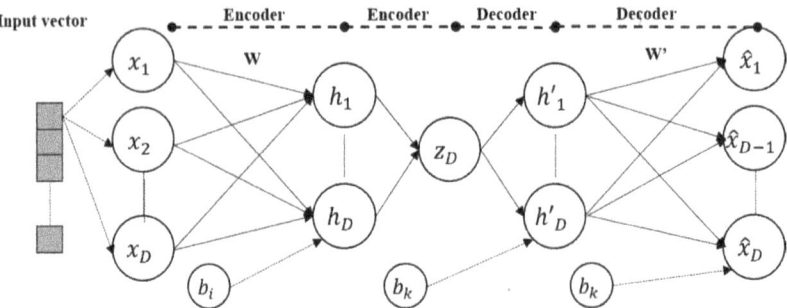

Figure 4. Structure of an SAE.

3. Proposed Method

3.1. Compression Method Using Wavelet Packet

Motion data of a horse rider consists of eight features that are compressed from 49,000 to 12,250 data samples using wavelet packets. The data that pass through the low-frequency filter are extracted as features, and the data that pass through the high-frequency filter are difficult to classify, because they are sparse in terms of their characteristics. High performance was obtained by selecting the two-layer wavelet feature. The computation for the generation of wavelet packets is simple when using an orthogonal wavelet. The sequence of functions shown as follows:

$$W_n(x), \ n = 0, \ 1, \ 2, \ \ldots \tag{12}$$

We have

$$W_{2n}(x) = \sqrt{2} \sum_{k=0}^{2N-1} h(k)W_n(2x - k) \tag{13}$$

$$W_{2n+1}(x) = \sqrt{2} \sum_{k=0}^{2N-1} g(k)W_n(2x - k) \tag{14}$$

in which $W_0(x) = \varphi(x)$ is the scaling function and $W_1(x) = \varphi(x)$ is the wavelet function. In this paper, $W_{3,0}$, $W_{2,0}$, and $W_{1,0}$ are used for input. Figure 5 shows the method of decomposition based on wavelet packets.

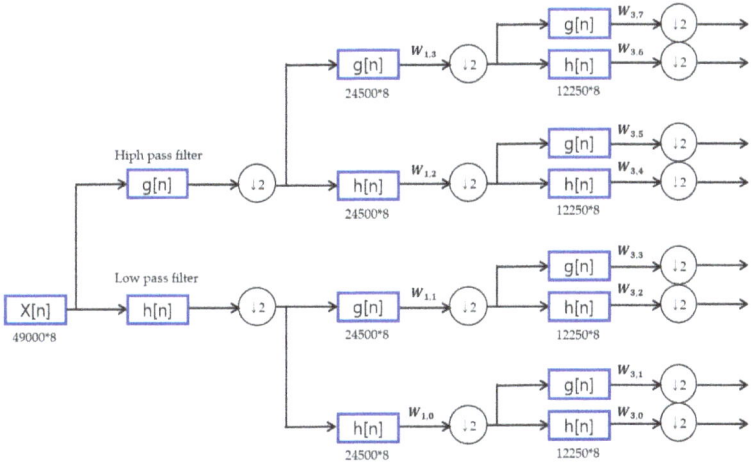

Figure 5. Method of decomposition based on wavelet packet.

3.2. Feature Extraction Based on Statistical Methods

In the previous study [26], we experimented with the elbow angle and the y-axis coordinate data of the hip, whereas in this study, we experimented with forty additional features. Feature values are extracted from 12,000 frames. The feature extraction method gradually extracts five feature values (average, maximum, minimum, variance, and standard deviation) from 1 to 20 frames like mask filter. The data are characterized by eight feature values: y-axis coordinates of the hip, backbone angle, right elbow angle, left elbow angle, right knee angle, left knee angle, elbow distance, and knee distance. We experimented sequentially with 10–100 frames; however, the best performance was achieved at 20 frames, and many features were obtained by taking advantage of the big data feature of AEs. Statistical values are a powerful tool for analyzing time series data. Eight features are generated from the sensor data, and five features are extracted for every frame based on eight features. Finally, 40 features are built. As a result, it is better to apply five feature values than to use a single minimum, a maximum, and an average value. Figure 6 shows a method of constructing statistical data.

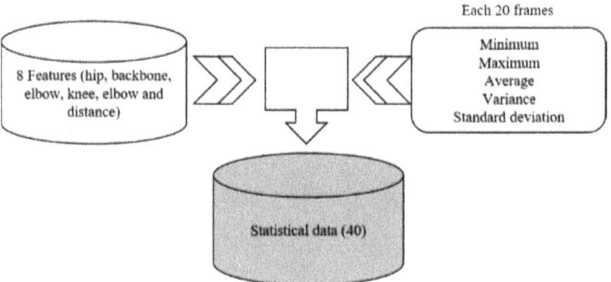

Figure 6. Method of constructing statistical data.

3.3. Ensemble Stacked Auto-Encoder

The softmax classifier provides us with the probabilities for each class label. It is more convenient for humans to interpret probabilities rather than margin scores of an SVM. The ESAE is constructed by building multiple SAE. By combining the probability values of the classifiers, an ensemble form is created. Thus, we could improve the classification performance by changing the structure. The softmax classifier is defined by Equation (16).

$$f_j(z) = \frac{e^{z_j}}{\sum_k e^z_k} \qquad (15)$$

The ensemble (sum) in the softmax classifier can be denoted as Equation (17). N is the number of SAE.

$$\text{Ensemble (sum)} = \frac{\sum_1^N f_j(z)}{N} \qquad (16)$$

The ensemble (product) in the softmax classifier can be denoted as Equation (18).

$$\text{Ensemble (product)} = f_1(z) \times f_2(z) \times \ldots f_j(z) \qquad (17)$$

In ML, ensemble methods use multiple learning algorithms to obtain better predictive performance than what is obtainable from any of the constituent learning algorithms alone. Unlike ML in statistical mechanics, which is usually infinite, an ML ensemble consists of only a concrete finite set of alternative types but typically allows a much more flexible structure to exist among those alternatives. For this purpose, this paper proposes an ESAE. An ESAE consists of two or more SAEs as feature extractors and improves the classification performance by averaging and multiplying the probability values extracted from the softmax classifier using the ensemble (sum) and ensemble (product) values. The performance of the data can be improved owing to the synergy. In this work, we changed the size of hidden nodes in (1) and (2) and experimented in ensemble form. Figure 7 shows the structure of ESAE. The training data is input to SAE (1), SAE (2), and SAE (3) respectively. Learning is performed in the ensemble form according to the change in the hidden layer. Finally, the probability values are obtained from the softmax. Performance is improved through sum and product methods are improved using probability values.

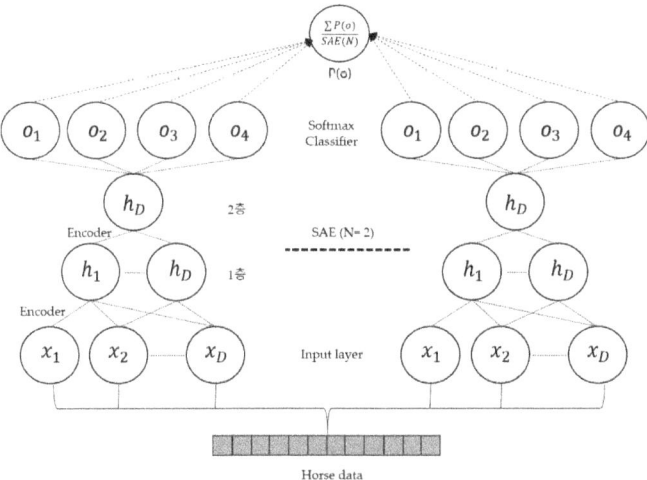

Figure 7. Structure of ESAE.

Pseudocode of Stacked Auto-Encoder

Procedure Ensemble Stacked Auto-Encoder
1.   Reading:
       Signal ← horse data
       K ← sampling factor
       N ← number of AE
       Target signal ← signal samples by K
2.   Initialization of variable:
       Hidden node ← select the number of layers of SAE
       Max-epochs ← select the number of layers of SAE
       Weight ← select the number of layers of SAE
       Sparsity Regularization ← select the number of layers of SAE
3.   Training the SAE network (minimum mean square error sense):

     (a)   Train the AE network of N $(2 - N)$
     (b)   Obtain probability from softmax classifier
     (c)   Fuse probability values for entire softmax classifier using average and product

4.   Testing

     (a)   Test the network using Signal against Target Signal
     (b)   Calculate the error as error ← predicted output—Target Signal

5.   Repeat until convergence

             if $|\text{error}| \geq$ tolerance level,
                   Go to Step.3
else
Finish training

---

4. Experiment and Database

4.1. Sensors of MVN Based Upon Miniature MEMS Inertial Sensor Technology

Sensors of Xsens are a camera-less 3D human motion measurement system. They are based on state-of-the-art MEMS inertial sensors, biomechanical models, and sensor fusion algorithms.

Sensors of Xsens are ambulatory and can be used indoors and outdoors regardless of lighting conditions. The results of the sensor trials require minimal post-processing, as there is no occlusion or lost markers. Results can easily be exported to other software applications. Figure 8 shows a horse rider with a motion capture suit.

Figure 8. Horse rider with motion capture suit.

4.2. Database

Acceleration sensors were used for data acquisition [26]. In order to obtain the reliability of data, data was additionally acquired. Accuracy can be trusted, because data is acquired over many days. There are four types of horse gaits in the database, including walk, sitting trot, rising trot, and canter. The database consists of 40 feature values with 2400 sizes. To describe the 40 feature values, there are eight features such as elbow angles (2), knee angles (2), elbow and knee distance (2), a backbone angle (1), and a hip-y-axis coordinates (1). The 40 features were extracted from 20 frames using mean, maximum, minimum, variance, and standard deviation with eight features. The size of the database for classification is 2400 × 40. The motion data of the horse rider is sized 49,500 × 1. It is projected into each data to obtain the angle data. Owing to the fact that it is an 8-angle value, data of 49,500 × 8 can be constructed. The dimension is reduced through a wavelet packet algorithm. A2 is selected to reduce the dimension, and statistical feature values are extracted through the reduced-size data.

Several AEs are modeled by using statistical feature values as input to the SAE, and by changing the hidden node. The average is obtained in an ensemble form through the probability values obtained from the respective softmax classifier. When all the results are collected, the final probability value can be obtained. To summarize, there are 4 data. The first data was applied to the AE without preprocessing the data acquired from the sensor to take advantage of the strength of the deep running. Performance is relatively low and excluded. The second data is the hip y data obtained from the sensor with the size 2800 × 70. The third is data using wavelet packet, and the size is 2800 × 18. The parameter settings are as follows. The weights of SAE are set to 0.0001, the max epoch is set to 3000, and the hidden size is 40. Finally, statistical features are extracted after the wavelet packet. The size is 2400 × 40. Training data and testing data is divided 50:50. Figure 9 shows the process of horse-riding coaching. Figure 10 shows hip y data with four kinds of horse gaits [26]. Hip y data with wavelet packet was experimented on in this paper.

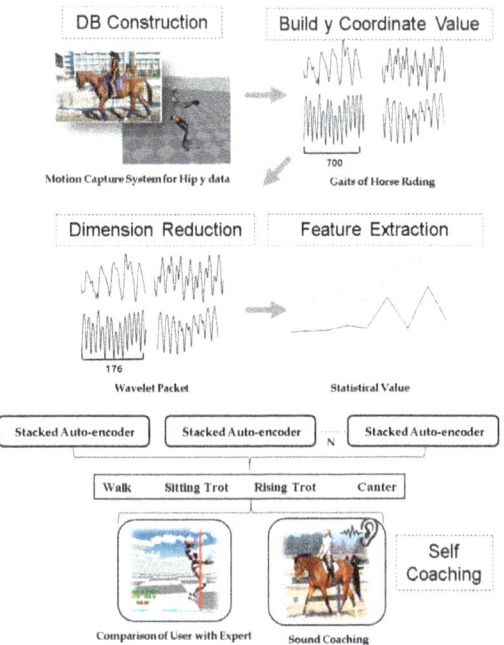

Figure 9. Process of horse-riding coaching.

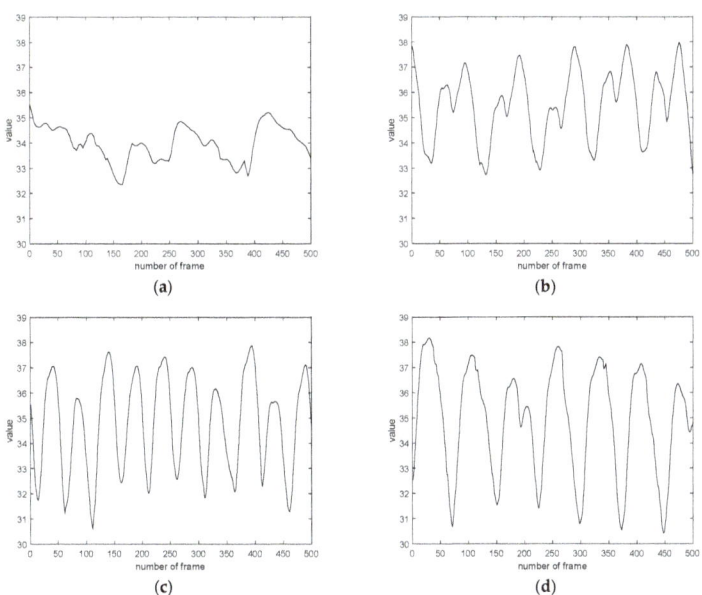

Figure 10. Hip y data with four kind of horse gaits. (**a**) Hip y data for walk, (**b**) hip y data for rising trot, (**c**) hip y data for sitting trot, and (**d**) hip y data for canter.

4.3. Environment

Motion data is acquired from an expert of horse riding who made one or two revolutions per gaits (walk, sitting, trot, rising trot, and canter) of an oval horse-riding course 20 m in length and 10 m in breadth while wearing a motion capture suit. Using the 3D motion capture suit based on Xsens inertial sensors, data were extracted in the order of Jeju (137 cm or less), Thoroughbred (160 cm), and Warm Blood (150–173 cm). It took 1 to 2 min to measure a file. Figure 11 shows the environment for data acquisition.

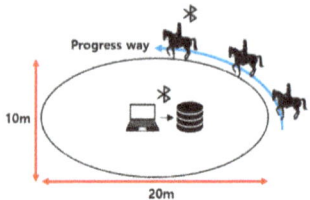

Figure 11. Environment for data acquisition.

4.4. Coaching for Horse Riding

There are various approaches to coaching horse riding, such as muscle utilization, posture, and tacit understanding with a horse. We develop a system that allows users to visualize their data and compare professional posture with amateur posture. Numerically, a range of the maximum value and the minimum value is visually expressed from each feature value. Self-coaching can be achieved by comparing feature values for each frame with an expert. Figure 12 shows a coaching system for horse riding, and Table 2 shows a numerical comparison of walk and canter.

Average Degree

	Elbow		Knee		Backbone Vertical	Distance			Unit : Degree
	Left	Right	Left	Right		Elbow	Knee	Elbow	
User	131	131	112	113	175	98	15	11	Backbone Vertical
Expert		135		107	177	92	131	131	Knee

Min Max Vluse

		Elbow		Knee		Backbone Vertical	Distance			Unit : Distance
		Left	Right	Left	Right		Elbow	Knee	Elbow	
User	Min	47	59	39	43	166	66	5	5	Knee
	Max	179	179	180	179	180	124	32	27	
Expert	Min		124		98	175	80	18	13	
	Max		146		109	180	105	20	15	

Figure 12. Coaching system for horse riding.

Table 2. Numerical comparison of walk and canter.

Feature	Minimum of Rising Trot	Maximum of Rising Trot	Minimum of Canter	Maximum of Canter
Hip y	32.08	38.79	31.87	38.31
Backbone	171.39	176.47	170.77	176.34
Angle of elbow	127.59	151.82	124.98	159.24
Angle of knee	123.92	172.20	119.50	135.80
Distance of elbow	23.02	27.02	25.87	25.78
Distance of knee	14.93	16.42	15.52	18.59

Characteristics of walk and canter are analyzed for the sake of coaching. Generally, the data value of horse riding has a cycle. By checking the frame-by-frame period, the user's motion can be recognized. Figure 13 shows a comparison of two feature values (hip y, an angle of backbone) by gaits. In the case of Figure 13, the change in the value controls the rider's hip height range. We can see it move more significantly in the canter than in the walk. The canter represents greater movement than the walk regarding elbow angle, knee angle, and backbone angle. Also, the cycle of canter is shorter than walk. Figure 14 shows a comparison of two feature values (an angle of right elbow, an angle of left elbow) by gaits. Figure 15 shows comparison of two feature values (an angle of right knee, an angle of left knee) by gaits. Figure 16 shows a comparison of two feature values (a distance of elbow, a distance of knee) by gaits.

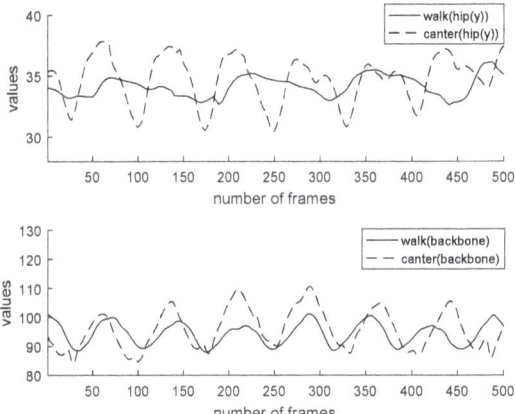

Figure 13. Comparison of two feature values (hip y, angle of backbone) by gaits.

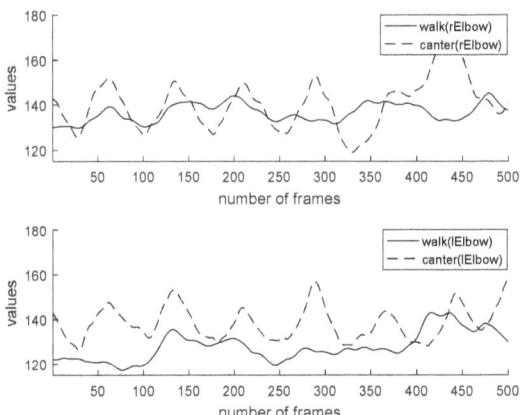

Figure 14. Comparison of two feature values (angle of right elbow, angle of left elbow) by gaits.

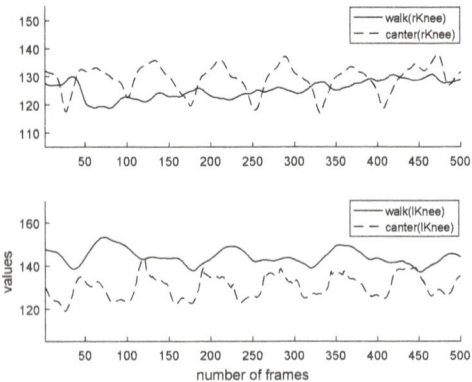

Figure 15. Comparison of two feature values (angle of right knee, angle of left knee) by gaits.

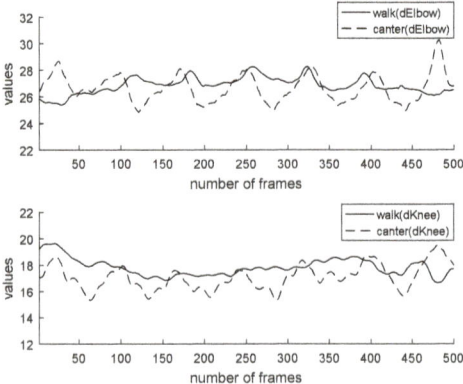

Figure 16. Comparison of two feature values (distance of elbow, distance of knee) by gaits.

4.5. Experiment and Result

This study focused on employing an ESAE to classify horse riding gaits to facilitate real-time coaching. To classify the actual horse-riding gaits, an ESAE with a higher classification rate and real-time posture coaching should be used. In summary, the ESAE exhibited the best performance for classification. According to classification results and the motion information such as the hip value, which is the main parameter for motion analysis and coaching, we can apply the proposed method to the coaching system, for each horse gaits, and for a rider under real or simulated environments. When three SAEs were used, the hidden size of ESAE was set to 30, 20, and 10, respectively, and when two SAEs were used, they were set to 46 and 15, respectively. When a single AE was applied, the average performance was 96.1%, and when a single SAE was applied, the performance was 96.8%. Two kinds of data were used: the hip value and eight characteristic values. The statistical data exhibited good performance in all algorithms, because it could separate all data characteristics well. Among them, the ESAE showed the best performance. Table 3 indicates a comparison of performance using hip y data with wavelet packet. Figure 17 shows the performance of ESAE for 40 feature data with wavelet packet. Figure 18 shows a comparison of performance using 40 feature data with wavelet packet.

Table 3. Comparison of performance using hip y data with wavelet packet.

Method	Performance
LDA	87.0
SVM	94.5
TREE	84.1
KNN	94.0
Ensemble Bagging	91.3
ELM (Sin)	91.8
ELM (Sig)	91.1
AE (Hidden Size = 15)	94.1
SAE (Hidden Size = 10, 15)	94.2
ESAE-1 (n = 3, sum)	94.2
ESAE-2 (n = 3, product)	94.3
ESAE-3 (n = 2, sum)	95.3
ESAE-4 (n = 2, product)	95.2

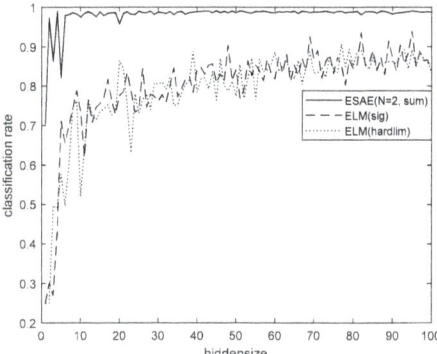

Figure 17. Performance of ESAE for 40 feature data with wavelet packet.

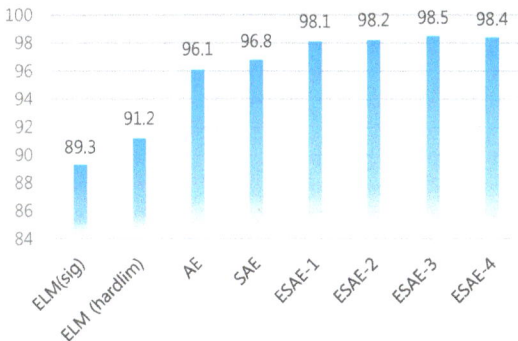

Figure 18. Comparison of performance using 40 feature data with wavelet packet.

4.6. Classification Performance

The accuracy Performance of SVM, TREE, KNN, and Ensemble Bagging in Table 3 is used as the ratio of correct classification to the number of total classified samples. The accuracy can be formulized as follows:

$$\text{Accuracy} = \frac{TP + TN}{TP + TN + FP + FN} \qquad (18)$$

TP is the number of correct predictions for positive samples, TN is the number of correct predictions for negative samples, FN is the number of incorrect predictions for positive samples, and FP is the number of incorrect predictions for negative samples. The performance of ESAE models presented in this study is obtained by softmax method described in Section 3.3.

5. Conclusions

This paper proposed the use of an ESAE to classify gaits of horse riding to facilitate real-time coaching. When the ESAE is used, the classification rate is 98.5%. According to classification results, and the motion information such as the hip value, which is the main parameter for motion analysis and coaching, we can apply the proposed method to the coaching system for each horse gait and for a rider under real or simulated environments. Moreover, ANFIS faced limitations in performance and time. Therefore, we could solve this problem by applying deep learning. Additionally, the ESAE is used for classifying horse gaits. AE can be converted into an ensemble form, which can have a synergistic effect on performance enhancement. As future work in the field of horse riding data analysis, we will study aspects of coaching in detail. Further, we plan to employ other body signals in the ESAE algorithm.

Author Contributions: J.-N.L. conceived and designed the research and experiments and contributed as the lead author of the article; Y.-H.B. performed database construction for the study; and K.-C.K. supervised the writing of the article, provided suggestions, and analyzed the data for the research.

Funding: This research was funded by the MIST (Ministry of Science & ICT), Korea, under the National Program for Excellence in SW supervised by the IITP (Institute for Information & communication Technology Promotion) (2017-0-00137).

Conflicts of Interest: The authors declare no conflict of interest.

References

1. Manic, M.; Rieger, C. Intelligent buildings of the future: Cyberaware, deep learning powered and human interacting. *Ind. Electron. Mag.* **2016**, *10*, 32–49. [CrossRef]
2. Deng, W.; Hu, J.; Guo, J. Transform-invariant PCA: A unified approach to fully automatic face alignment, representation, and recognition. *IEEE Trans. Pattern Anal. Mach. Intell.* **2014**, *36*, 1275–1284. [CrossRef] [PubMed]
3. Fourati, H.; Manamanni, N.; Afilal, L.; Handrich, Y. Complementary Observer for Body Segments Motion Capturing by Inertial and Magnetic Sensors. *IEEE/ASME Trans. Mechatron.* **2014**, *19*, 149–157. [CrossRef]
4. Fourati, H. Heterogeneous data fusion algorithm for pedestrian navigation via foot-mounted inertial measurement unit and complementary filter. *IEEE Trans. Instrum. Meas.* **2015**, *64*, 221–229. [CrossRef]
5. Zihajehzadeh, S.; Yoon, P.K.; Kang, B.S.; Park, E.J. UWB-Aided inertial motion capture for lower body 3-D dynamic activity and trajectory tracking. *IEEE Trans. Instrum. Meas.* **2015**, *64*, 3577–3587. [CrossRef]
6. Vartiainen, P.; Bragge, T.; Arokoski, J.P.; Karjalainen, P.A. Nonlinear state-space modeling of human motion using 2-D marker observations. *IEEE Trans. Biomed. Eng.* **2014**, *61*, 2167–2178. [CrossRef] [PubMed]
7. Wang, Y.; Hoai, M. Improving human action recognition by non-action classification. In Proceedings of the 2016 IEEE Conference on Computer Vision and Pattern Recognition (CVPR), Las Vegas, NV, USA, 27–30 June 2016; pp. 2698–2707.
8. Ligorio, G.; Sabatini, A.M. A novel kalman filter for human motion tracking with an inertial-based dynamic inclinometer. *IEEE Trans. Biomed. Eng.* **2015**, *62*, 2033–2043. [CrossRef] [PubMed]
9. Yilmaz, A.; Orhanli, T. Gait motion simulator for kinematic tests of above knee prostheses. *IET Sci. Meas. Technol.* **2015**, *9*, 250–258. [CrossRef]

10. Zhang, Y.; Chen, K.; Yi, J.; Liu, T.; Pan, Q. Whole-body pose estimation in human bicycle riding using a small set of wearable sensors. *IEEE/ASME Trans. Mechatron.* **2016**, *21*, 163–174. [CrossRef]

11. Villeneuve, E.; Harwin, W.; Holderbaum, W.; Sherratt, R.S. Signal quality and compactness of a dual-accelerometer system for gyro-free human motion analysis. *IEEE Sens. J.* **2016**, *16*, 6261–6269. [CrossRef]

12. Sarafianos, N.; Boteanu, B.; Ionescu, B.; Kakadiaris, A. 3D Human pose estimation: A review of the literature and analysis of covariates. *Comput. Vis. Image Underst.* **2016**, *152*, 1–20. [CrossRef]

13. Hasan, M.; Chowdhury, K.R. A continuous learning framework for activity recognition using deep hybrid feature models. *IEEE Trans. Multimedia* **2015**, *17*, 1909–1922. [CrossRef]

14. Hong, C.; Yu, J. Multimodal deep autoencoder for human pose recovery. *IEEE Trans. Image Process.* **2015**, *24*, 5659–5670. [CrossRef] [PubMed]

15. Liu, H.; Taniguchi, T. Feature extraction and pattern recognition for human motion by a deep sparse autoencoder. In Proceedings of the 2014 IEEE International Conference on Computer and Information Technology, Xi'an, China, 11–13 September 2014; pp. 174–181.

16. Yin, X.; Chen, Q. Deep metric learning autoencoder for nonlinear temporal alignment of human motion. In Proceedings of the 2016 IEEE International Conference on Robotics and Automation (ICRA), Stockholm, Sweden, 16–21 May 2016; pp. 2160–2166.

17. Hossein, R.; Ajmal, M.; Mubarak, S. Learning a deep model for human action recognition from novel viewpoints. *IEEE Trans. PAMI* **2018**, *40*, 667–681.

18. Potapov, A.; Potapova, V.; Peterson, M. A feasibility study of an autoencoder meta-model for improving generalization capabilities on training sets of small sizes. *Pattern Recognit. Lett.* **2016**, *80*, 24–29. [CrossRef]

19. Xu, C.; Liu, Q.; Ye, M. Age invariant face recognition and retrieval by coupled auto-encoder networks. *Neurocomputing* **2017**, *222*, 62–71. [CrossRef]

20. Kamyshanska, H.; Memisevic, R. The potential energy of an autoencoder. *IEEE Trans. Pattern Anal. Mach. Intell.* **2015**, *37*, 1261–1273. [CrossRef] [PubMed]

21. Geng, J.; Wang, H. Deep supervised and contractive neural network for SAR image classification. *IEEE Geosci. Remote Sens.* **2017**, *55*, 2442–2459. [CrossRef]

22. Zeng, K.; Yu, J. Coupled deep autoencoder for single image super-resolution. *IEEE Trans. Cybern.* **2017**, *47*, 27–37. [CrossRef] [PubMed]

23. Shi, B.; Chen, Y.; Zhang, P.; Smith, C.D. Nonlinear feature transformation and deep fusion for Alzheimer's disease staging analysis. *Pattern Recognit.* **2017**, *63*, 487–498. [CrossRef]

24. Saponara, S. Wearable biometric performance measurement system for combat sports. *IEEE Trans. Instrum. Meas.* **2017**, *66*, 2545–2555. [CrossRef]

25. Seshadri, D.R.; Drummond, C.; Craker, J.; Rowbottom, J.R. Wearable devices for sports: New integrated technologies allow coaches, physicians, and trainers to better understand the physical demands of athletes in real time. *IEEE Pulse* **2017**, *8*, 38–43. [CrossRef] [PubMed]

26. Lee, J.N.; Lee, M.W.; Byeon, Y.H.; Lee, W.S.; Kwak, K.C. Classification of horse gaits using FCM-based neuro-fuzzy classifier from the transformed data information of inertial sensor. *Sensors* **2016**, *16*. [CrossRef] [PubMed]

Article

Multi-axis Response of a Thermal Convection-based Accelerometer

Jae Keon Kim [1,2,†], Maeum Han [3,4,†], Shin-Won Kang [1,3], Seong Ho Kong [1,3,*] and Daewoong Jung [2,*]

1   Department of Sensor and Display Engineering, Kyungpook National University, Daegu 41566, Korea;
    jgk@knu.ac.kr (J.K.K); mehan@knu.ac.kr (S.-W.K.)
2   Aircraft System Technology Group, Korea Institute of Industrial Technology (KITECH), Daegu 42994, Korea
3   School of Electronics Engineering, College of IT Engineering, Kyungpook National University,
    Daegu 41566, Korea; mehan@kitech.re.kr
4   Construction Equipment R&D Group, Korea Institute of Industrial Technology (KITECH),
    Daegu 42994, Korea
*   Correspondence: shkong@knu.ac.kr (S.H.K.); dwjung@kitech.re.kr (D.J.);
    Tel.: +82-53-950-7579 (S.H.K.); +82-54-339-0624 (D.J.)
†   These authors contributed equally to this work.

Received: 20 May 2018; Accepted: 27 June 2018; Published: 29 June 2018

Abstract: A thermal convection-based accelerometer was fabricated, and its characteristics were analyzed in this study. To understand the thermal convection of the accelerometer, the Grashof and Prandtl number equations were analyzed. This study conducted experiments to improve not only the sensitivity, but also the frequency band. An accelerometer with a more voluminous cavity showed better sensitivity. In addition, when the accelerometer used a gas medium with a large density and small viscosity, its sensitivity also improved. On the other hand, the accelerometer with a narrow volume cavity that used a gas medium with a small density and large thermal diffusivity displayed a larger frequency band. In particular, this paper focused on a Z-axis response to extend the performance of the accelerometer.

Keywords: accelerometer; frequency; acceleration; heat convection

---

1. Introduction

An accelerometer is a device that measures the magnitude and direction of acceleration that is acting on a system. It is widely used in various areas. The airbag system of a vehicle and the suspension for posture control are typical examples of this device's application. Its application is currently increasing in terms of scope and frequency; thus, portable small electronics such as smart phones and tablet PCs contain accelerometers. As accelerometers are applied to advanced small electronic devices, the demand for small-sized accelerometers has been steadily increasing.

Since the 1990s, the development of microelectromechanical systems (MEMS) has microminiaturized devices that consist of mechanical or electrical components, which has also resulted in a considerable reduction in production cost and size. Sensor is one of the representative devices where MEMS has been concretely materialized and commercialized. Many types of measuring instruments have been microminiaturized, and the accelerometer is one of them. In the 1990s, Analog Device. Inc., a US company, developed and commercialized a capacitive-type subminiature accelerometer that achieved a drastic decrease in both size and price compared with existing accelerometers. Since then, many types of accelerometers have been developed and released to the market. Piezoresistive, piezoelectric, and capacitive types are examples of current commercial acceleration sensors [1–8]. Studies on a new type of accelerometer continue to advance.

Most of the traditional accelerometers use a solid proof mass to detect acceleration. On the other hand, a thermal convection-based accelerometer detects acceleration by utilizing the thermal convection in a sealed chamber [9–13]. This type of accelerometer possesses advantages and disadvantages compared with the traditional ones. The use of gas simplifies the internal structure of a sensor, which shortens the manufacturing process and reduces cost. In addition, the simpler shape, without a proof mass is more durable against impact or can withstand a larger impulse. A thermal convection-based accelerometer has an impulse-withstand value of 10,000 g or larger.

However, a thermal convection-based accelerometer suffers from the following disadvantages: it uses a heat source, it consumes more power because it uses the inertia of gas, and its bandwidth is lower than that of the existing accelerometers that use solids, which makes it unsuitable for detecting acceleration with a high frequency [14]. Therefore, the current study designed and fabricated an accelerometer using thermal convection and examined a method of improving its sensitivity and frequency band. We found that various environmental and structural parameters such as heater power, working gases, pressure, and cavity volume play an important role in the performance of a thermal convection-based accelerometer. In addition, more efforts have been spent to expand the sensing axis (Z-axis) as well as the planar axes (X- and Y-axes).

2. Materials and Methods

2.1. Device Structure and Working Principle

The proposed accelerometer consists of two main parts: top and bottom wafers. The bottom wafer contains temperature sensors and a microheater, which are necessary for the operation of the accelerometer. The top wafer secures the space for the gas that is used in the accelerometer and minimizes the effect of the external environment. Figure 1a shows the bottom wafer, which includes a heater and three pairs of temperature sensors. The bottom wafer was wet etched to form a 50 μm-thick membrane. The top and bottom wafers are joined to each other by epoxy resin. Figure 1b shows the top wafer of the proposed accelerometer, which creates the space by dry etching. Figure 1c shows a schematic diagram of the accelerometer, with the top and bottom wafers connected. The heater in the bottom wafer heats the gas in the space between the top and the bottom wafers, and the six temperature sensors that are located equidistant from the heater detect the temperature change in the space.

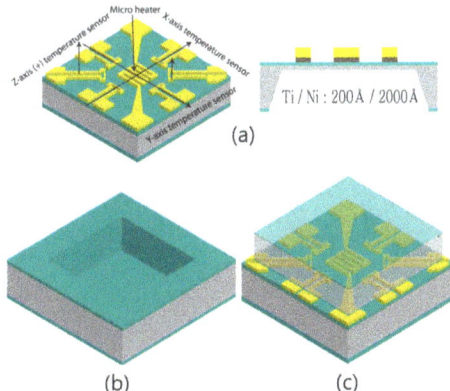

Figure 1. Proposed accelerometer with the (**a**) bottom and (**b**) top wafers and (**c**) bonded.

Figure 2 shows that the accelerometer operates when the gas is heated by the heater. Subsequently, the air convection around the heater produces a particular temperature distribution. The applied acceleration generates convection to its direction, which moves the gas inside the accelerometer. Acceleration is measured based on the change in the gas temperature that is detected by the temperature sensors.

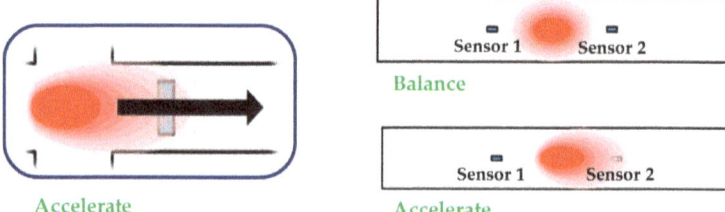

Figure 2. The principle of proposed convective sensor; the applied acceleration makes the deform temperature distribution inside the top wafer due to thermal convection, which gives an opposite movement of the temperature profile on both of the temperature sensors (i.e. temperature around sensor 1 decreases and that around sensor 2 increases).

2.2. Determination of Materials

The temperature sensors in the bottom wafer were fabricated by utilizing the property of metals, i.e., increases in resistivity as a function of temperature. A material with a higher temperature coefficient of resistivity (TCR) exhibits a larger change in resistivity with temperature, making it ideal for its use as a temperature sensor. In addition, if a material shows a linear change in its resistance with temperature, this property also demonstrates the material's suitability as a temperature sensor. Accordingly, the temperature sensors of the proposed accelerometer need to be made of a material with high TCR so that they can sensitively react to a slight temperature change. In addition, if a material shows a more linear change in resistance with temperature, it is more suitable for displaying the linearity of the sensor output.

Platinum (Pt) is a representative material for a temperature sensor [9]. Pt is highly resistant to corrosion and shows a stable and linear change in its resistance over a wide temperature range. On the other hand, nickel (Ni) has a narrower temperature range than Pt, but the TCR of Ni is $6.7 \times 10^{-3}\,°\mathrm{C}^{-1}$, which is approximately twice that of Pt. Moreover, Ni is less expensive; thus, the temperature sensors of the proposed accelerometer were made of Ni. The detailed fabrication process was described in Ref. [13]. Figure 3 shows the fabricated sensors on a coin and a printed circuit board (PCB) chip.

(a)   (b)

Figure 3. Thermal convection-based accelerometer on (**a**) coin and (**b**) PCB chip.

3. Results and Discussion

3.1. Characteristics of a Microheater and a Temperature Sensor

The microheater characteristics were investigated by measuring the temperature that was generated from the heater under the condition that the current was sequentially increased. The temperature was directly measured by using k-type thermocouple on the surface of the heater. Figure 4a shows that the temperature change, which occurred when current was applied to the heater to generate thermal convection, exhibited exponential function characteristics that were relative to the quantity of the applied current, because the electrical energy supplied to the heater was proportional to the square of the current.

The characteristics of the fabricated temperature sensor were determined by measuring the change in the resistance with temperature. The Ni temperature sensor showed a linear change following the temperature–resistance characteristic of the metal, as shown Figure 4b. A metal with a high TCR can be suitably used as a temperature sensor. The temperature sensors of the proposed accelerometer have a TCR value of approximately 5.1×10^{-3} ($°C^{-1}$). Although this value is slightly lower than 6.0×10^{-3} ($°C^{-1}$), which is the TCR of Ni, it is higher than 3.93×10^{-3} ($°C^{-1}$), which is the TCR of Pt. Consequently, the temperature sensors have good sensitivity.

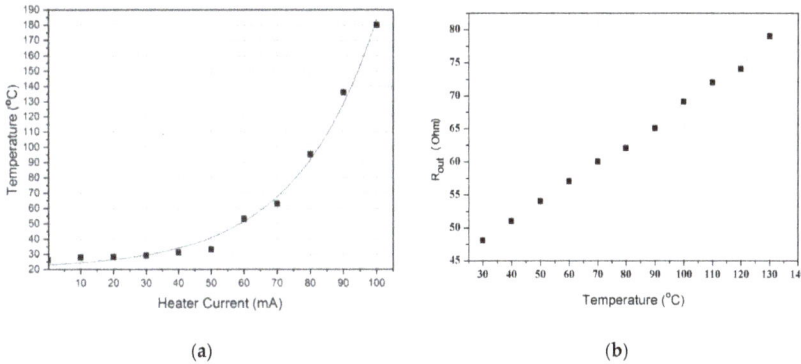

(a) (b)

Figure 4. (**a**) Current–temperature characteristics of the micro heater and (**b**) temperature-resistance characteristics of the micro temperature sensor.

3.2. Operating Principle

The governing equations that analyze the temperature profile of a thermal accelerometer are based on the principle of conservation of mass, momentum, and energy [15–17]. A continuity equation in physics describes the transport of a physical quantity being conserved. As mass, momentum, and energy are conversed quantities, numerous physical phenomena can be described by the continuity equations. In fluid mechanics, the continuity equation is a mathematical expression of the law of conversation of mass.

The performance of a thermal convection-based accelerometer is based on the heat transfer by natural convection. Therefore, analysis of natural-convection heat transfer is needed to analyze the operating process of a thermal convection-based accelerometer and to identify its unique characteristics. The heat transfer by natural convection is caused by the density gradient, due to a temperature difference. When a temperature difference occurs in an area where fluid exists, the density decreases in the part with a higher temperature and relatively increases in the other part with a lower temperature. As the high-density part moves along the acceleration direction, natural convection occurs because of the temperature difference.

As the governing equation of natural convection has no solid solution and ideal conditions should be given, a simplified degine was proposed to predict the performance of the thermal accelerometer [18]. The solutions to the equations of conitnity, mass, momentum, and energy are derived for the concentric sphere models. The solution is then derived from some non-dimensional numbers, Grashof number G_r, and Prandtl number P_r.

The use of these dimensionless numbers helps to predict and analyze the performance of the thermal accelerometers. G_r is a nondimensional parameter that is used in the correlation of heat and mass transfer due to thermally induced natural convection at a solid surface immersed in a fluid. The significance of the G_r is that it represents the ratio between the buoyancy force due to spatial variation in fluid density (caused by temperature differences) to the restraining force due to the viscosisty of the fluid [19]. The P_r characterizes the distribution of the velocities relative to the temperature distribution. It is a characteristic of thermal physics of fluid.

$$G_r = \frac{g\rho^2\beta L^3 \Delta T}{\mu^2} \tag{1}$$

$$P_r = \frac{\mu}{\alpha} \tag{2}$$

Here, g, ρ, β, L, ΔT, μ, and α are the applied acceleration, gas density, coefficient of volumetric expansion, characteristic size (generally denotes the cavity size), temperature difference between the heater and boundary of the sensor, kinematic viscosity, and thermal diffusivity, respectively [9,13,17,18].

To predict the performance of the thermal accelerometer, Gr and Pr numbers were calculated for the gas medium (using properties in Table 1) and are listed in Table 2. The calculation is based on atmospheric conditions, applied acceleration of 1g, characteristic size (L) of 400 μm, and temperature difference (ΔT) of 25 °C (assuming a heater current of 60 mA).

Table 1. The gas medium properties at 50 °C (adapted from [20]).

	Density (kg/m³)	Specific Heat (kJ/kg·K)	Kinematic Viscosity ($\times 10^{-6}$) (m²/s)	Thermal Diffusivity ($\times 10^{-4}$) (m²/s)	Thermal Conductivity (W/m·K)
Air	1.092	1.007	19.6	0.248	0.02735
N_2	1.0564	1.042	17.74	0.249	0.02746
CO_2	1.6597	0.8666	9.71	0.129	0.01858

Table 2. Calculated G_r and P_r numbers.

	Air	N_2	CO_2
G_r	7.44×10^{-3}	8.07×10^{-3}	4.24×10^{-2}
P_r	7.16×10^{-4}	6.46×10^{-4}	5.22×10^{-4}

3.3. Characteristics of the Accelerometer

To confirm how the characteristics of the accelerometer change according to the current input into the microheater, the current supplied to the temperature sensors was fixed at 10 mA, and the amount of current applied to the microheater was adjusted.

3.3.1. Effects of the Heating Power

Figure 5 shows the measurement results of the characteristics. The higher the current supplied to the micro heater was, the larger the heat generated by the heater was. As seen in Figure 4a, the temperature started to increase at 30 mA and rapidly rose from 50 mA. Thus, four currents (30, 50, 70, and 90 mA) were selected to examine the effect of the heating power on the sensitivity of the accelerometer. According to the results, the temperature increase in the microheater was accompanied by an increase in the voltage variation in the temperature sensor [21]. When the temperature of the microheater increased, the temperature difference (ΔT) between the temperature sensor increased, and thus, the sensitivity of the accelerometer increased according to G_r in Equation (1). A large electric power supply to the heater improves the sensitivity of the sensor.

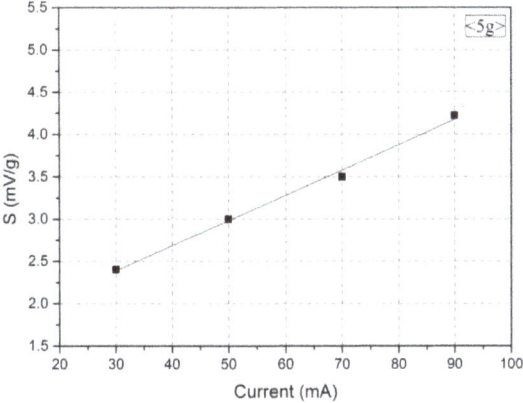

Figure 5. Sensitivity variation according to the current of the micro heater.

3.3.2. Effects of the Frequency

Figure 6 shows the measurement results that were obtained by fixing the current supply for the microheater at 70 mA and varying the frequency that was applied to the accelerometer. The purpose of the experiment was to determine how the accelerometer characteristics change according to the frequency variation that was applied to it. An acceleration of 5g was applied along the positive direction. Figure 6 shows that the results indicate that as the frequency increased, the variation in the output voltage of the temperature sensor decreased with the acceleration. The noise equivalent acceleration (NEA) is measured to be 0.25 mg RMS. When the value of the acceleration was fixed and only the magnitude of the frequency varied, the travel distance of the vibration shaker when accelerated became shorter, and thus, the temperature difference that was detected by the temperature sensor decreased [22]. Although the best sensitivity was measured at 1 Hz, the longest time to recover thermal equilibrium for the next measurement was also observed at this frequency. In other words, the sensitivity of the sensor and the frequency were inversely related. On this basis, we can predict that as Gr in Equation (1) increases, the sensitivity improves, but the frequency band decreases [12,13,15,16].

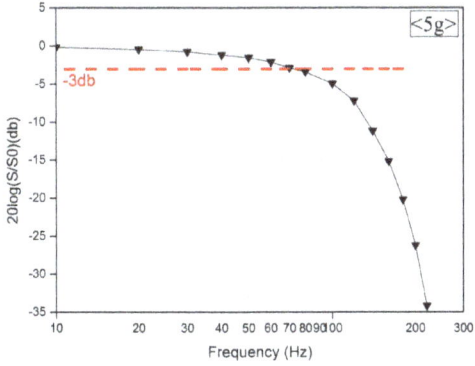

Figure 6. Frequency characteristics of the accelerometer.

3.3.3. Effects of the Medium Type

The sensitivity and frequency band were also measured using three different gas media to determine the effect of a gas medium on the accelerometer. Figure 7a clearly shows that the significant difference in the sensitivity was caused by different gas media. This result indicates that the characteristics of a gas medium greatly affect the thermal convection, which is the operating principle of the accelerometer. Gas media with large densities and small viscosities appeared to result in better sensitivity [23,24]. This result also agrees with Gr in Equation (1).

Figure 7b clearly shows that the relationship between the sensitivity and frequency according to the types of gas media produces the same result as that between the sensitivity and frequency in terms of the volume of the top wafer. Figure 7b shows that the gas media with smaller densities and larger thermal diffusivities have wider frequency bands [25,26]. The gases that have a smaller density can move faster than those with a larger density, giving a widened bandwidth.

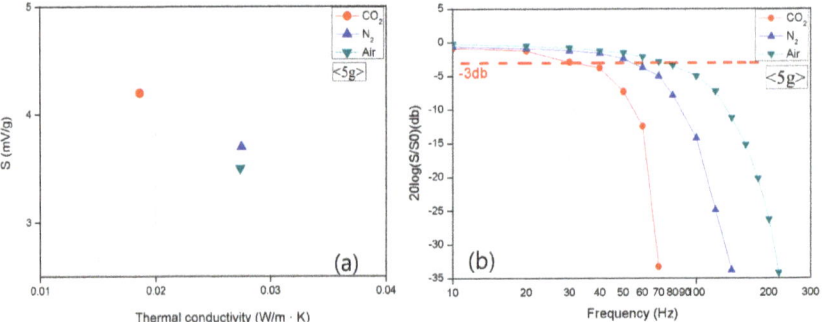

Figure 7. Output (**a**) sensitivity and (**b**) the frequency of the accelerometer according to gas medium.

3.3.4. Effects of the Gas Pressure

Figure 8a shows that an increase in pressure was accompanied by an improvement in sensitivity because the pressure increase led to the increase in the gas density, which increased Gr, thereby improving the sensitivity. This result is very significant as it indicates that high-pressure packaging could reduce energy consumption and improve sensitivity without any structural modification or additional increases in the heater power [23–27]. It is one of the great advantages that is introduced by using a gas medium instead of a liquid one in the proposed thermal convection-based accelerometer.

Figure 8b shows the variation in the frequency according to pressure. The result shows that an increase in the pressure was accompanied by a decrease in the frequency band. This result also confirmed that sensitivity and frequency were inversely related. When the pressure increased, the gas density increased, and its thermal diffusivity decreased. As is demonstrated by the frequency variation relative to the gas, the decrease in the thermal diffusivity narrowed the frequency band. As a result, when Gr increased, the sensitivity improved, but the frequency band became narrow. On the other hand, when Pr increased, the frequency band became wider, and the sensitivity improved with a smaller Pr. Consequently, in designing a thermal convection-based accelerometer, the use of an accelerometer must be carefully considered to determine the appropriate variables.

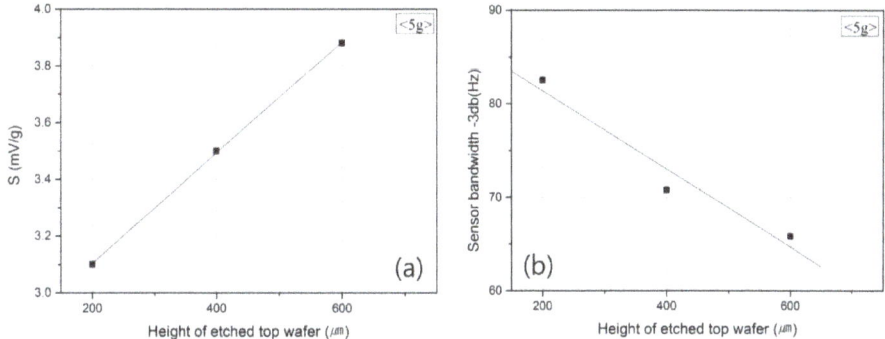

Figure 8. Output (**a**) sensitivity and (**b**) frequency of the accelerometer according to gas pressure.

3.3.5. Effects of the Cavity Volume

To investigate the effects of Gr and Pr on the sensitivity of the sensor and frequency band, the output of the accelerometer was measured by varying the volume of the top wafer where gas convection occurs, and by using other types of gas media.

Figure 9 shows the changes in the sensitivity of the accelerometer and the frequency band according to the volume of the top wafer. Figure 9a shows that an increase in the space where the medium can move was accompanied by an improvement in the sensitivity of the accelerometer [28] due to the increase in the length (L) of G_r. As the space volume increased, i.e., where the medium could move, the temperature difference between the heater inside the top wafer and that outside the top wafer also increased, which resulted in the improvement of the output characteristics [15,16].

However, as shown in Figure 9b, when the volume of the top wafer increased, the amount of medium that moved according to the acceleration value also increased, and the medium could not follow the fast movement of the sensor with the increase in frequency. Consequently, the frequency band that could be measured decreased. For this reason, when the volume of the top wafer is considered, a large volume needs to be selected for high sensitivity, and a small volume is appropriate for a large frequency band [12].

These results mean that a larger Pr in Equation (2) has a wider frequency band. To observe the effect of atmospheric pressure on the sensitivity of the sensor and frequency, an experiment was conducted by fabricating a chamber that could have its pressure controlled.

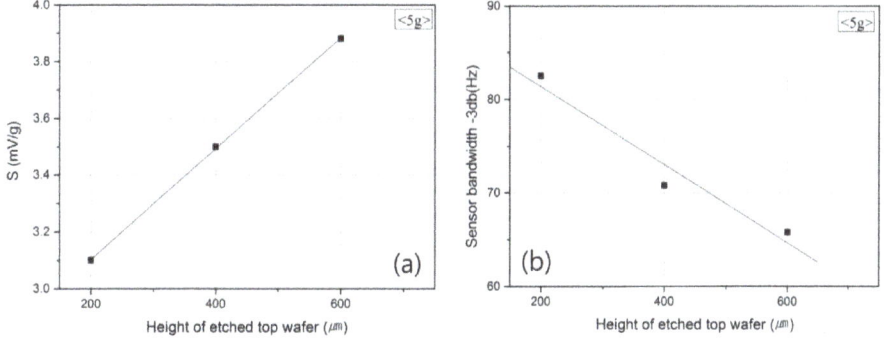

Figure 9. Output (**a**) sensitivity and (**b**) frequency of the accelerometer according to the height of the etched top wafer.

3.3.6. Z-axis Characteristics of the Accelerometer

The proposed thermal convection-based accelerometer can detect not only the x and y axes, but also the Z-axis. Figure 10 shows the measurement results for the three axes (X, Y, and Z). The X- and Y-axes showed almost the same level of sensitivity, and the output value in the upward (+) direction of the Z(+)-axis showed considerably lower sensitivity than that of the X- and Y- axes.

As shown in Figure 11, the measurement could be made in the upward positive (+) direction, but not in the downward negative (−) direction because the medium moved not to the left and right, but up and down. Because the gas near the heater had a high temperature and a low density, it rose upward. In this situation, when acceleration was applied in the upward (+) direction, the temperature distribution inside the accelerometer slightly increased. As the temperature sensor detected a decrease in temperature after the acceleration was applied, the output voltage decreased. Because the height of the top wafer was only 400 µm, the temperature distribution only slightly moved and the temperature sensor was placed on the surface of the bottom wafer, and the Z(+)-axis showed relatively lower sensitivity than that of the X- and Y- axes.

On the other hand, when the acceleration was applied in the downward (−) direction, the sensor moved downward, but the gas did not follow the sensor because of lower gas density and it providing no space to move. Accordingly, the same output value was observed, irrespective of the applied acceleration.

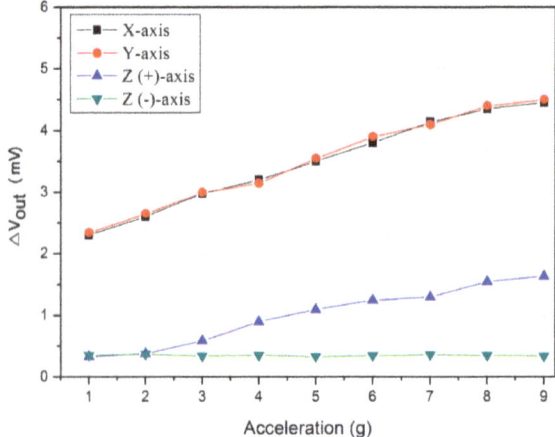

Figure 10. Output characteristic of the three-axis accelerometer.

To measure the acceleration in the negative direction on the Z-axis, the measurement was conducted by turning the sensor to the opposite direction. Figure 12 shows that when acceleration was applied in the positive direction, a constant output value was measured. On the other hand, when acceleration was applied in the negative direction, the output values showed linearity according to the magnitude of the acceleration. The reason is the same as that in the case of the positive direction in the Z-axis. In other words, when the sensor turned to the opposite direction, and when the acceleration was applied in the upward (+) direction, there was no place to move, and thus, even though the sensor moved upward, the same temperature distribution followed the movement. Consequently, the temperature sensor detected a constant temperature.

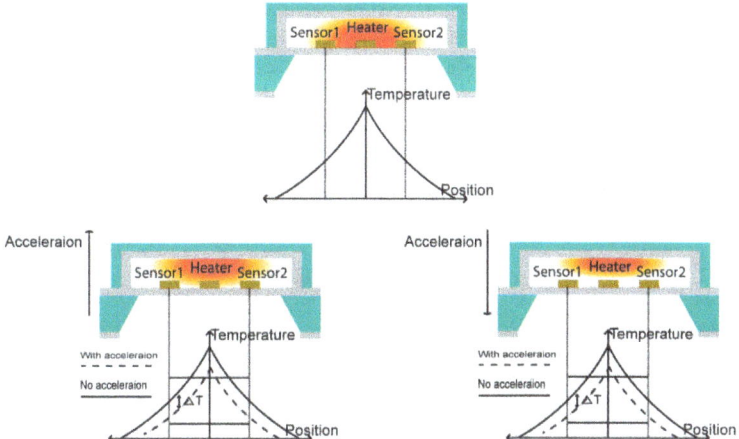

Figure 11. Schematic view of the Z-axis sensing principle.

To detect both directions of the Z-axis, two accelerometers were vertically attached, as shown in Figure 13. Figure 14 shows the outputs of the upper sensor for the positive direction in the Z-axis and those of the lower sensor for the negative direction in the Z-axis. The result shows that the values in the negative direction were always larger than those in the positive direction because the temperature distribution inside the sensor became more delicate when the sensor was turned upside down. The movement of the temperature profile in the turning sensor is limited due to the rising tendency of the hot air.

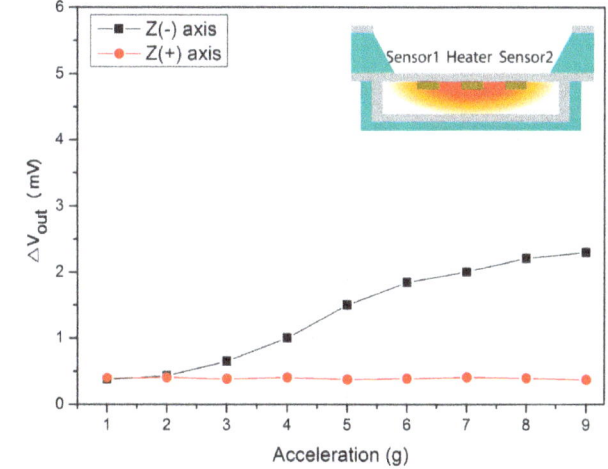

Figure 12. Output characteristic of the accelerometer as a function of the accelerometer on the Z-axis.

For an effective Z-axis measurement, we needed to install a temperature sensor in a cavity or to design a temperature sensor to be installed in the upper part of the top wafer.

The output values of the acceleration in the Z-axis directions were relatively low for the following reasons: the temperature difference between the upper and lower parts of the top wafer were much smaller than those between the left and right sides. Moreover, the measurements along the X-and

Y-axes represented the differences in the outputs between the temperature sensors on both sides, whereas the measurements along the Z-axis represented the output values of a single temperature sensor. To compensate for the output values of the acceleration in the Z-axis directions, an amplifier with an output that was larger than those of the X-and Y-axes may be included in the accelerometer, or the temperature sensors may be designed to be installed in the top wafer.

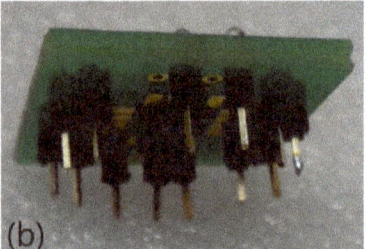

Figure 13. The structure of the accelerometer for the Z-axis.

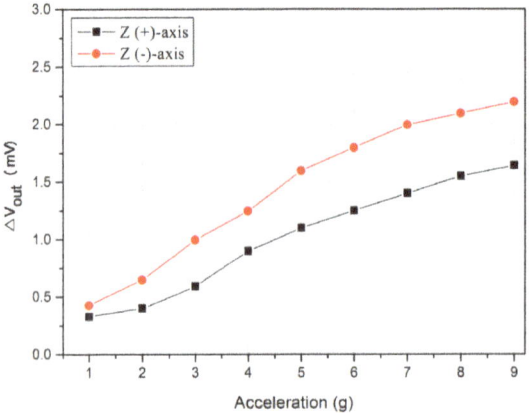

Figure 14. Output characteristic of the accelerometer on the Z-axis.

4. Conclusions

The MEMS technique was applied to fabricate a subminiature accelerometer. In addition, the problems that were associated with existing accelerometers that use a solid proof mass could be solved using gas as a medium. Because gas was used as a medium to measure the acceleration, the accelerometer achieved a great improvement in durability, which has not been possible when using solid proof mass accelerometers. Furthermore, the accelerometer was designed and experiments were conducted to improve the performance. The proposed accelerometer offered another advantage of a wide measurement range, from 1g to 9g. Many studies and trials have attempted to improve sensitivity, however sufficient attention has not been paid to the problem of a narrow frequency band, which is one of the disadvantages of the thermal convection-based accelerometers. Experimental results revealed that larger heating power increased the temperature difference (ΔT) between the temperature sensors, resulting in improved sensitivity of the accelerometer. Gases that have high densities and small viscosities show high sensitivity. In addition, an increase in the space where the medium can move was accompanied by an improvement in the sensitivity of the accelerometer. However, we found that the thermal convection-based accelerometer showed an inverse relationship between frequency and sensitivity. Gases that have a small density and large thermal diffusivity have a wider bandwidth. Smaller cavities showed a better frequency response than larger ones. Moreover, the Z-axis response was characterized to extend the performance of the accelerometer. When the acceleration was applied to an upward direction, the temperature profile rose along with the applied direction, resulting in lowered temperatures around the temperature sensor. Owing to its sensing mechanism and its structural design however, the same value was output irrespective of the applied downward direction of the acceleration. To solve the half detection of the Z-axis, two accelerometers were vertically attached.

Author Contributions: M.H., J.K.K., S.-W.K., and D.J. conceived the idea and designed the experiment. M.H., and J.K.K. fabricated the experimental sensor. J.K.K. analyzed the experimental results. M.H., S.H.K., and D.J. wrote the paper and discussed the contents.

Acknowledgments: This study was supported by the BK21 Plus project that was funded by the Ministry of Education, Korea (21A20131600011). This study has been conducted with the support of the Korea Institute of Industrial Technology.

Conflicts of Interest: The authors declare no conflict of interest.

References

1. He, J.; Xie, J.; He, X.; Du, L.; Zhou, W. Analytical study and compensation for temperature drift of a bulk silicon MEMS capacitive accelerometer. *Sens. Actuators A* **2016**, *239*, 174–184. [CrossRef]
2. Yu, H.; Guo, B.; Haridas, K.; Lin, T.-H.; Cheong, J.H.; Tsai, M.L.; Yee, T.B. Capacitive micromachined ultrasonic transducer based tilt sensing. *Appl. Phys. Lett.* **2012**, *101*, 153502. [CrossRef]
3. Zhang, L.; Lu, J.; Kurashima, Y.; Takagi, H.; Maeda, R. Gase studies of a planar piezoresistive vibration sensor: Measuring transient time history signal waves. *Microelectron. Eng.* **2016**, *165*, 27–31. [CrossRef]
4. Li, Y.; Zheng, Q.; Hu, Y.; Young, J. Micromachined piezoresistive accelerometers based on an asymmetrically gapped cantilever. *IEEE/ASME J. Microelectromech. Syst.* **2011**, *20*, 83–94. [CrossRef]
5. Zhu, R.; Ding, H.; Su, Y.; Zhou, Z. Micromachined gas inertial sensor based on convection heat transfer. *Sens. Actuators A* **2006**, *130–131*, 68–74. [CrossRef]
6. Liu, S.; Zhu, R. Mcromachined fluid inertial sensors. *Sensors* **2017**, *17*, 367. [CrossRef] [PubMed]
7. Dinh, T.; Phan, H.P.; Qamar, A.; Woodfield, P.; Nguyen, N.T.; Dao, D.V. Thermoresistive effect for advanced thermal sensor: Fundamentals, design consideration, and applications. *J. Microelectromech. Syst.* **2017**, *26*, 966–986. [CrossRef]
8. Mailly, F.; Martinez, A.; Giani, A.; Pascal-Delannoy, F.; Boyer, A. Design of a micromachined thermal accelerometer: Thermal simulation and experimental results. *Microelectron. J.* **2003**, *34*, 275–280. [CrossRef]
9. Mezghani, B.; Tounsi, F.; Masmoudi, M. Development of an accurate heat conduction model for micromachined convective accelerometers. *Microsyst. Technol.* **2015**, *21*, 345–353. [CrossRef]
10. Goustouridis, D.; Kaltsas, G.; Nassiopoulou, A.G. A silicon thermal accelerometer without solid proof mass using porous silicon thermal isolation. *IEEE Sensors J.* **2007**, *7*, 983–989. [CrossRef]
11. Luo, X.B.; Li, Z.X.; Guo, Z.Y.; Yang, Y.Z. Theraml optimization on micromachined convective accelerometer. *Heat Mass Transfer* **2001**, *38*, 705–712. [CrossRef]
12. Luo, X.B.; Yang, Y.J.; Zheng, F.; Li, Z.X.; Guo, Z.Y. An optimized micromachined convective accelerometer with no proof mass. *J. Micromech. Microeng.* **2001**, *11*, 504–508. [CrossRef]
13. Han, M.; Kim, J.K.; Park, J.H.; Kim, W.; Kang, S.W.; Kong, S.H.; Jung, D. Sensitivity and frequency response improvement of a thermal convection-based accelerometer. *Sensors* **2017**, *17*, 1765. [CrossRef] [PubMed]
14. Courteaud, J.; Crespy, N.; Combette, P.; Sorli, B.; Giani, A. Studies and optimization of the frequency response of a micromachined thermal accelerometer. *Sens. Actuators A* **2008**, *147*, 75–82. [CrossRef]
15. Hodnett, P. Natural convection between horizontal heat concentric circular cylinders. *J. Appl. Math. Phys.* **1973**, *24*, 507–516. [CrossRef]
16. Chaehoi, A.; Mailly, F.; Latorre, L.; Nouet, P. Experimental and finite-element study of convective accelerometer on CMOS. *Sens. Actuators A* **2006**, *132*, 78–84. [CrossRef]
17. Mukherjee, R.; Basu, J.; Mandal, P.; Guha, P.K. A revew of micromachined thermal accelerometers. *J. Micromech. Microeng.* **2017**, *27*, 123002. [CrossRef]

18. Garraud, A.; Giani, A.; Combette, P.; Charlot, B.; Richard, M. A dual axis CMOS micromachined convective thermal accelerometer. *Sens. Actuators A* **2011**, *170*, 44–50. [CrossRef]

19. Sanders, C.J.; Holman, J.P. Franz Grashof and the Grashof number. *Int. J. Heat Mass Transfer.* **1972**, *15*, 562–563. [CrossRef]

20. Singal, R.K. *Refrigeration and psychrometric charts with property tables (S.I. units), Appendix 1 Property tables and charts (SI units)*; S.K.KATARIA & SONS: Delhi, India, 2006.

21. Kaltsas, G.; Goustouridis, D.; Nassiopoulou, A.G. A thermal convective accelerometer system based on a silicon sensor-study and packaging. *Sens. Actuators A* **2006**, *132*, 147–153. [CrossRef]

22. Bahari, J.; Jones, J.D.; Leung, A.M. Sensitivity improvement of micromachined convective accelerometers. *J. Microelectromech. Syst.* **2012**, *21*, 646–655. [CrossRef]

23. Courteaud, J.; Combette, P.; Crespy, N.; Cathebras, G.; Giani, A. Thermal simulation and experimental results of a micromachined thermal inclinometer. *Sens. Actuators A* **2008**, *141*, 307–313. [CrossRef]

24. Zhao, K.; Dalton, P.; Yang, G.C.; Scherer, P.W. Numerical modeling of turbulent and laminar airflow and odorant transport during sniffing in the human and rat nose. *Chem. Senses.* **2006**, *31*, 107–118. [CrossRef] [PubMed]

25. Garraud, A.; Combette, P.; Courteaud, J.; Giani, A. Effect of the detector width and gas pressure on the frequency response of a micromachined thermal accelerometer. *Micromachines* **2011**, *2*, 167–178. [CrossRef]

26. Garraud, A.; Combette, P.; Pichot, F.; Courteaud, J.; Charlot, B.; Giani, A. Frequency response analysis of an accelerometer based on thermal convection. *J. Micromech. Microeng.* **2011**, *21*, 035017. [CrossRef]

27. Mailly, F.; Martinez, A.; Giani, A.; Pascal-Delannoy, F.; Boyer, A. Effect of gas pressure on the sensitivity of a micromachined thermal accelerometer. *Sens. Actuators A* **2003**, *109*, 88–94. [CrossRef]

28. Van, T.; Dzung, V.D.; Susumu, S. A 2–DOF convective micro accelerometer with a low thermal stress sensing element. *Smart Mater. Struct.* **2007**, *16*, 2308–2314.

Article

Design and Performance Test of an Ocean Turbulent Kinetic Energy Dissipation Rate Measurement Probe

Bian Tian [1], Huafeng Li [1], Hua Yang [2,*], Yulong Zhao [1], Pei Chen [3] and Dalei Song [4]

[1]   State Key Laboratory for Manufacturing Systems Engineering, Xi'an Jiaotong University, Xi'an 710049,
       China; t.b12@mail.xjtu.edu.cn (B.T.); lihuafeng@stu.xjtu.edu.cn (H.L.); zhaoyulong@mail.xjtu.edu.cn (Y.Z.)
[2]   College of Information Science and Engineering, Ocean University of China, Qingdao 266100, China
[3]   School of Construction Machinery, Chang'an University, Xi'an 710064, China; chdchenpei@chd.edu.cn
[4]   College of Engineering, Ocean University of China, Qingdao 266100, China; songdalei@ouc.edu.cn
*   Correspondence: hyang@ouc.edu.cn; Tel.: +86-532-6678-2926

Received: 13 May 2018; Accepted: 13 June 2018; Published: 20 June 2018

Abstract: Ocean turbulent kinetic energy dissipation rate is an essential parameter in marine environmental monitoring. Numerous probes have been designed to measure the turbulent kinetic energy dissipation rate in the past, and most of them utilize piezoelectric ceramics as the sensing element. In this paper, an ocean turbulent kinetic energy dissipation rate measurement probe utilizing a microelectromechanical systems (MEMS) piezoresistor as the sensing element has been designed and tested. The triangle cantilever beam and piezoresistive sensor chip are the core components of the designed probe. The triangle cantilever beam acts as a velocity-force signal transfer element, the piezoresistive sensor chip acts as a force-electrical signal transfer element, and the piezoresistive sensor chip is bonded on the triangle cantilever beam. One end of the triangle cantilever beam is a nylon sensing head which contacts with fluid directly, and the other end of it is a printed circuit board which processes the electrical signal. A finite element method has been used to study the effect of the cantilever beam on probe performance. The Taguchi optimization methodology is applied to optimize the structure parameters of the cantilever beam. An orthogonal array, signal-to-noise ratio, and analysis of variance are studied to analyze the effect of these parameters. Through the use of the designed probe, we can acquire the fluid flow velocity, and to obtain the ocean turbulent dissipation rate, an attached signal processing system has been designed. To verify the performance of the designed probe, tests in the laboratory and in the Bohai Sea are designed and implemented. The test results show that the designed probe has a measurement range of 10^{-8}–10^{-4} W/kg and a sensitivity of 3.91×10^{-4} (Vms²)/kg. The power spectrum calculated from the measured velocities shows good agreement with the Nasmyth spectrum. The comparative analysis between the designed probe in this paper and the commonly used PNS probe has also been completed. The designed probe can be a strong candidate in marine environmental monitoring.

Keywords: turbulent kinetic energy dissipation rate; probe; microelectromechanical systems (MEMS) piezoresistive sensor chip; Taguchi method; marine environmental monitoring

---

1. Introduction

The accurate measurement of the ocean turbulent dissipation rate is a key point in marine environmental monitoring. The turbulent kinetic energy dissipation rate represents the rate at which turbulent kinetic energy is converted into molecular thermal kinetic energy under the action of molecular viscosity, which can be used to build mathematical models to simulate the macroscopic motions of the ocean. Through the study of the turbulent kinetic energy dissipation rate, the ocean turbulent dissipation process can be constructed, which is of great importance to improve the physical model of the ocean and to study the law of mixing in the ocean. Ocean turbulence contributes significantly to the transport of momentum, heat, and mass in the ocean and has significant effects on the velocity, temperature characteristics, and distribution of dissolved and granular substances in the ocean.

Broadly used probes designed to measure the ocean turbulent kinetic energy dissipation rate utilize piezoelectric ceramics as the sensitive element. The first probe based on piezoelectric ceramics was designed by Ribner and Siddon [1,2], which was designed for atmospheric environmental monitoring. Osborn [3,4], at the University of British Columbia, studied the airfoil probe, which was a suitable and useful velocity sensor for oceanic turbulence. The probe had advantages over heated anemometry due to its rugged nature, lower susceptibility to fouling, and inherent linearity. However, improvements were also needed in the construction technique, so he improved the design of the former probe and tested it in Howe Sound near Vancouver, British Columbia. He concluded that the airfoil probe combined with a free-fall instrument housing was ideal for studying the vertical current shear in the ocean. The resolution and sensitivity were sufficient to provide estimates of the energy dissipation directly. This offers a classical approach to measuring the ocean turbulent dissipation rate. Wolk et al. [5] evaluated the performance of a free-falling microstructure profiler-TurboMAP, and the probe used on TurboMAP was almost exactly the same size and shape as the so-called Osborn probe. By assuming a universal form of the turbulence spectrum, turbulent kinetic energy dissipation rates below 5×10^{-4} W/kg can be estimated. Moum et al. [6], at Oregon State University, designed a shear sensor that utilizes piezobimorph ceramic as the sensing element which they named OSU, and they compared almost 1000 microstructure profiles obtained from different shear probes. Measurements of ocean microstructure were made in the turbulent Faroe Bank Channel overflow using a turbulence instrument equipped with SPM-38 turbulence shear probe [7]. The dissipation rate measurement results revealed that the lowest detection level was as low as 5×10^{-11} W/kg, which was comparable to the best available vertical microstructure profilers. Tianjin University in China designed a series of ocean turbulent dissipation rate measurement probes, such as TMR1, TJUA, and TJUB [8–11]. The TMR1 was the first ocean turbulence sensor probe they designed, and after that, they proposed the TJUA and the TJUB. Both the TMR1 and TJUA used piezoelectric ceramics as the sensitive material, and the TJUB used carbon fiber reinforced polymer plastic as the sensitive material, which lead to a high sensitivity of 2.52×10^{-4} (Vms^2)/kg. The PNS series probes developed by Prandke at ISW Wassermesstechnik have been broadly used in marine environmental monitoring. PNS shear probes are airfoil-type microstructure velocity fluctuation sensors designed for a microstructure profiler. PNS-93 was specially designed for use in operational microstructure measuring systems. The use of a cantilever which transmits the lift force on the probe tip to the piezoceramic beam in the interior of the sensor was a huge step forward. The sensitivity of PNS-93 shear probe varied in a range from 1.5×10^{-4} to 2.1×10^{-4} (Vms^2)/kg [12]. Then, a series of PNS probes were developed, such as PNS-98, PNS-03, and PNS-06. Cisewski et al. [13] investigated the mixing regime of the upper 180 m of a mesoscale eddy in the vicinity of the Antarctic Polar Front at 47° S and 21° E using the MSS profiler equipped with a microstructure shear sensor PNS-98. PNS-03 has an airfoil diameter and length of 3 mm and 4 mm, respectively, and PNS-06 has an airfoil diameter and length of 6 mm and 10 mm, respectively. PNS-03 and PNS-06 shear probes are available in a compact version and a version with thread M10. The sensitivities are in the order of 1×10^{-4} (Vms^2)/kg for the PNS-03 and 4×10^{-4} (Vms^2)/kg for the PNS-06 [14]. Now, the PNS series probes are the basic configurations of MSS-series microstructure turbulence profilers. Table 1 lists the detailed parameters of five series of probes and of the probe designed in this paper.

The aforementioned probes are almost based on the piezoelectric effect, and the probes based on piezoresistive effect are rarely mentioned. The signal processing system of piezoelectric-effect-based probes is complicated, but the signal processing system of piezoresistive-effect-based probes is simple. It is difficult for piezoelectric-effect-based probes to measure static signals, and the calibration also can be difficult. For the financial cost, the piezoresistive chip is suitable for mass production and micromation, but piezoelectric ceramics are sputtered and of high difficulty and high cost. In this paper, we designed an ocean turbulent kinetic energy dissipation rate measurement probe based on the piezoresistive effect. The triangle cantilever beam and piezoresistive sensor chip are the core components of the designed probe. An attached signal processing system has been designed and connected with a printed circuit board in the probe. Tests in the lab and in the ocean have been conducted separately. Test results reveal that the probe has good performance both in measurement range and sensitivity. The comparisons between the designed probe and the PNS probe have also been made.

Table 1. The detailed parameters of probes.

Probe	Dimensions (mm)	Withstand Pressure (mm)	Spatial Resolution (mm)	Sensitivity (Vms²/kg)
Osborn's	6.3 in diameter - in length	230	1	4×10^{-4}
SPM-38	9.5 in diameter 127 in length	1000	10	0.57×10^{-4}
TJUB	10 in diameter 63.3 in length	1000	5	2.52×10^{-4}
PNS-03	8 in diameter 77 in length	1000	-	1×10^{-4}
PNS-06	8 in diameter 77 in length	1000	-	4×10^{-4}
This paper	8 in diameter 57 in length	1000	-	3.91×10^{-4}

2. Structure Design and Working Principle

The schematic diagram of the proposed ocean turbulent kinetic energy dissipation rate measurement probe is shown in Figure 1. The proposed probe consisted of a nylon sensing head, a piezoresistive sensor chip, a printed circuit board, a stainless-steel triangle cantilever beam, a half-cylinder gasket, and a stainless-steel shell. The nylon sensing head contacts with fluid flow mass directly and transforms the velocity signal to a force signal. The piezoresistive sensor chip was fabricated with an SOI (silicon-on-insulator) wafer, which consists of two layers of silicon and a silicon oxide in-between. The piezoresistive sensor chip was attached to the stainless-steel triangle cantilever beam, and a printed circuit board was also attached to it. The piezoresistive sensor chip was wire-bonded to the printed circuit board and further connected to a signal processing system. The piezoresistive sensor chip, printed circuit board, and stainless-steel triangle cantilever beam were packaged using a half-cylinder gasket and a stainless-steel shell, and the nylon sensing head was fixed on the free end of the stainless-steel triangle cantilever beam.

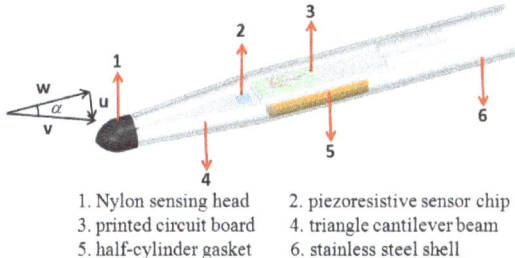

1. Nylon sensing head   2. piezoresistive sensor chip
3. printed circuit board   4. triangle cantilever beam
5. half-cylinder gasket   6. stainless steel shell

Figure 1. The structure of the designed ocean turbulent kinetic energy dissipation rate measurement probe.

When there is a kind of fluid flow through the probe, according to the theory of fluid dynamics, assuming the fluid is nonviscous, the force produced by the fluid flow per unit length of the probe is [15]:

$$f_P = \frac{1}{2}\rho v^2 \frac{dA}{dx} \sin 2\alpha \tag{1}$$

where f_P is the force per unit length of probe, ρ is the density of the fluid, v is the velocity of the fluid flow, dA/dx is the rate of change in the cross-section area in the axial direction, and α is the angle of attack, as shown in Figure 1. In fact, the viscous effect also plays a role in terms of the kinetic energy dissipation, but the viscosity coefficient of water is very small and thus the viscous force produced by water is also very small [4]. The nonviscous fluid is considered as the flow fluid in this paper is to simplify the mathematical model. Then, the overall force on the probe is the integration from the bottom to the top of the probe:

$$F = \int f_P dx = \frac{1}{2}\rho Av^2 \sin 2\alpha = \rho A(v \sin\alpha)(v \cos\alpha) = \rho Auw \tag{2}$$

where u is the cross velocity and w is the axial velocity.

When the fluid flows through the probe, the nylon sensing head produces deformation and thus the stainless-steel triangle cantilever beam produces deformation too. Then, the strain of the piezoresistive sensor chip changes and the variation of resistance of the piezoresistor is

$$\frac{\Delta R}{R} = \pi F \tag{3}$$

where ΔR is the variation of resistance of the piezoresistor, R is the original value of the piezoresistor, and π is the piezoresistive coefficient of silicon. Then, the output voltage of piezoresistive sensor chip is

$$U_o = U_i \frac{\Delta R}{R} = U_i \pi F = U_i \pi \rho Auw \tag{4}$$

where U_o is the output voltage of the piezoresistive sensor chip and U_i is the input voltage. From Equation (4), we know that the output voltage of the piezoresistive sensor chip is directly proportional to the cross velocity of fluid.

Defining S as the sensitivity of the designed probe, the sensitivity can be expressed as

$$S = \frac{U_o}{u} = U_i \pi \rho Aw. \tag{5}$$

Equation (5) can be rewritten as

$$u = \frac{U_o}{S} \tag{6}$$

then, the variation of cross velocity can be obtained as

$$\frac{\partial u}{\partial t} = \frac{1}{S}\frac{\partial U_o}{\partial t}. \tag{7}$$

According to the Taylor frozen theory:

$$\frac{\partial u}{\partial x} = \frac{1}{Sw}\frac{\partial U_o}{\partial t}. \tag{8}$$

Then, the turbulent kinetic energy dissipation rate can be calculated as shown in Figure 2. This is done by first deleting the singular value and calculating the shear frequency spectrum $\varphi(f)$ in the frequency domain, followed by calculating the shear wavenumber spectrum $\psi(k)$ in the wavenumber domain. Then, the probe response correction and motion compensation correction is completed, as shown in Figure 2. Finally, by confirming the integral cutoff wavenumber k_{max} of the shear spectrum, the turbulent kinetic energy dissipation rate can be expressed as

$$\varepsilon = 7.5\gamma \int_{k_{min}}^{k_{max}} \psi(k)dk \tag{9}$$

where ε is the turbulent kinetic energy dissipation rate, γ is the kinematic viscosity coefficient, and $\psi(k)$ is the power spectrum of the shear velocity.

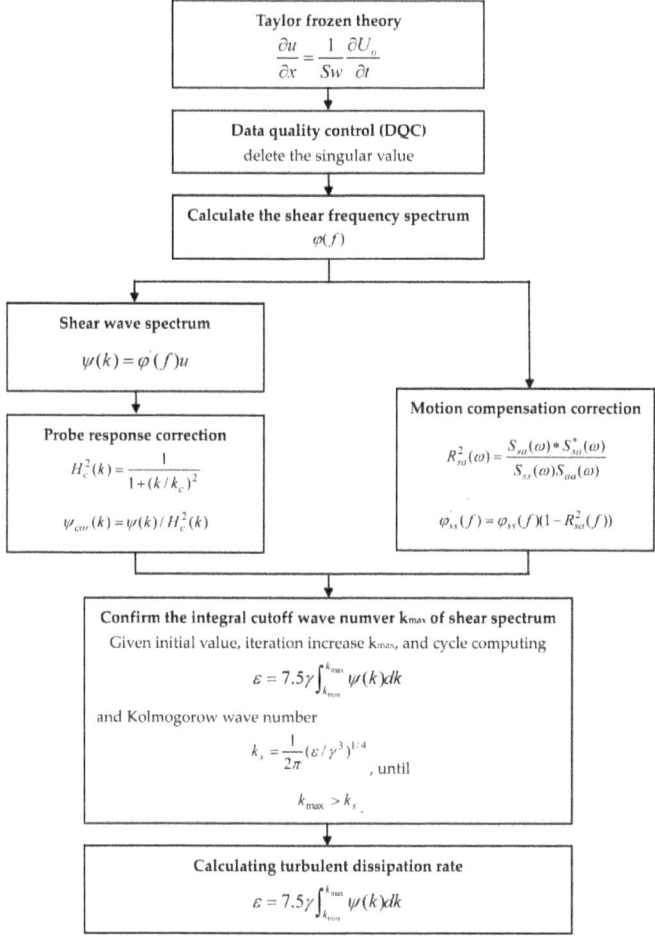

Figure 2. The flow chart of calculating turbulent kinetic energy dissipation rate.

A signal processing system was designed and connected with the probe. The signal processing system main consisted of signal amplification, signal filtering, AD sample, reference source, single chip microcomputer, and power source. Further, some guard blocks were also designed, such as a protective guard, a fixing cap, a seal ring, and a water-tight joint. The overall diagram of the probe and the signal processing system is shown in Figure 3.

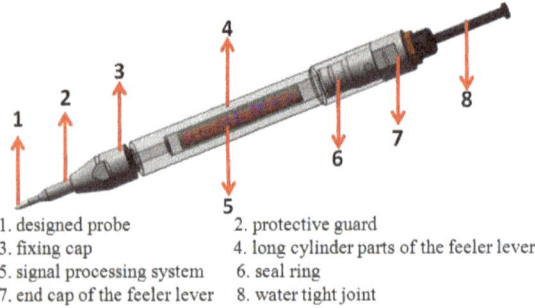

1. designed probe 2. protective guard
3. fixing cap 4. long cylinder parts of the feeler lever
5. signal processing system 6. seal ring
7. end cap of the feeler lever 8. water tight joint

Figure 3. The overall diagram of the probe and signal processing system.

3. Finite Element Simulation and Optimal Design

According to the structure design and the working principle of the probe, the triangle cantilever beam has great influence on the performance of the probe. High sensitivity and high natural frequency are needed when the probe is working. Thus, the confirmation of the cantilever beam dimensions is an essential step in probe design. The Taguchi method was used to analyze and determine the dimensions of the cantilever beam. Taguchi uses a simple design of orthogonal array to study the entire parameter space with only a small number of experiments [16]. The greatest advantage of this method is that it saves effort in conducting experiments by reducing the experimental time, reducing the cost, and accelerating the pace at which significant factors are discovered [17]. In this paper, three parameters (height, thickness, and width of cantilever beam) at five levels were designed, and the fractional factorial design used was $L_{25}(5^3)$ orthogonal array, as shown in Table 2.

We focused on the sensitivity of the probe, so the stress and deflection of the triangle cantilever beam when it was placed in the fluid flow was studied. Since we were concerned with the robustness of the probe, the natural frequency of the triangle cantilever beam had to be analyzed. The COMSOL Multiphysics (Version 5.3a, COMSOL Inc., Stockholm, Sweden) was used to simulate the results. In the simulation, water was used as the flow fluid. The triangle cantilever beam was made from 316L stainless steel, and the density, Young modulus, and Poisson's ratio were 7850 kg/s, 2×10^{11} Pa, and 0.33, respectively. The sensing head was made from nylon material, and the density, Young modulus, and Poisson's ratio were 1150 kg/s, 2×10^9 Pa, and 0.4, respectively. The input flow velocity was set as 1 m/s, and the output condition was set as 0 Pa. Different simulation conditions lead to different results, so the conditions such as material parameters, input and output direction, input and output velocity, boundary condition, area of flow field, and meshing influenced the simulation results. The simulation results are listed in Table 3. Orthogonal arrays of Taguchi, the signal-to-noise (*S/N*) ratio, and the analysis of variance (ANOVA) were employed to find the optimal levels and to analyze the effect of the cantilever beam structure parameters on probe performance.

Table 2. Cantilever beam parameters and their levels.

Symbol	Parameter	Unit	Level 1	Level 2	Level 3	Level 4	Level 5
A	Height	mm	16	18	20	22	24
B	Thickness	mm	0.2	0.25	0.3	0.35	0.4
C	Width	mm	4	5	6	7	8

The range analysis was aimed at illuminating the significant levels of different influencing parameters on the performance of probe. Thus, the most significant parameter could be disclosed according to the results of the range analysis. The range analysis results of stress, deflection, and frequency are shown in Figures 4a, 5a and 6a, respectively. The larger the stress and deflection is, the higher the sensitivity is. Thus, we can see in Figures 4a and 5a that thickness is the most significant parameter influencing sensitivity among the three parameters, and the thinner the thickness is, the higher the sensitivity is. Also, the greater the frequency is, the stronger the robustness is. Thus, we can know from Figure 6a that thickness is the most significant parameter influencing robustness among the three parameters, and the thicker the thickness is, the higher the robustness is.

Table 3. Experimental layout and results using an $L_{25}(5^3)$ orthogonal array.

Experiment Number	A	B	C	Stress	Deflection	Frequency
1	16	0.2	4	6.86	49.4	232.68
2	16	0.25	5	3.38	20.5	360.31
3	16	0.3	6	2.07	9.93	513.37
4	16	0.35	7	1.27	5.39	690.19
5	16	0.4	8	0.81	3.16	888.59
6	18	0.2	5	5.99	54.3	218.5
7	18	0.25	6	3.23	23.4	331.03
8	18	0.3	7	2	11.7	464.05
9	18	0.35	8	1.25	6.46	616.23
10	18	0.4	4	1.88	8.59	544.03
11	20	0.2	6	5.69	60.4	204.39
12	20	0.25	7	3.13	26.7	304.65
13	20	0.3	8	1.86	13.6	421.86
14	20	0.35	4	2.71	17.2	383.35
15	20	0.4	5	1.71	9.12	516.99
16	22	0.2	7	5.35	67	190.92
17	22	0.25	8	3.08	30.3	281.05
18	22	0.3	4	4.25	35.2	265.61
19	22	0.35	5	2.39	17.8	369.4
20	22	0.4	6	1.53	9.87	487.13
21	24	0.2	8	5.01	74.3	178.4
22	24	0.25	4	6.82	77.1	178.63
23	24	0.3	5	3.62	35.8	259.16
24	24	0.35	6	2.24	18.9	352.26
25	24	0.4	7	1.58	11.6	456.95

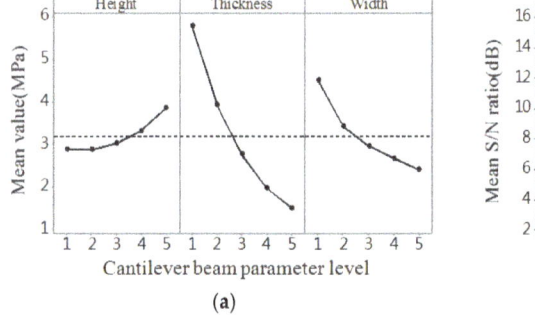
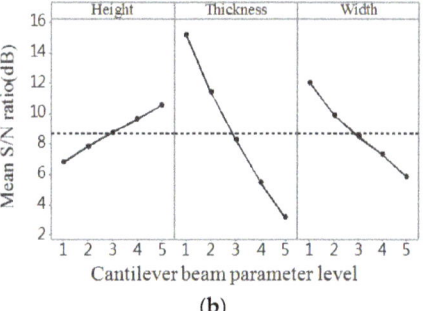

Figure 4. The result graph for stress. (**a**) Mean value; (**b**) Mean *S/N* ratio.

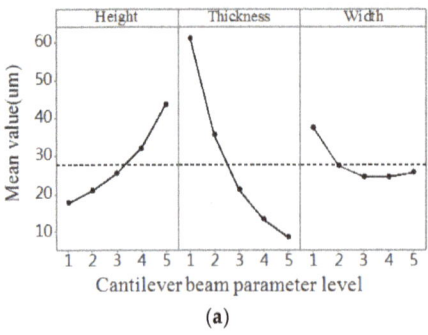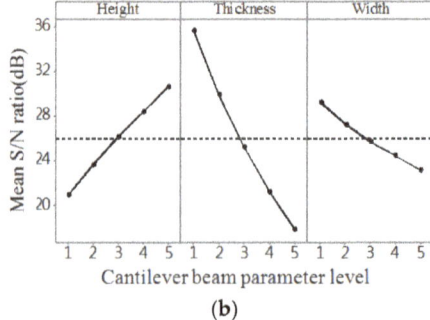

Figure 5. The result graph for deflection. (**a**) Mean value; (**b**) Mean *S/N* ratio.

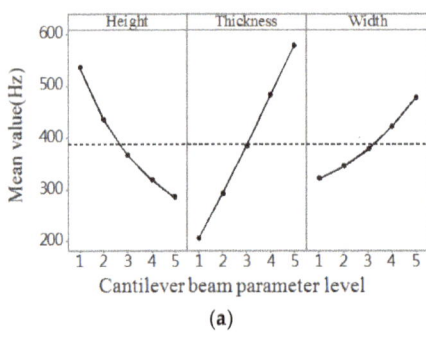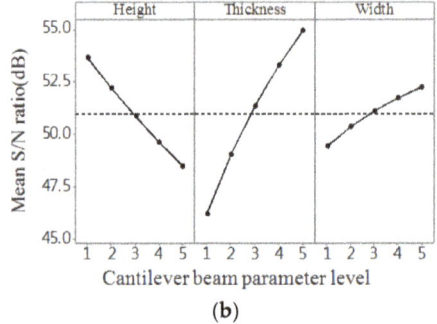

Figure 6. The result graph for frequency. (**a**) Mean value; (**b**) Mean *S/N* ratio.

Taguchi used *S/N* ratio as the quality characteristic of choice. There are three categories of *S/N* ratio characteristics when the characteristic is continuous [17–19]. First, larger is better:

$$S/N = -10\log\frac{1}{n}\left(\sum\frac{1}{y^2}\right); \tag{10}$$

second, nominal is the best:

$$S/N = 10\log\frac{\bar{y}}{s_{\bar{y}}^2}; \tag{11}$$

and third, smaller is better:

$$S/N = -10\log\frac{1}{n}\left(\sum y^2\right) \tag{12}$$

where n is the number of observations, y is the observed data, $\bar{y}$ is the average of observed data, and $s_{\bar{y}}^2$ is the variance of y. Large stress and deflection generate high sensitivity, and high natural frequency leads to strong robustness. For all the types of characteristics with the above *S/N* ratio transformation, the higher the *S/N* ratio, the better the result. The *S/N* ratio analysis results of stress, deflection, and frequency are shown in Figures 4b, 5b and 6b, respectively. Similar results can be concluded from the range analysis, that is, the thickness is the most significant parameter influencing sensitivity and robustness among the three parameters, and the thinner the thickness, the higher the sensitivity, and the thicker the thickness, the higher the robustness. Figure 7 shows the frequency domain response of 25 combinations. According to the results from the range analysis and *S/N* ratio analysis, B1C1A5 is the optimal combination of the structure parameters for sensitivity, and B5A1C5 is the optimal combination of the structure parameters for robustness.

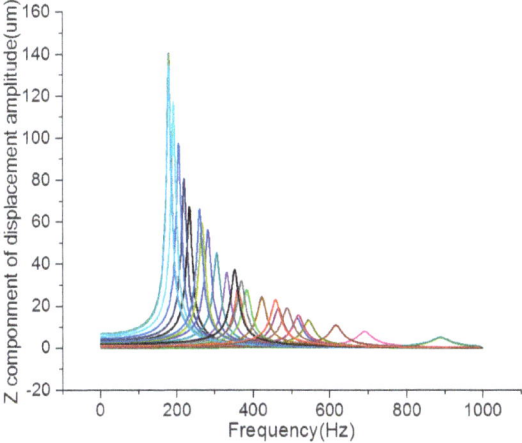

Figure 7. The frequency domain response of 25 combinations.

The purpose of the ANOVA was to investigate the design parameters that significantly affected the quality characteristic [20]. The total sum of square SS_T from the S/N ratio η can be calculated as [21]

$$SS_T = \sum_{i=1}^{m} (\eta_i - \bar{\eta})^2 \tag{13}$$

where m is the number of the experiment, η_i is the mean S/N ratio for the ith experiment, and η is the mean S/N ratio. The sum of squares from the tested parameter SS_P can be calculated as

$$SS_P = \sum_{j=1}^{t} \frac{(S\eta_j)^2}{t} - \frac{1}{m}\left(\sum_{i=1}^{m} \eta_i\right)^2 \tag{14}$$

where P denotes one of the parameters, j is the level number of this parameter P, t is the repetition of each level of the parameter P, and $S\eta_j$ is the sum of the S/N ratio involving this parameter P and level j. The sum of square from error SS_E is

$$SS_E = SS_T - SS_A - SS_B - SS_C. \tag{15}$$

The total degrees of freedom D_T is $D_T = m - 1$, and the degree of freedom from tested parameters D_P is $D_P = t - 1$. Thus, the degree of freedom from error D_E is $D_E = D_T - D_A - D_B - D_C$. The variance of the tested parameters V_P is $V_P = SS_P/D_P$, and the variance from the error V_E is $V_E = SS_E/D_E$. Then, the F value for each design parameter is simply the ratio of the mean-of-square deviation to the mean-of-square error:

$$F_P = V_P/V_E. \tag{16}$$

The corrected sum of square S_P can be calculated as

$$S_P = SS_P - D_P V_E. \tag{17}$$

Then, the percentage contribution ρ_P can be calculated as

$$\rho_P = S_P/SS_T. \tag{18}$$

Table 4 shows the results of ANOVA for stress, which shows that the thickness parameter is the most significant structure parameter affecting the stress. The width parameter also has a significant effect on stress, and the height parameter has an insignificant effect on stress. The contributions for the stress of the three structure parameters height, thickness, and width are 3.21%, 73.67%, and 16.21%, respectively. Table 5 shows the results of the ANOVA for deflection, which shows that the thickness parameter is the most significant structure parameter affecting the deflection. The height parameter also has a significant effect on deflection, and the width parameter has an insignificant effect on deflection. The contributions for the deflection of the three structure parameters height, thickness, and width are 14.77%, 69.9%, and 2.75%, respectively. Table 6 shows the results of the ANOVA for frequency, which shows that the thickness parameter is the most-significant structure parameter affecting the frequency. The height parameter and width parameter also have a significant effect on frequency. The contributions for the frequency of the three structure parameters height, thickness, and width are 26.2%, 58.53%, and 9.54%, respectively.

From the results of the Taguchi method, we conclude that high sensitivity derives from the thin thickness of the cantilever beam and strong robustness derives from the thick thickness of the cantilever beam. This results in a trade-off between sensitivity and robustness. In this paper, we finally chose a height of 20 mm, a thickness of 0.25 mm, and a width of 6 mm in consideration of the comprehensive performance of the designed probe.

Table 4. Results of the ANOVA for stress.

Parameter	DOF	Sum of Squares	Variance	F Value	p Value	Contribution(%)
Height	4	3.432	0.8580	3.79	0.032	3.21
Thickness	4	58.852	14.7131	64.95	0.000	73.67
Width	4	13.654	3.4134	15.07	0.000	16.21
Error	12	2.718	0.2265			6.91
Total	24	78.657				100

Table 5. Results of the ANOVA for deflection.

Parameter	DOF	Sum of Squares	Variance	F Value	p Value	Contribution(%)
Height	4	2108.0	527.00	8.04	0.002	14.77
Thickness	4	8998.2	2249.55	34.32	0.000	69.9
Width	4	605.0	151.26	2.31	0.118	2.75
Error	12	786.7	65.56			12.58
Total	24	12,497.9				100

Table 6. Results of the ANOVA for frequency.

Parameter	DOF	Sum of Squares	Variance	F Value	p Value	Contribution(%)
Height	4	201,236	50,309	28.44	0.000	26.2
Thickness	4	440,802	110,200	62.30	0.000	58.53
Width	4	77,740	19,435	10.99	0.001	9.54
Error	12	21,227	1769			5.73
Total	24	741,004				100

4. Fabrication and Encapsulation

The fabrication of the designed probe main contained three parts: first, the fabrication of the piezoresistive sensor chip; second, the fabrication of the printed circuit board and signal processing circuit board; and third, the fabrication of the stainless-steel parts, such as the triangle cantilever beam and protective guard.

The fabrication of the piezoresistive sensor chip utilized microelectromechanical systems (MEMS) technology. A very thin active layer SOI wafer was used as the starting material. The thickness of the top silicon layer and the buried silicon oxide layer were about 5 μm and 0.4 μm, respectively. The front side of the SOI wafer was implanted by boron ions by means of reactive ion etching (RIE) and the piezoresistors were formed. After that, a high dose of boron ions was diffused and the wafer was annealed. To protect the piezoresistors on the front side of SOI wafer from being corroded, a silicon nitride layer was deposited by low pressure chemical vapor deposition (LPCVD). Then, the metal wire was etched by RIE and contact pads were also formed. The fabricated piezoresistive sensor chip is shown in Figure 8d.

The fabrication of the printed circuit board and signal processing circuit board utilized integrated circuit (IC) technology and microelectronics technology. The fabrication of stainless-steel parts utilized a line-cutting process. To package the piezoresistive sensor chip, some preparatory work had to be completed. First, ultrasonic cleaning of the nylon sensing head, the triangle cantilever beam, the half-cylinder gasket, and the stainless-steel shell was performed using a KH3200DB CNC ultrasonic cleaner. The piezoresistive sensor chip was cleaned using acetone. Then, all of the components were dried on a hot plate. Second, the piezoresistive sensor chip was adhered onto the triangle cantilever beam using M-Bond 610 glue. Also, the printed circuit board was adhered onto the triangle cantilever beam. The fully adhered finished probe is shown in Figure 8c. Third, the gold wire was soldered between the piezoresistive sensor chip contact pad and the printed circuit board contact pad. Then, silica gel was coated on the probe to protect the piezoresistive sensor chip and gold wire. After that, the wire was soldered on the printed circuit board to connect it to the signal processing system. Finally, the triangle cantilever beam and half-cylinder gasket were configured into the stainless-steel shell, and the components were affixed by AB glue. To guarantee that the the probe was leakproof, the shell was filled with silica gel. The packaged probe is shown in Figure 8b. The fabrication and encapsulation of the signal processing system was similar to the fabrication and encapsulation of the probe, and the fabricated probe and signal processing system is shown in Figure 8a.

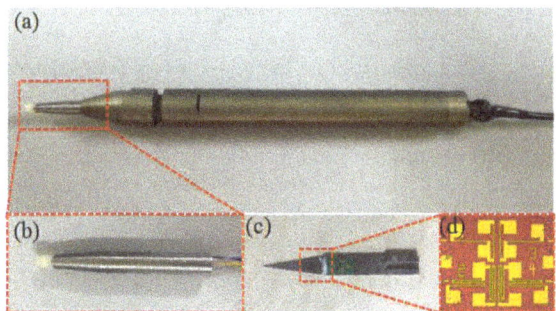

Figure 8. The fabricated probe and MEMS piezoresistive sensor chip. (**a**) The overall of measurement system; (**b**) The fabricated probe; (**c**) The cantilever beam with piezoresistive sensor chip and PCB board on it; (**d**) The picture of piezoresistive sensor chip.

5. Experiments and Results

To verify the performance of designed probe, tests in the laboratory and in the Bohai Sea were designed and conducted. A flowing cycling experiment system was purposely designed in the laboratory, as shown in Figure 9. The system consisted of top and bottom sinks, an overflow gap, a water inlet and water outlet, spin equipment, a jet orifice, turbulence compensation, and an emptying valve. The probe was installed in the experiment system, and the fluid flow was set under a constant velocity with attack angles from −10° to 10° with a step of 2°. The experimental setup in the laboratory is shown in Figure 10.

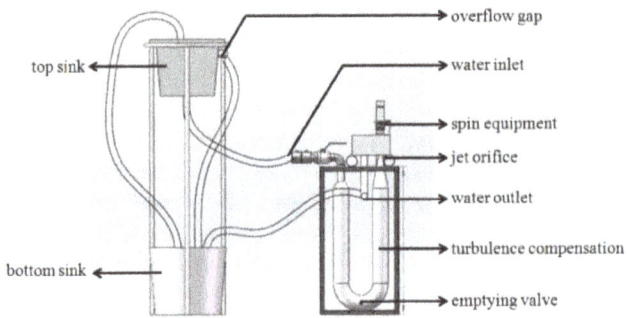

Figure 9. The schematic diagram of flowing cycling experiment system.

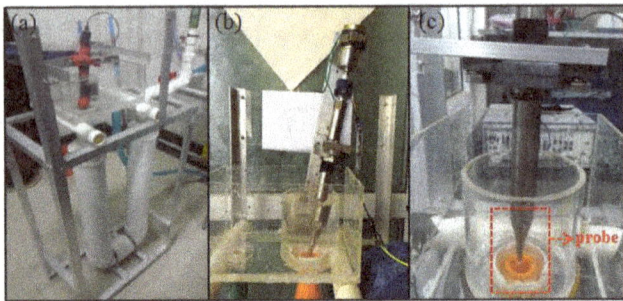

Figure 10. The schematic diagram of experimental setup in the laboratory. (**a**) The flowing cycling experiment system; (**b**) The measurement system in the experiment; (**c**) The designed probe.

We obtained 11 sets of data in each experiment, and every set of data includes the angle of attack, the value of sin 2α, the output voltage, and the value of $U_o/\rho v^2$. The sensitivity was calculated using Equation (5), and the experiment results are listed in Table 7. From the experiment results, the relationship between the value of $U_o/\rho v^2$ and sin 2α can be described, which is shown in Figure 11, where the cubic polynomial fitting is used to describe the relationship. The first order coefficient of the polynomial is the sensitivity of probe, which is 3.91×10^{-4} (Vms^2)/kg. This sensitivity is larger than that of SPM-38, TJUB, and PNS-03, and close to that of Osborn's and PNS-06, as listed in Table 1. As a result, we can conclude that the probe designed in this paper can be a probable choice in ocean turbulent kinetic energy dissipation rate measurements.

Table 7. The experimental results in the flowing cycling experiment system.

α	Sin 2α	U_o	$U_o/\rho v^2$
-10	-0.342	-1.0261	-1.6367
-8	-0.276	-0.8001	-1.2773
-6	-0.208	-0.6037	-0.9653
-4	-0.139	-0.3727	-0.5982
-2	-0.070	-0.1842	-0.2984
0	0.000	0.0457	0.0669
2	0.070	0.2241	0.3504
4	0.139	0.3388	0.5328
6	0.208	0.4427	0.6980
8	0.276	0.7183	1.1359
10	0.342	0.9579	1.5186

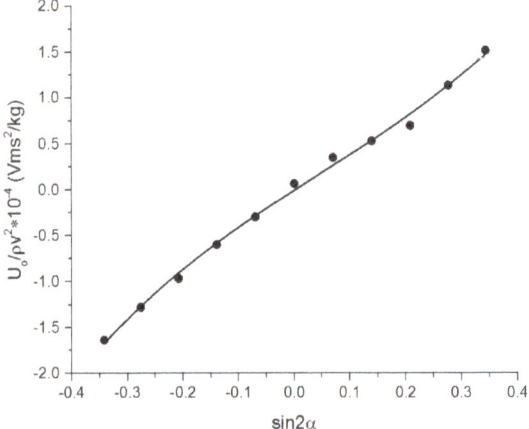

Figure 11. The relationship between the value of $U_o/\rho v^2$ and sin 2α.

The experiment in the Bohai Sea was designed to verify the practical performance of the designed probe. The probe was carried by an ocean vertical profiler and every profiler carried two probes which were designed in this paper and two PNS-series probes, as shown in Figure 12. A gallows was used for the profiler's release and recovery. The release velocity was about 1 m/s and the profiler was released at a constant speed. Figure 13a shows the velocity shear data of MEMS and PNS in the experiment, respectively. Note that normalization processing was used to make the comparison between MEMS and PNS. The data collected by MEMS and PNS in the experiment are similar to each other. Figure 13b shows the power spectrum of the velocity shear of MEMS and PNS, respectively. From Figure 13, we can see that the probe designed in this paper has comparable performance with PNS series probes.

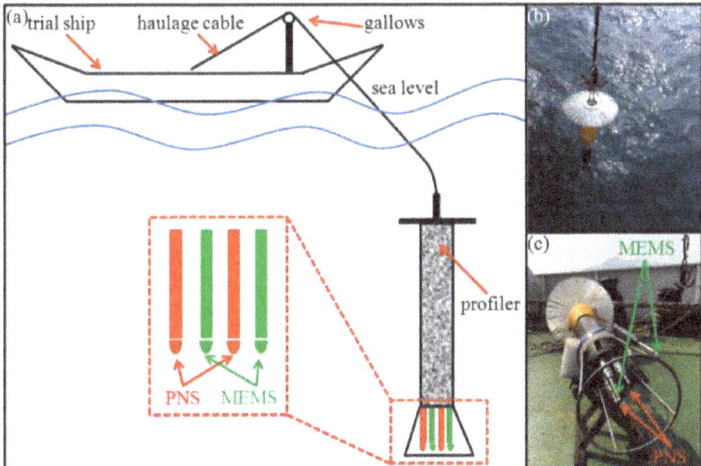

Figure 12. The configuration of the ocean test. MEMS denotes the probe designed in this paper. PNS denotes the PNS series probes. (**a**) The schematic diagram of the ocean test; (**b**) The release of the profiler; (**c**) The configuration of the probes.

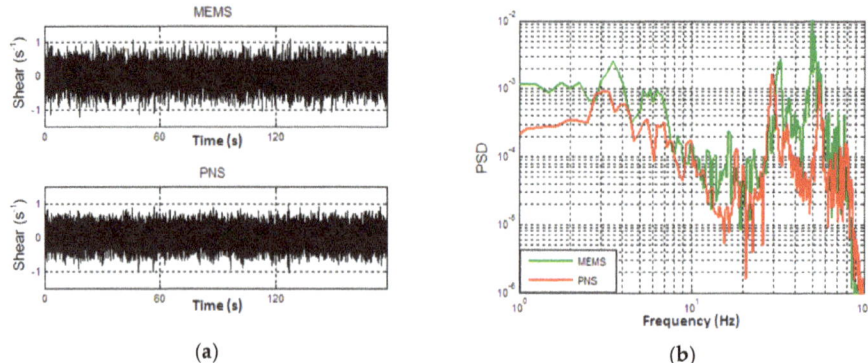

Figure 13. The response results of MEMS and PNS. (**a**) The velocity shear data of MEMS and PNS; (**b**) The power spectrum of velocity shear of MEMS and PNS.

The measured power spectrum is routinely compared to the empirical turbulence spectrum, which was measured by Nasmyth [22], as shown in Figure 14a, and the black dashed curve represents the Nasmyth spectrum. The Nasmyth spectrum is an ideal curve, but in the actual experiments, the results were influenced by noise and vibration, so the experiment results do not match with Nasmyth spectrum strictly. The probes were carried by an ocean vertical profiler, and the release velocity of the profiler and the vibration of the profiler caused by the ship affected the experimental results, so the velocity measurements have spurious contributions from the high-frequency vibrations of towed vehicle. Further, the roll and heave of the ship included large variations of the speeds and depths of the towed vehicles. However, according to Figure 14a, we know that the experiment results both from MEMS and PNS show good agreement with the Nasmyth spectrum. The cut-off frequency is about 100 cpm, and the ocean turbulent kinetic energy dissipation rate of MEMS and PNS are about 3.34×10^{-7} W/kg and 1.25×10^{-7} W/kg, respectively. The measured ocean turbulent kinetic energy dissipation rates are coincident with the estimated value in this ocean area.

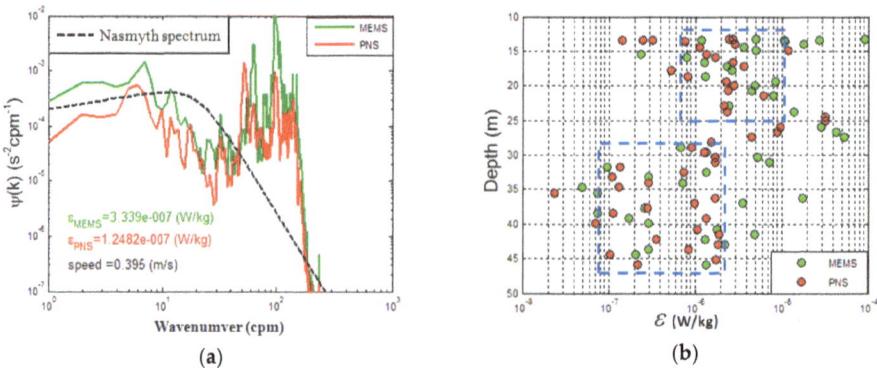

Figure 14. The experiment results of ocean test. (**a**) The power spectrum compared with Nasmyth spectrum. The black dashed curve represents the Nasmyth spectrum; (**b**) The ocean turbulent dissipation rate of MEMS and PNS.

The ocean turbulent kinetic energy dissipation rate was calculated using Equation (9) and the experiment results are shown in Figure 14b. The ocean turbulent kinetic energy dissipation rates measured by MEMS and PNS in the 10–50-m upper mixing area are similar and mainly between 10^{-8} W/kg to 10^{-4} W/kg. From Figure 14b, we know that the ocean turbulent kinetic energy dissipation rate decreases with the increase of depth. The ocean turbulent kinetic energy dissipation rate is centered on 10^{-6}–10^{-5} W/kg in the 10–25-m upper mixing area and centered on 10^{-7}–10^{-6} W/kg in the 30–50-m upper mixing area, as shown in Figure 14b, marked by the blue dashed box. The experiment results indicate that the probe has similar performance with PNS-series probes once again.

6. Conclusions

This paper introduced an ocean turbulent kinetic energy dissipation rate measurement probe. Different from numerous probes that utilize piezoelectric ceramics as the sensing element, the probe designed in this paper utilizes a MEMS piezoresistor as the sensitive element. The structure design and working principle have been introduced, and a signal processing system also been designed and connected with the probe. The Taguchi method has been used to study the influence of the cantilever beam structure parameters on the probe's performance. Range analysis, signal-to-noise ratio analysis, and analysis of variance were studied. Fluid flowing cycling experiments in the lab revealed that the probe has a sensitivity of 3.91×10^{-4} $(Vms^2)/kg$. The experiments in the Bohai Sea revealed that the probe has a measurement range between 10^{-8}–10^{-4} W/kg. The comparative analysis between the designed probe and the commonly used PNS-series probe shows that the designed probe has equivalent performance with the PNS-series probe. The designed probe can be a strong candidate in marine environmental monitoring.

Author Contributions: B.T. designed and fabricated the probe; Y.Z. and P.C. conceived and designed the experiments; H.Y. and D.S. performed the experiments and analyzed the data; H.L. simulated and optimized the probe and wrote the paper.

Funding: This research received funding from National High Technology Research and Development of China (863 Program, No. 2014AA093404), National Natural Science Foundation of China (No. 91748207, No. 51720105016), and China National Heavy Machinery Research Institute Co., Ltd. (No. 20170521).

Acknowledgments: This work was supported by Collaborative Innovation Center of Suzhou Nano Science and Technology.

Conflicts of Interest: The authors declare no conflict of interest.

References

1. Ribner, H.S.; Siddon, T.E. An aerofoil probe for measuring the transverse component of turbulence. *J. Am. Inst. Aeronaut. Astron.* **1965**, *3*, 747–749. [CrossRef]
2. Siddon, T.E. *A Turbulence Probe Utilizing Aerodynamic Lift*; UTIAS Technical Note No. 88; University of Toronto: Toronto, ON, Canada, 1965.
3. Osborn, T.R. Vertical profiling of velocity microstructure. *J. Phys. Oceanogr.* **1974**, *4*, 109–115. [CrossRef]
4. Osborn, T.R.; Crawford, W.R. *Turbulent Velocity Measurement with an Airfoil Probe*; IOUBC Manuscript Report No. 31; The University of British Columbia: Vancouver, BC, Canada, 1977.
5. Wolk, F.; Yamazaki, H.; Seurony, L.; Lueck, R.G. A new free-falling profiler for measuring biophysical microstructure. *J. Atmos. Ocean. Technol.* **2002**, *19*, 780–793. [CrossRef]
6. Moum, J.N.; Gregg, M.C.; Lien, R.C.; Carr, M.E. Comparison of turbulence kinetic energy dissipation rate estimates from two ocean microstructure profilers. *J. Atmos. Ocean. Technol.* **1995**, *12*, 346–366. [CrossRef]
7. Fer, I.; Peterson, A.K.; Ullgren, J.E. Microstructure measurement from an underwater glider in the turbulent Faroe Bank Channel overflow. *J. Atmos. Ocean. Technol.* **2014**, *31*, 1128–1150. [CrossRef]
8. Gu, L.; Liu, Y.H.; Wang, Z.L.; Wang, Y. Design and Experiments of Shear Probe for Ocean Turbulence. *J. Tianjin Univ.* **2009**, *42*, 733–738.
9. Gu, L.; Liu, Y.H.; Wang, Y.; Wang, Z.L. Optimization Design and Simulation of Shear Probe for Ocean. *Piezoelectr. Acoustoopt.* **2010**, *32*, 268–270.
10. Gu, L.; Wang, S.X.; Wang, Y.; Wang, Z.L.; Liu, Y.H. Optimization Design of Shear Probe for Measurement of Ocean Turbulence. In Proceedings of the IEEE International Conference on Measuring Technology and Mechatronics Automation, Zhangjiajie, China, 11–12 April 2009; Volume 2, pp. 857–860.

11. Wang, Y.H.; Xu, T.Y.; Wu, Z.L.; Liu, Y.H.; Wang, S.X. Structure Optimal Design and Performance Test of Airfoil Shear Probes. *IEEE Sens. J.* **2015**, *15*, 27–36. [CrossRef]
12. Prandke, H.; Pfeiffer, K. Shear Probe for Use in Operational Microstructure Measuring Systems. In Proceedings of the IEEE Oceans Engineering for Today's Technology and Tomorrow's Preservation, Brest, France, 13–16 September 1994; Volume 1, pp. I414–I418.
13. Cisewski, B.; Strass, V.H.; Prandke, H. Upper-ocean vertical mixing in the Antarctic Polar Front Zone. *Deep Sea Res. II* **2005**, *52*, 1087–1108. [CrossRef]
14. PNS03/06 Shear Probes for Microstructure Measurements. Available online: http://www.isw-wasser.com/prandke/images/pdf/shear_sensor.pdf (accessed on 28 March 2018).
15. Macoun, P.; Lueck, R. Modeling the Spatial Response of the Airfoil Shear Probe Using Different Sized Probes. *J. Atmos. Ocean. Technol.* **2004**, *21*, 284–297. [CrossRef]
16. Yang, W.H.; Tarng, Y.S. Design optimization of cutting parameters for turning operations based on the Taguchi method. *J. Mater. Process. Technol.* **1998**, *84*, 122–129. [CrossRef]
17. Bagci, E.; Aykut, S. A study of Taguchi optimization method for identifying optimum surface roughness in CNC face milling of cobalt-based alloy (stellite 6). *Int. J. Adv. Manuf. Technol.* **2006**, *29*, 940–947. [CrossRef]
18. Ghani, J.A.; Choudhury, I.A.; Hassan, H.H. Application of Taguchi method in the optimization of end milling parameters. *J. Mater. Process. Technol.* **2004**, *145*, 84–92. [CrossRef]
19. Asilturk, I.; Akkus, H. Determining the effect of cutting parameters on surface roughness in hard turning using the Taguchi method. *Measurement* **2011**, *44*, 1697–1704. [CrossRef]
20. Nalbant, M.; Gokkaya, H.; Sur, G. Application of Taguchi method in the optimization of cutting parameters for surface roughness in turning. *Mater. Des.* **2007**, *28*, 1379–1385. [CrossRef]
21. Lin, T.R. Experimental design and performance analysis of TiN-coated carbide tool in face milling stainless steel. *J. Mater. Process. Technol.* **2002**, *127*, 1–7. [CrossRef]
22. Nasmyth, P.W. Ocean Turbulence. Ph.D. Thesis, The University of British Columbia, Vancouver, BC, Canada, 1970.

Article

Design and Application of a High-G Piezoresistive Acceleration Sensor for High-Impact Application

Xiaodong Hu [1,*], Piotr Mackowiak [2], Manuel Bäuscher [1,2], Oswin Ehrmann [1,2], Klaus-Dieter Lang [1,2], Martin Schneider-Ramelow [2], Stefan Linke [3] and Ha-Duong Ngo [2,4,*]

[1] Department Electrical Engineering, Technische Universität Berlin, Gustav-Meyer-Allee 25, 13355 Berlin, Germany; Manuel.Baeuscher@izm.fraunhofer.de (M.B.); Oswin.Ehrmann@izm.fraunhofer.de (O.E.); kdlang@izm.fraunhofer.de (K.-D.L.)
[2] Wafer Level System Integration, Fraunhofer Institute for Reliability and Microintegration, Gustav-Meyer-Allee 25, 13355 Berlin, Germany; Piotr.Mackowiak@izm.fraunhofer.de (P.M.); martin.schneider-ramelow@izm.fraunhofer.de (M.S.-R.)
[3] Department Development, TE Connectivity GmbH, Hauert 13, 44227 Dortmund, Germany; Stephan.Linke@te.com
[4] Microsystems Engineering, University of Applied Sciences Berlin, Wilhelminenhofstraße 75A, 12459 Berlin, Germany
* Correspondence: xiaodonghu2010@gmail.com or hu@mat.ee.tu-berlin.de (X.H.); ha-duong.ngo@izm.fraunhofer.de (H.-D.N.); Tel.: +49-030-464-037981(X.H.); +49-030-464-03188 (H.-D.N.)

Received: 17 April 2018; Accepted: 24 May 2018; Published: 28 May 2018

Abstract: In this paper, we present our work developing a family of silicon-on-insulator (SOI)–based high-g micro-electro-mechanical systems (MEMS) piezoresistive sensors for measurement of accelerations up to 60,000 g. This paper presents the design, simulation, and manufacturing stages. The high-acceleration sensor is realized with one double-clamped beam carrying one transversal and one longitudinal piezoresistor on each end of the beam. The four piezoresistors are connected to a Wheatstone bridge. The piezoresistors are defined to 4400 Ω, which results in a width-to-depth geometry of the pn-junction of 14 μm $\times$ 1.8 μm. A finite element method (FEM) simulation model is used to determine the beam length, which complies with the resonance frequency and sensitivity. The geometry of the realized high-g sensor element is $3 \times 2 \times 1$ mm^3. To demonstrate the performance of the sensor, a shock wave bar is used to test the sensor, and a Polytec vibrometer is used as an acceleration reference. The sensor wave form tracks the laser signal very well up to 60,000 g. The sensor can be utilized in aerospace applications or in the control and detection of impact levels.

Keywords: high acceleration sensor; piezoresistive effect; MEMS; micro machining

1. Introduction

Nowadays, high-g sensors have become an important measurement unit in technological applications. The areas where sensors are most commonly applied include aerospace technologies, military and security systems, and renewable energy technologies [1–3]. It is important to reduce the size of the sensor to extend the field of applications. A silicon microfabrication technique makes it possible to reduce both the size of the sensor and the production cost through batch fabrication, making it suitable for mass production. In state-of-the-art technology and research, high-g acceleration sensors measure the acceleration in one- or three-axis with proof masses [3–7]. Therefore, the electrical variation from stress influence is an important parameter. Another important aspect, relevant to the accelerometer sensitivity, is the maximum displacement of the system. Obtaining a high sensitivity is the goal of recent research. Hence, new sensing mechanisms, like silicon nanowires, will be developed. These mechanisms, with their new materials, are difficult to manufacture and not yet economically-feasible. For this reason, research and development still focuses on geometry and design optimization [8]. Therefore, a novel high-g sensor with double-clamped beam was developed at Fraunhofer IZM with a measurement range of up to 60,000 g. This paper describes the concept, the simulation, and the process flow of the sensor, with device characterization at the end. For this high-g sensor, the piezoresistive effect is used. It is a stable and well-known state-of-the-art method, with a simple evaluation unit and precise accuracy. Important aspects of this developed sensor are its high robustness and its resolution. The production process of this sensor family aims to be precise and low-cost to fulfill economic requirements and make it accessible for a variety of new applications.

2. Sensor Design

The developed sensor design contains a silicon beam with four integrated piezoresistors. The geometry of the sensor is optimized for a high acceleration range up to 60,000 g. Figure 1 shows a sketch of the top view of the sensor design.

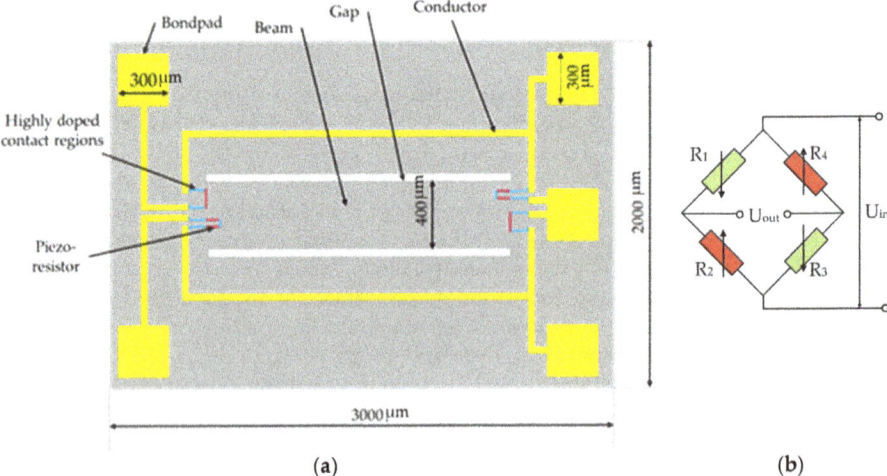

(a)   (b)

Figure 1. (**a**) Sketch of top view of sensor design, featuring one double-clamped beam carrying one transversal and one longitudinal piezoresistor on each end of the beam; (**b**) the Wheatstone bridge for equivalent circuit.

Piezoresistors and conductors are connected to form an open full Wheatstone bridge. It contains one double-clamped beam and carries one transversal and one longitudinal piezoresistor on each end of the beam. Highly doped contact regions are employed to connect the piezoresistors to conductors. The overall chip size is 3 mm × 2 mm in length and width, respectively. According to Equation (1), a change of mechanical strain on the Wheatstone bridge is transformed into a change of output voltage of the piezoresistive acceleration sensor.

$$\Delta U = U_0 \cdot \varepsilon \cdot k \tag{1}$$

where ΔU represents the change of the output voltage and U_0 the supply voltage of the Wheatstone bridge. To increase the sensitivity, both the piezoresistive gauge factor k and the mechanical strain ε can be increased.

The gauge factor of silicon depends on the dopant concentration and is typically limited to values below 100. Piezoresistors with a high gauge factor are also more sensitive to changes of temperature, i.e., they have a higher temperature coefficient of resistance (TCR). To obtain a high sensitivity, a dopant concentration of $2 \times 10^{18}/\text{cm}^3$ after annealing is implanted to form the piezoresistors, which should have a gauge factor of at least 90. The high influence of temperature on such piezoresistors is accepted to obtain a very high sensitivity without coming close to the yield strain of silicon. For the design of the dimensions of the piezoresistors, the joule heating effect should also be considered, which is one of the biggest influencing factors for the resistors' performance. For this reason, the current density within each resistor should be kept low. The increase of electrical resistance and of the piezoresistors' cross-section helps to reduce the current density. Therefore, the values of resistance are defined as 4400 Ω and the cross section of the piezoresistors is designed as 14 µm × 1.8 µm (width × depth of pn-junction, respectively).

$$\varepsilon = \frac{\rho \cdot f \cdot g}{2 \cdot E \cdot t} \cdot l^2 \qquad (2)$$

The mechanical strain can be calculated with Equation (2), which is based on the beam theory for a double-clamped beam. Where the density ρ and Young's modulus E are fixed material parameters, and load f and gravity g are determined by external factors, only the geometrical parameters of beam thickness t and of beam length l can be modified to increase the strain. To provide high sensitivity as well as sufficient mechanical strength to tolerate a maximal acceleration overload of twice the specified full range acceleration, the beam thickness is set to 20 µm [2,9]. Therefore, the only mechanical design parameter changeable to provide a higher strain and thus a higher sensitivity is the beam length l. The FEM model can be used to determine the beam length, which complies with the resonance frequency and sensitivity. While the sensitivity of the sensor is increased by increasing the beam length, its resonance frequency decreases (stability). Sensor design is therefore always a compromise between high sensitivity and high resonance frequency. Another parameter that should be considered is the beam width, because all four of the piezoresistors should be placed on the beam ends to form a full Wheatstone bridge configuration within the areas of high strain on the beam ends. However, according to Equation (2), the beam width has no influence on the strain the piezoresistors are subjected to. In this project, a beam with a width of 400 µm was set for the double-clamped beam.

To analyze the mechanical behavior of the novel double clamped beam structure and to determine the length of the beam length, a simplified model was built, as shown in Figure 2. Here, a quarter of the model is shown, for symmetry of the sensor design can be exploited the reduce computational time. The influence of the different beam lengths on the resonance frequency and sensitivity of piezoresistive sensors were calculated and simulated. A static mechanical analysis using ANSYS mechanical was generated. The element type used was Solid64, as it defines eight knots and three degrees of freedom. As border conditions for this simulation, the surface of the sensor was defined free of displacement. The sensor's symmetry was exploited, and therefore, the symmetrical borders were set free of displacement in the normal direction of symmetry.

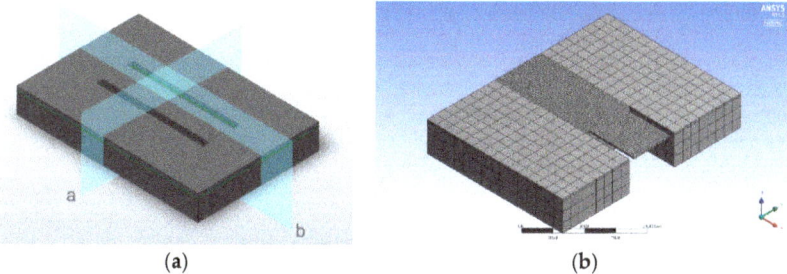

(a)   (b)

Figure 2. (**a**) Symmetrical axis of acceleration sensor; (**b**) mashed model half of used sensor. Meshing is homogenous over the surface with element sizes of 20 µm in the beam area and 100 µm around the beam area on the bulk material.

The resonance frequency is a direct result of the modal analysis carried out for each beam configuration. Equation (3) can be used to estimate the first mode resonance frequency of a double-clamped beam.

$$f_1 = \frac{\frac{e_1^2 \cdot t}{l^2}\sqrt{\frac{E}{12 \cdot \rho}}}{2 \cdot \Pi} \tag{3}$$

where the first resonance frequency f_1, beam length l, Density ρ, and Young's modulus E. Among them, ρ and E are fixed material parameters, and f_1 will decrease with increasing l. The influence of the different beam lengths on the sensitivity of piezoresistive sensors to acceleration S_a is calculated by Equation (4).

$$S_a = \frac{\Delta U}{\Delta a \cdot U_0} \tag{4}$$

Using a gauge factor of 90, the required strain to obtain the specified sensitivity can be calculated. By extracting the strain values for each beam configuration at the location of the piezoresistors from the simulation result files, the average strain of the piezoresistors were evaluated. The simulation values for resonance frequency and strain were plotted over the beam length for the acceleration ranges of 60,000 g in Figure 3a, and sensitivity at full scale, calculated from the simulated average effective strain with SOI thickness of 20 μm and k-factors of 70 and 90, were plotted in Figure 3b.

The simulated values given in Figure 3a were divided by the required values for resonance frequency and strain. The intersection points of required and simulated strain and frequency allow the reduction of the beam length, complying with the given specifications. At an acceleration of 60,000 g, a beam of not more than 600 μm length will comply with the frequency specification. To obtain the required strain, a beam length of more than 14,000 μm is required at 60,000 g. Based on the simulation results presented here, the selection of specific sensor geometries, fulfilling at least one of the sensor specifications, is possible. The simulations clearly show that no single sensor design exists to fulfill both the resonance frequency and sensitivity specification for the acceleration range of 60,000 g. Thus, the establishment of a beam length is a compromise between resonance frequency and average effective strain.

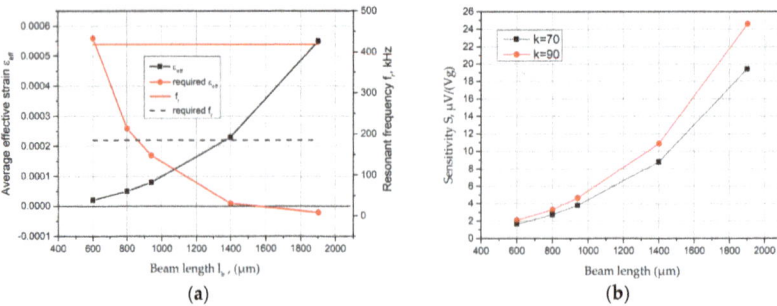

Figure 3. (**a**) Simulated average effective strain and resonance frequency over beam length at an acceleration load of 60,000 g; (**b**) sensitivity at full scale, calculated from the simulated average effective strain with an SOI thickness of 20 μm and k-factors of 70 and 90, respectively.

3. Fabrication Flow of the Sensor Wafer

The sensors were produced from SOI wafers with a device layer of 20 μm. Table 1 lists the thickness of each layer during the sensor process. A handle-wafer thickness from 300 μm to 325 μm, a buried-oxide thickness from 0.2 μm to 0.4 μm, and a base dopant concentration of approximately 1×10^{15} phosphorus ions/cm^3 are recommended.

The fabrication process of the sensor, which is based on the micromachine technology, is shown in Figure 4. The first processing step was aimed at etching the alignment marks into the device layer of the wafers (see Figure 4a). For this purpose, a standard photolithographic process flow composed of photoresist spin coating, prebake, exposure, development, and postbake was implemented. The second step was to implant the contact region and the piezoresistors. Before the implantation, a layer of approximately 50-nm-thick silicon oxide was grown in a furnace at 1000 °C (stray oxide). Then the implantation of contact regions and piezoresistors was accomplished with the standard photolithographic process described above. A dopant dose of 3×10^{15} and one of 1.1×10^{14} boron ions/cm² were used for the contact regions and piezoresistors, respectively. An implantation energy of 60 keV and a maximum beam current of 100 µA were used for the implantation. After the implantation, the wafers were cleaned and prepared for the annealing and oxidation process. Annealing and oxidation took place simultaneously in a furnace at 1000 °C. Furthermore, SUPREM simulations indicated that an annealing time of eight hours was necessary to establish the ion concentration of approximately 2×10^{18}/cm³ required to obtain a gauge factor of 90 and to realize a pn-junction depth of 1.8 µm. At the beginning of the annealing, oxygen was added to the furnace gas to grow a layer of insulating silicon oxide. After the annealing and oxidation, a silicon nitride layer was deposited through a low-pressure chemical vapor deposition (CVD) process. The next step was to establish electrical contact between the metallization layer and the contact regions. In order to form conductors and bond pads, the standard photolithographic process and dry etching process were employed to open the contact areas. As demonstrated in Figure 4c, a thin AlSiCu layer was then sputtered onto the surface and into the contact holes of the wafer. After structuring the AlSiCu layer by chemical wet etching and removing the photoresist, the metal layer was annealed in a forming gas atmosphere at 450 °C to establish ohmic contact between the metallization layer and the implanted contact regions.

Table 1. Thickness of each layer during the sensor fabrication.

Layers	Thickness (µm)
Handle wafer	300–325
BOX (buried oxide)	0.2–0.4
Device layer	20
Stray oxide	0.05
Insulation oxide	0.1
Silicon nitride	0.1
Pn-junction depth	1.8

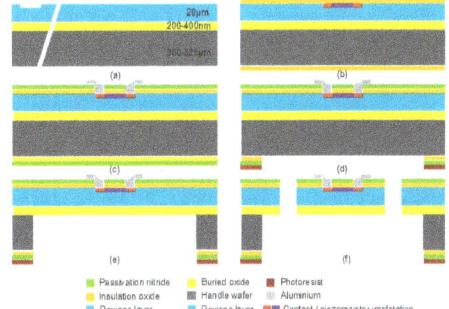

Figure 4. Fabrication flow of a high-g acceleration sensor. (**a**) Cross-section of SOI wafer after dry etching of the alignment marks; (**b**) after contact and piezoresistor implantation and photoresist removal; (**c**) after structuring of the metallization layer. (**d**) After structuring of the backside nitride and oxide layer; (**e**) after etching of the handling wafer from the back side; (**f**) after release of the beam structure by wet etching.

To realize the mechanical sensor structure, the process started by coating the back side of the wafer with a thick photoresist layer (see Figure 4d). This layer served as an etch mask for deep silicon etching, and therefore it is typically spun at a low rotational speed to obtain a thickness of over 4 μm. The photoresist edge bead was removed with acetone on a spin coater after the prebake to ensure good wafer clamping during the deep silicon etching. After exposure, development, and postbake, the back side photoresist mask featured openings for the back side cavities. Before deep silicon etching was conducted, the passivation and insulation layers needed to be removed locally. Removal of the silicon nitride was conducted by dry etching, and wet etching with the BOE was used to remove the silicon oxide. Etching of the front side AlSiCu metallization by the BOE was prevented by applying a protective foil on the wafer front side before wet etching. The silicon oxide insulation was etched within a few minutes. After the wafer was rinsed and dried, the protective foil was removed manually. Alternatively, protection of the aluminum was possible by coating the wafer top side with photoresist. The wafer was etched from its back side with the Bosch process until the BOX layer was reached and exposed to the entire cavity bottom, as seen in Figure 4e. The photoresist mask was removed subsequently. To form a double-clamped beam, two trenches per chip needed to be etched through the device layer, where deep silicon etching was also applied (see Figure 4f). Moreover, a standard photolithographic process was utilized, and after the postbake, the nitride and the oxide layers were removed locally by dry and BOE etching, respectively. A short, deep silicon etching process was sufficient to structure the device layer down to the BOX layer. After the release of the beam structure and the removal of the photoresist, the chips were finished and tested. Figure 5 displays a realized MEMS high-g-sensor element ($3 \times 2 \times 1$ mm³) and its package system.

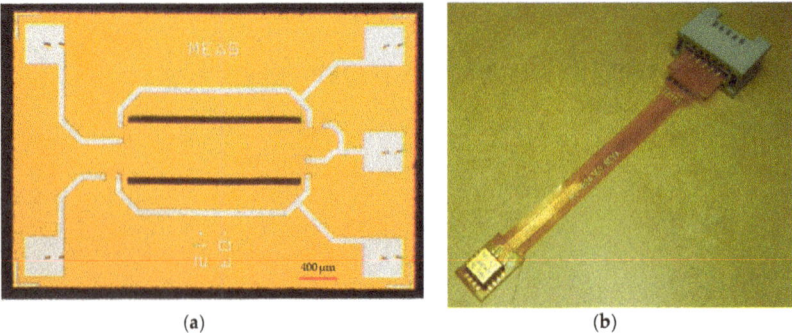

(a) (b)

Figure 5. (a) A realized MEMS high-g-sensor element ($3 \times 2 \times 1$ mm³). On top, the silicon beam with integrated piezoresistors, the leads, and the bond pads can be clearly seen; (b) packaged high-g-sensor system.

4. Device Characterization

In order to generate the necessary accelerations, a shock wave bar (Figure 6) and a Polytec vibrometer were employed as an acceleration reference. The MEMS sensor element was attached to a stainless-steel test fixture using the standard die attach method. The die was wire-bonded to the Printed-Circuit-Board (PCB) where the output wires were soldered. Furthermore, a cover was added for protection. The test fixture was mounted to the end of the shock wave bar. It also served as the reference for the laser. Figure 7a shows the time domain by excitation at about 60,000 g. Note that the left scale in m/s² is for the reference laser sensor, and the right scale in mV is for the tested sensor. The sensor wave form represents the laser signal. When applying higher accelerations, the time domain signal becomes more distorted because of structural ringing at the bar and test fixture interface. The linearity was also calculated from the maxima of the vibrometer and accelerometer signal. A typical output signal is demonstrated in Figure 7b. The sensor wave form tracks the laser signal very well.

Figure 6. Infrastructure (Shock Wave Bar) used to characterize the sensors.

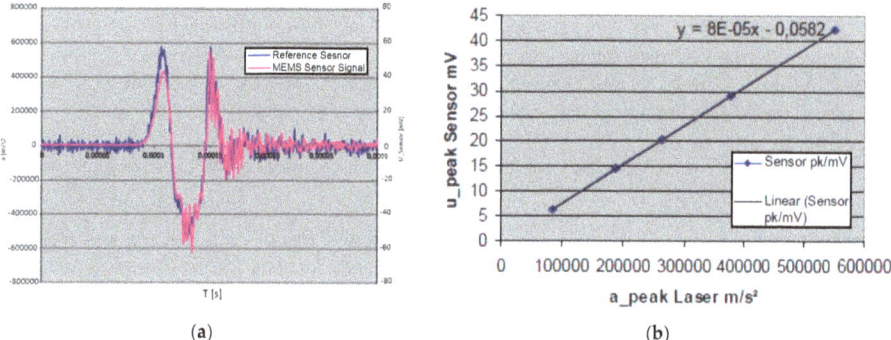

(a) (b)

Figure 7. (**a**) Test results of the fabricated 60,000 g sensor. Magenta: MEMS sensor signal. Blue: reference sensor; (**b**) the calculated linearity of the sensor.

5. Conclusions

In this paper, a novel method for a high-impact sensor with an impact up to 60,000 g is proposed. The sensor is realized with a double-clamped beam form and has been successfully fabricated using silicon micromachining and diffusion techniques. The fabricated sensor was also tested with a shock test, and the measurement results reveal that the fabricated devices exhibit a linear response. In addition, with the time domain by excitation up to 60,000 g, the sensor wave form tracks the laser signal very well. The sensors can be utilized in aerospace applications or in the control and detection of impact levels.

Author Contributions: X.H., H.-D.N., P.M. and S.L. conceived, designed and simulated the sensors. H.-D.N., P.M. and X.H. fabricated the sensors in clean room. P.M., S.L., X.H. and H.-D.N. developed the electronics, the packaging for the sensors and characterized the sensors. P.M., S.L., X.H., H.-D.N., M.B., O.E., K.-D.L. and M.S.-R. analyzed the sensor data. H.-D.N., X.H., P.M. and M.B. wrote the paper.

Funding: This work was funded by Measurement Specialities.

Acknowledgments: The author gratefully acknowledges the financial and technical support for this work provided by Fraunhofer Institute for Reliability and Microintegration in Berlin as well as Technische Universität Berlin and University of Applied Sciences Berlin for their ongoing support. Authors would also like to thank Measurement Specialities for financial support of the work.

Conflicts of Interest: The authors declare no conflict of interest.

Micromachines **2018**, *9*, 266

References

1. Ning, Y.; Loke, Y.; McKinnon, G. Fabrication and characterization of high g-force, silicon piezoresistive accelerometers. *Sens. Actuators A Phys.* **1995**, *48*, 55–61. [CrossRef]
2. Mackowiak, P.; Mukhopadhyay, B.; Hu, X.; Ehrmann, O.; Lang, K.D.; Linke, S.; Chu, A.; Ngo, H.D. Development and fabrication of a very High-g sensor for very high impact applications. *J. Phys. Conf. Ser.* **2016**, *757*, 012016. [CrossRef]
3. Jung, H.-I.; Kwon, D.-S.; Kim, J. Fabrication and characterization of monolithic piezoresistive high-g three-axis accelerometer. *Micro Nano Syst. Lett.* **2017**, *5*, 7. [CrossRef]
4. Davis, B.S.; Denison, T.; Kuang, J. A Monolithic High-G SOI-MEMS Accelerometer for Measuring Projectile Launch and Flight Accelerations. *Shock Vib.* **2006**, *13*, 127–135. [CrossRef]
5. Narasimhan, V.; Li, H.; Tan, C.S. Monolithic CMOS-MEMS integration for high-g accelerometers. In *Emerging Technologies in Security and Defence II; and Quantum-Physics-Based Information Security III*; SPIE: Bellingham, WA, USA, 2014; p. 10.
6. Wang, J.; Li, X. A High-Performance Dual-Cantilever High-Shock Accelerometer Single-Sided Micromachined in (111) Silicon Wafers. *J. Microelectromech. Syst.* **2010**, *19*, 1515–1520. [CrossRef]
7. Dong, P.; Li, X.; Yang, H.; Bao, H.; Zhou, W.; Li, S.; Feng, S. High-performance monolithic triaxial piezoresistive shock accelerometers. *Sens. Actuators A Phys.* **2008**, *141*, 339–346. [CrossRef]
8. Xu, Y.; Zhao, L.; Jiang, Z.; Ding, J.; Peng, N.; Zhao, Y. A Novel Piezoresistive Accelerometer with SPBs to Improve the Tradeoff between the Sensitivity and the Resonant Frequency. *Sensors* **2016**, *16*, 210. [CrossRef] [PubMed]
9. Buder, U. *High-Acceleration Sensors Final Report*; TU Berlin Microsensor and Actuator Technology Center: Berlin, Germany, 2008.

Article

Method of Measuring the Mismatch of Parasitic Capacitance in MEMS Accelerometer Based on Regulating Electrostatic Stiffness

Xianshan Dong [1], Shaohua Yang [1,2], Junhua Zhu [1], Yunfei En [1] and Qinwen Huang [1,*]

[1] Science and Technology on Reliability Physics and Application of Electronic Component Laboratory, No.5 Electronics Research Institute of the Ministry of Industry and Information Technology, Guangzhou 510610, China; dongxs@pku.edu.cn (X.D.); yangsh@ceprei.com (S.Y.); zhujunhua@ceprei.com (J.Z.); enyf@ceprei.com (Y.E.)
[2] College of Physics and Optoelectronic Engineering, Guangdong University of Technology, Guangzhou 510006, China
* Correspondence: huangqinwen@ceprei.com; Tel.: +86-020-8723-6477

Received: 29 January 2018; Accepted: 2 March 2018; Published: 15 March 2018

Abstract: For the MEMS capacitive accelerometer, parasitic capacitance is a serious problem. Its mismatch will deteriorate the performance of accelerometer. Obtaining the mismatch of the parasitic capacitance precisely is helpful for improving the performance of bias and scale. Currently, the method of measuring the mismatch is limited in the direct measuring using the instrument. This traditional method has low accuracy for it would lead in extra parasitic capacitive and have other problems. This paper presents a novel method based on the mechanism of a closed-loop accelerometer. The strongly linear relationship between the output of electric force and the square of pre-load voltage is obtained through theoretical derivation and validated by experiment. Based on this relationship, the mismatch of parasitic capacitance can be obtained precisely through regulating electrostatic stiffness without other equipment. The results can be applied in the design of decreasing the mismatch and electrical adjusting for eliminating the influence of the mismatch.

Keywords: MEMS accelerometer; mismatch of parasitic capacitance; electrostatic stiffness

---

1. Introduction

An accelerometer is a key device in inertial navigation and control systems for measuring the acceleration information of a carrier. With the progress of MEMS technology, the MEMS accelerometer has been rapidly developed and is widely used in military, industry, medicine, and consumer electronics fields for its small volume, light weight, small power consumption, and low cost. Among MEMS accelerometers, the closed-loop capacitive accelerometer based on electrostatic force balance is an important form for its relatively good performance [1,2].

The MEMS capacitive accelerometer measures the acceleration through electrically detecting the changed differential capacitance of sensor caused by the movement of proof-mass under acceleration. As is known to all, parasitic capacitance is a serious problem in MEMS capacitive accelerometers [3–5]. Its mismatch between electrodes including in the sensor, package, and circuit would produce an offset and deteriorate the performance of bias and scale. The mismatch of effective capacitance due to process variation during sensor fabrication can be eliminated by the closed-loop system, but the mismatch of parasitic capacitance remains. Some research has been carried out for eliminating the influence of the parasitic capacitance [6–8], but these methods are either unsolved completely or lead to extra questions. Reducing the mismatch of parasitic capacitance is more direct and effective, and another solution is compensating the mismatch through electrical adjusting or adding an extra capacitor which is widely used [9,10]. Either reducing or compensating the mismatch of parasitic capacitance should be measured accurately.

Currently, the method of measuring the parasitic capacitance is limited in the direct measuring using the instrument or the capacitive measuring circuit [11,12]. This method has low accuracy for it would lead to extra parasitic capacitance and the measuring result is the state of off-power, moreover, some equivalent parasitic capacitance cannot be obtained and it cannot be implemented in some occasions. This paper proposes a novel method of measuring the mismatch of parasitic capacitance in MEMS accelerometer based on the mechanism of a closed-loop system. Through regulating the electrostatic negative stiffness and obtaining the curve between the output of electric force and the square of pre-load voltage, the mismatch can be obtained according to the coefficient of linear fitting. This method can be applied in the design for reducing the mismatch and electrical adjusting for eliminating the influence of mismatch, and the research for the characteristics of the mismatch influenced by the temperature and the self-calibrating technique of eliminating the mismatch can be further studied with this method.

2. Method of Measuring the Mismatch of Parasitic Capacitance

2.1. Influence of Parasitic Capacitance

Figure 1 shows the schematic of effective and parasitic capacitances in MEMS capacitive accelerometer interfaced with a C/V converting circuit. Obviously, there are several parasitic capacitances and the mismatch of parasitic capacitances ΔC_{m1} between C_{p1} and C_{p2}—including in the sensor, package, and circuit—will confuse the differential effective capacitances ΔC between C_{top} and C_{bottom} that would produce an offset. The mismatch ΔC_{m2} between C_{p3} and C_{p4} will also have an influence on the output. Besides, the parasitic capacitances, C_{p5} and C_{p6}, can affect the influence of ΔC_{m1} and ΔC_{m2} on the output.

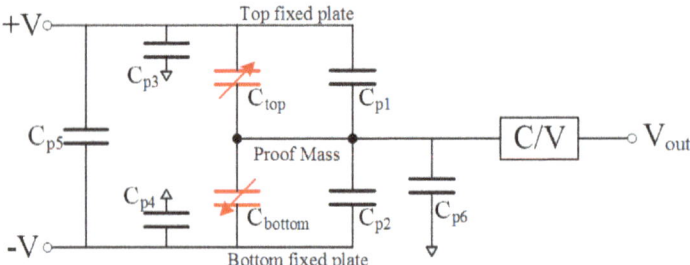

Figure 1. Schematic of capacitance in system of MEMS accelerometer.

Generally, the sensitivity of effective capacitance is about 100 fF/g or even smaller and the mismatch of parasitic capacitance can be up to 100 fF that will result in an offset of 1 g. This large offset would severely deteriorate the performance of the accelerometer. Therefore, it is necessary to study the mismatch and do some work for reducing the influence. Measuring the mismatch accurately is a basic step. Though there are many discrete parasitic capacitances, we only need to obtain the total equivalent mismatch.

2.2. Theory of Measuring the Mismatch

In the closed-loop system of a MEMS capacitance accelerometer, there is electrostatic force between fixed plates and proof mass that balances the inertial force caused by acceleration [13], and the proof mass is not at the geometrical center for the mismatch of parasitic capacitance. Figure 2 shows a working diagram of the sensor.

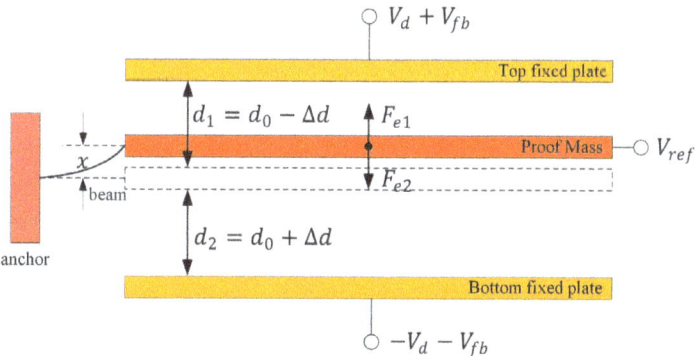

Figure 2. Sensor working diagram of electrostatic force balance.

Considering the process variation and parasitic capacitance, the electrostatic force F_e of the proof mass is:

$$F_e = F_{e1} - F_{e2} = \frac{\varepsilon_r \varepsilon_0 A \times \left(V_d + V_{fb} - V_{ref}\right)^2}{2(d_0 - \Delta d - x)^2} - \frac{\varepsilon_r \varepsilon_0 A \times \left(-V_d - V_{fb} - V_{ref}\right)^2}{2(d_0 + \Delta d + x)^2} \tag{1}$$

where ε_r and ε_0 are the relative and absolute dielectric constant respectively, A is the overlapped area of capacitance, V_d is the modulated voltage, V_{fb} is the feedback voltage, V_{ref} is the pre-load voltage, d_0 is the average gap between electrodes, Δd is the gap deviation due to process variation, and x is the bending value of the beam due to the mismatch of effective and parasitic capacitance.

In general, x and Δd are far smaller than d_0, and then, Equation (1) can be simplified to:

$$F_e = \frac{2\varepsilon_r \varepsilon_0 A \times V_{ref} V_{fb}}{d_0^2} - \frac{2\varepsilon_r \varepsilon_0 A \times \left(V_{ref}^2 + V_{fb}^2 + V_d^2\right)}{d_0^3} \times (x + \Delta d) \tag{2}$$

where the bending value x consists of x_1 brought by the mismatch of effective capacitance and x_2 brought by the mismatch of parasitic capacitance, so $x = x_1 + x_2 = -\Delta d + x_2$. Substituting this equation to Equation (2), the electrostatic force F_e can be expressed as:

$$F_e = \frac{2\varepsilon_r \varepsilon_0 A \times V_{ref} V_{fb}}{d_0^2} - \frac{2\varepsilon_r \varepsilon_0 A \times \left(V_{ref}^2 + V_{fb}^2 + V_d^2\right)}{d_0^3} \times x_2 \tag{3}$$

where $2\varepsilon_r \varepsilon_0 A \times \left(V_{ref}^2 + V_{fb}^2 + V_d^2\right)/d_0^3 = k_e$ is called electrostatic stiffness.

In the closed-loop system, there is the force balance for the proof mass:

$$F_e + kx + ma + F_s = 0 \tag{4}$$

where k is the stiffness of the beam, m is the inertial mass of the proof mass, a is the external acceleration, and F_s is the residual stress. Replacing Equation (3) into Equation (4), the formula of force balance can be expressed as:

$$\frac{2\varepsilon_r \varepsilon_0 A \times V_{ref} V_{fb}}{d_0^2} - \frac{2\varepsilon_r \varepsilon_0 A \times \left(V_{ref}^2 + V_{fb}^2\right)}{d_0^3} \times x_2 = B_0 \tag{5}$$

where $B_0 = 2\varepsilon_r\varepsilon_0 A \times V_d^2 \times x_2/d_0^3 - kx - ma - F_s$. When the input acceleration is unchanged, the parameter B_0 can be considered as a fixed value. When the input acceleration and offset are small, V_{fb}^2 is far smaller than V_{ref}^2, so Equation (5) can be simplified to:

$$\frac{2\varepsilon_r\varepsilon_0 A \times V_{ref}V_{fb}}{d_0^2} = \frac{2\varepsilon_r\varepsilon_0 A \times x_2}{d_0^3} \times V_{ref}^2 + B_0 \qquad (6)$$

For the digital acquisition system, the left portion in Equation (6) can be transformed to $F_e' = 2\varepsilon_r\varepsilon_0 A \times V_{ref}V_{fb}/d_0^2 = U_{out}/K_1 \times m \times g_L$ where U_{out} is digital output which unit is LSB, K_1 is the scale of accelerometer which unit is LSB/g and g_L is local gravity acceleration. Then, Equation (6) can be transformed to:

$$\frac{U_{out}}{K_1} \times m \times g_L = \frac{2\varepsilon_r\varepsilon_0 A \times x_2}{d_0^3} \times V_{ref}^2 + B_0 \qquad (7)$$

Equation (7) can be transformed to:

$$Y = B_1 \times X + B_0 \qquad (8)$$

where $Y = U_{out}/K_1 \times m \times g_L$ is dependent variable, $X = V_{ref}^2$ is independent variable, $B_1 = 2\varepsilon_r\varepsilon_0 A \times x_2/d_0^3$ is linear coefficient and B_0 is intercept which is a fixed value.

Equation (8) shows that the relationship between output of electrostatic force $F_e' = U_{out}/K_1 \times m \times g_L$ and the square of pre-load voltage V_{ref}^2 is linear. Thus, we can make a curve with F_e' as y-axis and V_{ref}^2 as x-axis, and then, a linear fitting of the curve is made. Lastly, the mismatch of the parasitic capacitance can be obtained from the linear coefficient B_1 through the equation:

$$\Delta C_p = \frac{\varepsilon_r\varepsilon_0 A}{d_0 - x_2} - \frac{\varepsilon_r\varepsilon_0 A}{d_0 + x_2} \approx \frac{2\varepsilon_r\varepsilon_0 A \times x_2}{d_0^3} \times d_0 = B_1 \times d_0, \qquad (9)$$

where d_0 can be calculated through the obtained scale of the closed-loop system. Meanwhile, we can get the offset and the deviation from geometrical center due to the mismatch of parasitic capacitance.

3. Measurement Results and Discussion

Measuring tests have been done with closed-loop MEMS accelerometer to verify this novel method and two applications with this method are present. The measuring work were implemented on a printed circuit board (PCB) with discrete component, interfaced with a packaged sensor using ceramic shell and bond wire. The senor is fabricated with bulk silicon process and the structure is comb finger. The control system is achieved by analogue circuit and the analogue output is digitally acquired through Analog to Digital Convert (ADC) and Field Programmable Gate Array (FPGA) chip. The full-scale range of the accelerometer is 30 g, and the noise is $10\mu g/\sqrt{Hz}$. In this system, the parasitic capacitances originate from the sensor, the ceramic shell, the bond wire and the PCB circuit. In our designed accelerometer, this mismatch commonly leads in an offset of several hundred mg that severely deteriorates the performance of accelerometer.

3.1. Measurement Results

3.1.1. Verification Experiment and Results

In the verification experiment, the accelerometer is placed on the marble platform and the input acceleration is about 0 g which purpose is to make the external acceleration stable and the output very small. This step can improve the accuracy of the measurement. Because the pre-load voltage goes through voltage follower and resistance, and then reaches the node of proof-mass, so, the pre-load voltage does not directly connect to this C/V node. We draw out a line from the node of pre-load voltage that did not change the output. Then, the pre-loaded voltage of the accelerometer is changed, and the scale is tested through turning the accelerometer. The changed pre-loaded voltage, the digital output and the scale are record. Table 1 contains the measuring data with different pre-loaded voltage.

Table 1. The measuring data with different V_{ref}.

V_{ref} (V)	U_{out} (LSB)	K_1 (LSB/g)	V_{ref}^2 (V²)	F_e' (N)
1.00	5058	137,837	1.00	7.33×10^{-8}
2.00	1526	68,051	4.00	1.96×10^{-8}
3.00	−49	45,092	9.00	-8.48×10^{-8}
4.00	−1079	33,768	16.00	-2.29×10^{-7}
5.00	−1946	26,993	25.00	-4.09×10^{-7}
6.00	−2651	22,491	36.00	-6.22×10^{-7}
7.00	−3329	19,253	49.00	-8.73×10^{-7}
8.00	−4043	16,811	64.00	-1.18×10^{-6}
9.00	−4682	14,942	81.00	-1.51×10^{-6}

Using these recorded data, we make a figure by taking V_{ref}^2 as x-axis and F_e' as y-axis as shown in Figure 3, and a linear fitting of the curve is made.

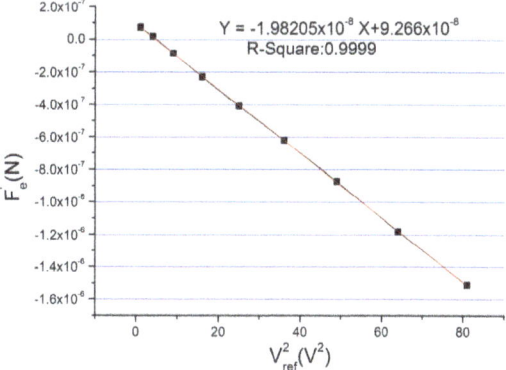

Figure 3. Relationship between V_{ref}^2 and F_e'.

The R^2 of the linear fitting is 0.9999 which shows highly linear correlation between V_{ref}^2 and F_e'. The strong linear relationship validates the theory of formula deduction. From the linear fitting formula, the linear coefficient can be obtained which is -1.98205×10^{-8}. Through calculation according to this number, the bending value x_2 of the beam owing to the mismatch of parasitic capacitance which is also the deviation from the geometrical center is −13.48 nm. It should be noted that the bending value of the beam is a vector. That is to say it can be positive or negative. The bending direction of the beam depends on the sum of x_1 and x_2, and the minus sign of this x_2 indicates that the beam bends to the bottom plate, owing to the mismatch of parasitic capacitance. Correspondingly, the mismatch of parasitic capacitance is −69.372 fF and the offset caused by the mismatch is 219 mg.

3.1.2. Applications and Results

The charge amplifier and diode ring are the common used C/V converting circuit. Because the charge amplifier is based on current measurement, the parasitic capacitance Cp3 and Cp4 in figure1 has little influence on the output of charge amplifier. However, in our design the diode ring detecting circuit is adopted for its simple structure. In diode ring detecting circuit, the principle of C/V converting is based on charge-discharge of capacitance. The capacitance Cp3 and Cp4 would affect the charge–discharge process of demodulating capacitance, so, it has an effect on the output. We carried out an experimental test to study the influence on output of capacitance to ground (GND) previously. A 1 pF difference between C_{p3} and C_{p4} was made in MEMS accelerometer using diode ring detecting circuit and a change of 0.5 g on output was observed, so it is necessary to study the influence of the parasitic capacitance between the fixed plate and GND. It should be noted that the effect of this equivalent mismatch on output is not equal to the effective differential capacitance, so its equivalent mismatch cannot be measured using the direct measuring method. The experiment for measuring the equivalent mismatch of the parasitic capacitance between the fixed plate and GND is carried out.

A chip capacitor of 1 pF is intentionally added between the top fixed plate and GND. Because this operation changes the bias of accelerometer which is equivalent to changing the equivalent mismatch. Then, the total equivalent mismatch is measured using this novel method before and after adding this capacitor. Figure 4 is the testing results which show the influence of this mismatch.

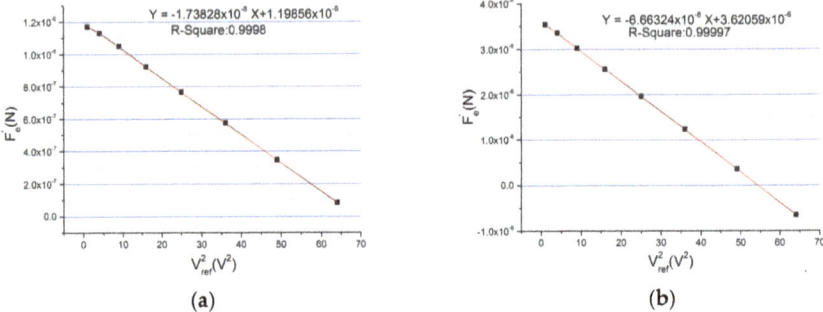

Figure 4. Measuring the equivalent mismatch between fixed plate and GND: (a) the initial state; (b) state of adding a capacitance of 1 pF.

The linear coefficient after adding the chip capacitor is much bigger than the one of initial state that indicates the parasitic capacitance between the fixed plate and GND can seriously affect the output. Through calculating, the initial mismatch of the accelerometer is -60.840 fF and the mismatch after adding the 1 pF capacitance is -233.213 fF. So, the equivalent mismatch of the 1 pF capacitance between the fixed plate and GND is 172.373 fF.

Another application using this method is improving the design of circuit to reduce the mismatch of parasitic capacitive. Table 2 shows the mismatch of parasitic capacitive for different sensors on same circuit board. For these six sensors, the average bending value x_2 is -11.0 nm and the average mismatch is -56.44 fF, which causes an offset of 179 mg. It can be seen that the values of the mismatch are near that indicates the mismatch is mainly from the circuit board for the mismatch of different sensors would have large discreteness.

Table 2. Mismatch of different sensors on same board.

Sensor	X_2 (m)	Mismatch/fF
1	-1.12×10^{-8}	-57.64
2	-0.99×10^{-8}	-50.95
3	-0.98×10^{-8}	-50.43
4	-1.14×10^{-8}	-58.67
5	-1.17×10^{-8}	-60.21
6	-1.18×10^{-8}	-60.73
average	-1.10×10^{-8}	-56.44

The design of the circuit should be improved to reduce the mismatch of parasitic capacitance on the circuit board. An improved circuit was fabricated and the mismatch is measured with the same sensor welded on different circuit boards. Figure 5 is the contrast of mismatch on different circuit boards. The mismatch of parasitic capacitance is −69.372 fF on the before-optimization circuit board, and it is +22.332 fF on the after-optimization circuit board. It can be seen that through optimizing the circuit design, the mismatch of parasitic capacitance is reduced by 69% and the sign of the mismatch is changed.

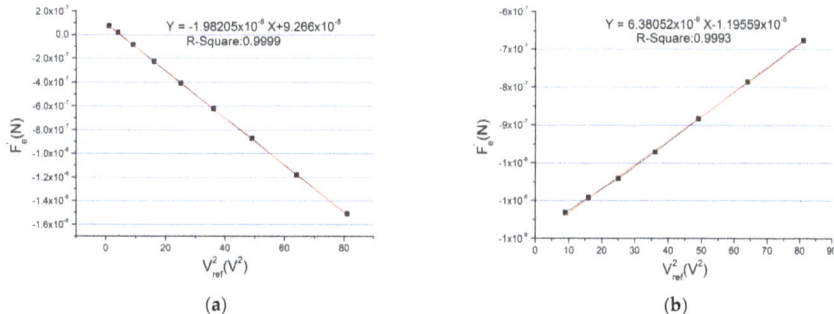

Figure 5. Mismatch of different circuit design: (**a**) result of before-optimization circuit; (**b**) result of after-optimization circuit.

3.2. Discussion

The linear relationship between output of electrostatic force and the square of pre-load voltage is validated by the experiment. In an ideal system with no mismatch, the force F'_e is a fixed value for the feedback and pre-load voltage are changed at inverse proportions. However, due to the existence of the mismatch of parasitic capacitance in real system, the force F'_e will be changed in proportion to x_2 following the changed force $k_e x_2$ when regulating the electrostatic stiffness through changing the pre-load voltage. The novel method exploits this characteristic to obtain the mismatch of parasitic capacitance.

It should be pointed out that the curve deviates from the straight line when the pre-load voltage is small, especially when the mismatch is small. This is because the force $k_e x_2$ has little change with a small pre-load voltage or a small mismatch that makes the linear relationship disturbed by the feedback voltage. Nevertheless, the mismatch of parasitic capacitance can be obtained precisely through regulating electrostatic stiffness with relatively high pre-loaded voltage.

The measured results show the mismatch of capacitance parasitic is fF level. The mismatch is so small that requires testing equipment of very high precision. Different from the traditional methods, in this novel method a line is just drawn out from the pre-loaded node which does not interfere with any electrical node of the C/V frond-end circuit, so it does not introduce additional parasitic capacitance. Moreover, the measured result is the equivalent mismatch of all parasitic capacitance when the accelerometer is in an operating state. Therefore, the mismatch result is that we want.

4. Conclusions

This paper describes a novel method for measuring the mismatch of parasitic capacitance in MEMS capacitive accelerometer. The strong linear relationship between output of electrostatic force and the square of pre-load voltage is validated by the theory and experiment. The total equivalent mismatch of parasitic capacitance can be obtained precisely and conveniently through regulating electrostatic stiffness with changing the pre-loaded voltage. The results can be used in the design and electrical adjusting for decreasing the influence of the mismatch that is helpful for improving the performance of accelerometer, and the temperature characteristics of the mismatch and the self-calibrating technique of eliminating the mismatch can be further studied with this method.

Acknowledgments: The authors would like to thank the grants supported by National Natural Science Foundation of China (Grant No. 51505089), Natural Science Foundation of Guangdong Province (Grant No. 2016A030313672), and Generality Research Foundation of Component (China JAD1628200).

Author Contributions: Xianshan Dong and Qinwen Huang conceived the method, deduced the formula, and designed the measurement; Shaohua Yang performed the measuring experiments; Junhua Zhu analyzed the data; Shaohua Yang and Yunfei En contributed the sample and measuring tools; Xianshan Dong wrote the paper.

Conflicts of Interest: The authors declare no conflict of interest. The founding sponsors had no role in the design of the study; in the collection, analyses, or interpretation of data; in the writing of the manuscript, or in the decision to publish the results.

References

1. Dong, Y.; Zwahlen, P.; Nguyen, A.M.; Frosio, R.; Rudolf, F. Ultra-high precision MEMS accelerometer. In Proceedings of the 16th International Conference on Solid-State Sensors, Actuators and Microsystems, Beijing, China, 5–9 June 2011.
2. Ullah, P.; Ragot, V.; Zwahlen, P.; Rudolf, F. A new high performance sigma-delta MEMS accelerometer for inertial navigation. In Proceedings of the 2015 Dgon Inertial Sensors and Systems Symposium, Karlsruhe, Germany, 22–23 September 2015.
3. Yazdi, N.; Kulah, H.; Najafi, K. Precision readout circuits for capacitive microaccelerometers. In Proceedings of the IEEE Sensors 2004, Vienna, Austria, 24–27 October 2004.
4. Zhu, Z.Y.; Liu, Y.D.; Jin, Z.H. The parasitic capacitance's influence on noise in a MEMS accelerometer sensor. *Chin. J. Sens. Actuators* **2013**, *26*, 17–20.
5. He, J.B.; Xie, J.; He, X.P.; Du, L.M.; Zhou, W. Research on nonlinear error of micro-accelerometers considering the fringe and parasitic capacitance. *Mech. Sci. Technol. Aerosp. Eng.* **2016**, *35*, 752–757.
6. Jeong, Y.; Ayazi, F. A novel offset calibration method to suppress capacitive mismatch in MEMS accelerometer. In Proceedings of the Samsung Electro-Mechanics Best Paper Award, Seoul, Korea, 13 November 2013.
7. Lajevardi, P.; Petkov, V.P.; Murmann, B. A delta sigma interface for MEMS accelerometers using electrostatic spring constant modulation for cancellation of bondwire capacitance drift. *IEEE J. Solid-State Circ.* **2016**, *48*, 265–275. [CrossRef]
8. Xiong, X.X.; Wu, Y.L.; Jone, W.B. Control circuitry for self-repairable MEMS accelerometers. In *Technological Developments in Education and Automation*, 1st ed.; Iskander, M., Kapila, V., Karim, M.A., Eds.; Springer: Berlin, Germany, 2010; pp. 265–270.
9. Ko, H. Highly configurable capacitive interface circuit for tri-axial MEMS microaccelerometer. *Int. J. Electron.* **2012**, *99*, 945–955. [CrossRef]
10. Liu, M.J.; Dong, J.X. Compensation for bias in capacitive micro accelerometer. *J. Chin. Inert. Technol.* **2008**, *16*, 86–89.
11. Zhou, H.Y.; Shen, T.Y.; Li, L.R.; Xu, F.Y.; Hu, J.W.; Xie, Y.H. Study on the measurement of micro-capacitance based on RLC series resonance and increment. In Proceedings of the International Conference on Mechanical, Electronic and Information Technology Engineering, Chongqing, China, 21–22 May 2016.
12. Dascher, D.J. Measuring parasitic capacitance and inductance using TDR. *Hewlett-Packard J.* **1996**, *47*, 83–96.
13. Li, J.; Dong, J.X.; Liu, Y.F.; Wu, T.Z. Effects of preload voltage on performance of force-rebalance micro silicon accelerometer. *J. Transducer Technol.* **2004**, *23*, 35–40.

Article

Design, Fabrication, and Performance Characterization of LTCC-Based Capacitive Accelerometers

Huan Liu [1], Runiu Fang [1], Min Miao [2,*], Yichuan Zhang [2], Yingzhan Yan [3], Xiaoping Tang [3], Huixiang Lu [3] and Yufeng Jin [1,4]

[1]   National Key Laboratory of Science and Technology on Micro/Nano Fabrication, Peking University, Beijing 100871, China; liuhuan169@pku.edu.cn (H.L.); fangruniu@pku.edu.cn (R.F.); yfjin@pku.edu.cn (Y.J.)

[2]   Institute of Information Microsystem, Beijing Information Science and Technology University, Beijing 100085, China; zhangyichuan@rails.cn

[3]   China Electronics Technology Group Corporation No. 54 Research Institute, Hebei 050081, China; yyz712@gmail.com (Y.Y.); xptang54@gmail.com (X.T.); luhuixiang54@gmail.com (H.L.)

[4]   Shenzhen Graduate School of Peking University, Shenzhen 518055, China

*   Correspondence: miaomin@bistu.edu.cn; Tel.: +86-010-6275-2536

Received: 28 December 2017; Accepted: 7 March 2018; Published: 9 March 2018

Abstract: In this paper, two versions of capacitive accelerometers based on low-temperature co-fired ceramic (LTCC) technology are developed, different with respect to the detection technique, as well as the mechanical structure. Fabrication of the key structure, a heavy proof mass with thin beams embedded in a large cavity, which is extremely difficult for the conventional LTCC process, is successfully completed by the optimized process. The LC resonant accelerometer, using coupling resonance frequency sensing which is first applied to LTCC accelerometer and may facilitate application in harsh environments, demonstrates a sensitivity of 375 KHz/g over the full scale range 1 g, with nonlinearity less than 6%, and the telemetry distance is 5 mm. The differential capacitive accelerometer adopting differential capacitive sensing presents a larger full scale range 10 g and lower nonlinearity less than 1%, and the sensitivity is 30.27 mV/g.

Keywords: low-temperature co-fired ceramic (LTCC); capacitive accelerometer; wireless; process optimization; performance characterization

---

1. Introduction

Low-temperature co-fired ceramic (LTCC) technology, which was initially applied for RF applications, is one of the integration techniques for microelectronic systems. Due to the ability to embed integrated passive devices into substrates and good electrical properties, such as low dielectric loss and high-speed transmission thanks to the usage of low dielectric ceramic and highly-conductive Ag/Pd/Au conductors, LTCC technology is widely used in the field of microwave circuits and highly-reliable electronic military components [1,2]. LTCC technology enables the fabrication of 3D structures by micromachining perforated features into individual green tape and then laminating and sintering the multilayer stack to form the compact integrated substrate/interposer. The merit is soon exploited by various applications, including biomedical devices, electrochemical devices, microfluidic devices, pressure sensors, and temperature sensors [3–7]. In the field of micro-accelerometers, silicon-based accelerometers have been widely used in inertial measurement, aerial navigation, and gravity gradient measurement [8–10]. LTCC-based accelerometers utilizing different sensing principles were also reported. Neubert et al. [11] reported the first LTCC accelerometer, which uses piezoresistors and measures the voltage gap in the bridge circuit to determine the acceleration. Subsequently, Jurkow et al. [12] proposed an LTCC accelerometer utilizing the piezoelectric effect. A patented PZT film was applied to the surface of LTCC membrane as the acceleration-sensing component. Moreover, a triaxial LTCC accelerometer using piezoresistors is proposed as the follow-up [13]. An early exploration of a LTCC capacitive accelerometer was conducted in [14], which mainly focused on simulation, and a fabrication process based on sacrificial material was conceived.

Due to the hermeticity, chemical inactivity, and high-temperature stability of the LTCC material, one major advantage of LTCC-based sensors over their silicon counterparts is the resistance to harsh environments [15], which facilitates the application of LTCC-based sensors in harsh environments where silicon-based sensors cannot be deployed. Additionally, LTCC-based sensors can be easily integrated in multi-component modules (MCMs) which usually use LTCC substrates as a platform to achieve a compact-sized system [3]. Compared to LTCC accelerometers based on piezoresistive and piezoelectric principles, which may introduce materials incompatible with LTCC and instability at high temperature, the capacitive accelerometer is more suitable for high-temperature applications. In this paper, the LTCC-based capacitive accelerometers are designed, fabricated, and characterized. The conventional LTCC process is unable to fabricate the key structure for the capacitive accelerometer, namely, a heavy proof mass with thin beams embedded in a large cavity, because the structure would collapse during co-firing. Thus, an optimized LTCC process flow is developed to solve the problem. Based on the acceleration-sensitive structure, two signal processing methods are applied to capacitive accelerometers: one is telemetry of the resonance frequency between the sensor and the readout unit by inductive coupling, and the other is translating the differential capacitive input into a voltage output using a commercial readout chip. The performances of accelerometers are confirmed by experiments, which demonstrate good wireless acceleration-input transmission for the LC resonant accelerometer, and stable performances for the differential capacitive accelerometer.

2. Structure Design and Process Optimization

The structure of the LTCC accelerometer is shown in Figure 1, which consists of three parts. In the middle part, the proof mass is suspended by four symmetrical beams. By screen-printing metal on the proof mass, it acts as a movable electrode, thereby forming a variable capacitor with top or bottom electrodes. Movement of the proof mass due to an out-of-plane acceleration causes changes in capacitance.

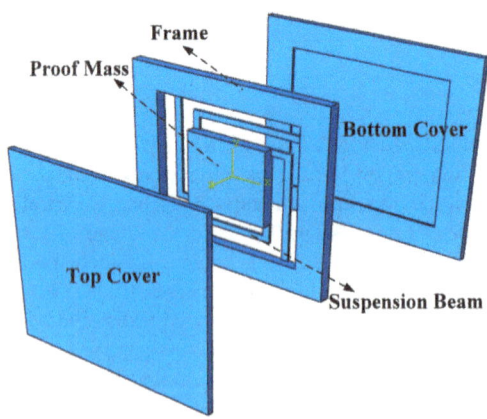

Figure 1. Schematic of the LTCC accelerometer.

The seismic middle part is sandwiched between top and bottom covers, which protect the sensitive structure and form a cavity for the proof mass to vibrate. In this paper, we applied two signal processing methods to the accelerometer structure. One method is the coupling resonant frequency sensing, which embeds a variable capacitor and an inductor in the accelerometer, and measures the resonance frequency of the LC circuit by a remote reader coil. The other method is differential capacitive sensing, which embeds a pair of differential capacitors into the accelerometer, and then translates differential capacitive input into voltage output using a commercial readout chip. Figure 2 shows the profile of the LTCC accelerometers. For the LC resonant accelerometer, a variable plate capacitor is formed between the top cover and the proof mass by screen printing electrodes on them. A spiral inductor is printed on the surface of the top cover, and wired to the capacitor's electrodes with vertical interconnection vias and horizontal interconnections. The differential capacitive accelerometer has a similar profile, and will be discussed in detail in Section 3.2. The beam-mass structure has a significant effect on the performance of the accelerometer, such as the measuring range and sensitivity, therefore, two types of beams, L-shapedd beams and Z-shapedd beams, are designed and fabricated as shown in Figure 3, and the location vias are designed for precise alignment of different parts in fabrication. Since coupling resonance frequency sensing is more sensitive to noise, the L-shapedd beam, which is easier to deform, is used to guarantee a high sensitivity. Additionally, differential capacitive sensing is more stable, and the Z-shaped beam is used to achieve a large measuring range.

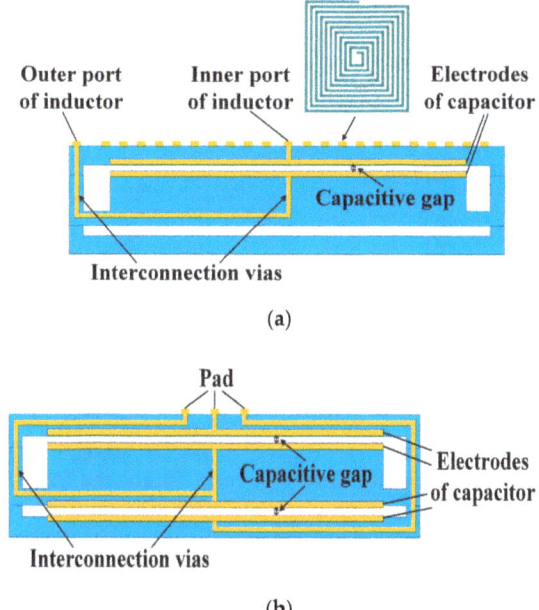

Figure 2. Profile of the accelerometers: (a) LC resonant accelerometer; and (b) differential capacitive accelerometer.

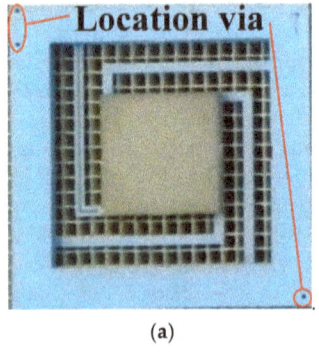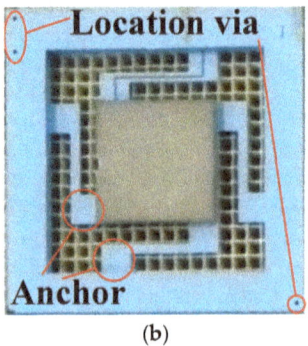

(a)   (b)

Figure 3. Beam-mass structure: (**a**) L-shapedd beams used in the LC resonant accelerometer; and (**b**) Z-shapedd beams used in the differential capacitive accelerometer.

Table 1 lists the designed physical dimensions of the accelerometers. The overall dimension of the accelerometer is 30 mm × 30 mm × 2.3 mm.

Table 1. Physical dimensions of the accelerometer.

Dimension Parameters	Value
Middle frame/top cover/bottom cover	30 mm × 30 mm × 1.5 mm/0.4 mm/0.4 mm
Cavity	22 mm × 22 mm × 1.7 mm
Proof mass	12 mm × 12 mm × 1.5 mm
L-shapedd beam	16 mm × 1 mm × 0.3 mm
Z-shapedd beam	6 mm × 1 mm × 0.3 mm
Anchor of Z-shapedd beam	3 mm × 3 mm × 0.3 mm
Capacitive gap	0.1 mm

Numerical simulations were performed to obtain the mechanical behavior of the beam-mass structure using FEM (finite element method) software ANSYS (ANSYS Inc., Canonsburg, PA, USA). The material properties are referred to in [16]. Due to its longer effective beams, the L-shaped beam-mass structure demonstrates much higher sensitivity than that of the Z-shaped beam-mass structure, which is 2.99 μm/g compared with 0.321 μm/g. However, the trade-off between sensitivity and bandwidth results in a lower resonance frequency for the L-shaped beam-mass structure. The first resonance frequency of the two beam-mass structure is 291 Hz and 885 Hz, corresponding to a vibration of the proof-mass in the Z direction. The next two modes following the first mode are torsional vibration around the two diagonal lines of the proof-mass, respectively. The simulation results are listed in Table 2. It is noted that the accelerometers are rotationally symmetrical about the center of the proof-mass, and the angle of rotational symmetry is 90 degrees, so the second and third resonance frequencies are the same due to equal stiffness and moments around the X-axis and Y-axis for these two modes.

Table 2. Simulated mechanical properties of beam-mass structure.

Parameters	L-Shaped Beams	Z-Shaped Beams
Displacement sensitivity	2.99 μm/g	0.321 μm/g
First resonance frequency	291 Hz	885 Hz
Second resonance frequency	634 Hz	1549 Hz
Third resonance frequency	634 Hz	1549 Hz

The accelerometers were fabricated with LTCC technology, but the traditional LTCC process has difficulties in fabricating the high-quality large cavity and beam-mass structures of accelerometers. To solve this problem, an optimized process, as shown in Figure 4, is proposed.

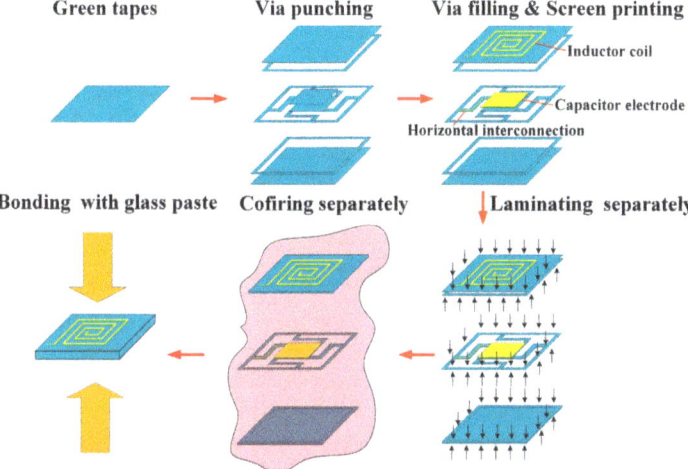

Figure 4. Optimized process flow for LTCC accelerometers.

Green tapes, which consist of alumina ceramic-filled glass systems mixed with an organic vehicle, are a basic material of LTCC technology. They are available from commercial suppliers at different thickness, and Dupont 951PT green tapes with 100 µm thickness were adopted in the fabrication. Depending on the thickness, the accelerometer needs 23 layers of green tape in total, of which four layers are for the top cover, 15 layers are for the middle frame, and four layers are for the bottom cover.

After preparing green tapes, the process moved to the via punching step to fabricate signal interconnection vias, location vias, and cavities. The programmable punching machine is controlled by a document which records the patterns of the green tapes.

Then the interconnection vias were filled with metal paste (Dupont Ag) using screen printing techniques, and the inductor, capacitor electrodes, and horizontal interconnection lines were also screen-printed.

The major difference between the proposed process and traditional LTCC process are subsequent steps. For the traditional LTCC process, after the previous steps, all green tapes will be laminated and co-fired together. However, the features of the accelerometer, particularly the large cavity with dimensions of 22 mm × 22 mm × 1.7 mm embedded in the structure and enormous difference in the mass of the beams and proof mass (the mass ratio of the proof mass to beams is 45 for L-shaped beams and 30 for Z-shaped beams), imposed a great challenge to fabrication because unfired green tapes were in a relatively soft state, the movable thin beams could not support the heavy proof mass structure, and they would collapse in the cavity.

In most cases, sacrificial layers that are easy to burn out, such as graphite powder-based paste, can be applied to solve this problem [17,18]. This method is typically used to fabricate cavities and channels free of deformation. The sacrificial layer supports the three-dimensional structure up to the burnout temperature during co-firing and, when the structure is stiff enough, it is burned out into gas and escapes from the intrinsic pores in green tapes, which is followed by densification and elimination of the pores of the LTCC tapes. Control of the burnout characteristics of the sacrificial layer is critical for this method. If the sacrificial layer starts to burn out before the tapes become stiff, the embedded structure will sag or even collapse. If the burnout of the sacrificial layer is not complete after the tapes' densification, the gas generated afterward will swell the tapes. Sagging and swelling problems also have a negative effect on the interconnections located on the surface of tapes. For the designed accelerometer, fabrication of the cavity is a challenge, and the existence of the suspension proof-mass makes it more difficult because neither of the two covers of the accelerometer could touch the proof-mass and the space is only 100 μm. Therefore, the next steps were optimized for the accelerometer, where the three parts of the accelerometer were laminated and co-fired separately and then bonded together with glass paste.

In step 4, the green tapes of each part were stacked, and these three parts were laminated separately in a laminating machine which adopts isostatic pressing in heated water. The process setting is isostatic hydraulic pressure of 20 MPa in 70 °C water for ten minutes. The parts were vacuum sealed in a plastic bag to prevent the water from coming into contact with them.

Then the three parts were co-fired separately. The temperature profile of co-firing is as follows: 20–400 °C for 5 h to volatize the organic particles; 400–600 °C for 6 h for structure formation, and the green tapes started to harden around 500 °C; then, 600–900 °C for 5 h for complete densification; and 900–20 °C for 3 h for cooling down.

The final step is bonding with glass paste. First, the three parts were aligned precisely with the help of a computer aided vision system, images of location vias on the surface of LTCC tapes to be aligned were captured by a CCD (charge-coupled device) camera (Sony, Tokyo, Japan), and alignment is accomplished by adjusting the images until they coincide. Then the stack was sintered at 600 °C, and a good bonding strength can be achieved because both glass and LTCC are isotropic materials.

The optimized LTCC process flow is very useful to fabricate movable structures in LTCC substrates, where cavities can be avoided during co-firing and, thus, more control on movable structures during fabrication can be obtained. With the process optimization, the LTCC accelerometers were fabricated successfully, as shown in Figure 5, and the X-ray inspection image proved the structural integrity.

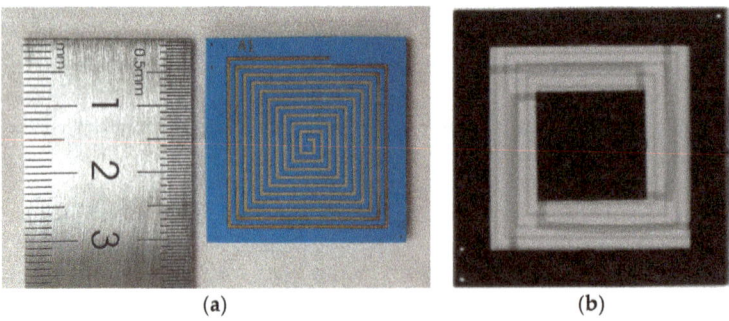

(a)   (b)

Figure 5. Fabricated LTCC accelerometers: (**a**) optical image; and (**b**) X-ray inspection image.

3. Signal Processing Methods

In this section, the two different signal processing methods applied on the designed accelerometers are discussed in detail, which are coupling resonance frequency sensing and differential capacitive sensing. By using the coupling resonance frequency sensing, the accelerometer can be easily deployed in harsh environments for its separated sensing circuits and reader antenna, but the involved signal processing is complicated, while the differential capacitive sensing is more stable because of its fully-developed interface circuit.

3.1. Coupling Resonance Frequency Sensing

Due to the mechanical stability of the LTCC material, LTCC-based sensors can be deployed in harsh environments. However, the signal readout and processing unit still need to be in a safe environment. One solution is the telemetry between the sensor and readout unit by inductive coupling, which has been applied to LTCC-based pressure sensors [19,20] and temperature sensors [21,22]. Wireless readout of the acceleration is first introduced to the field of LTCC accelerometers. The principle is shown in Figure 6. The acceleration signal is translated into resonance frequency changes by a variable capacitor, which is then detected through the coupling between the reader coil and the sensor coil.

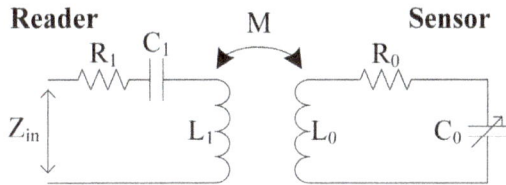

Figure 6. Equivalent circuit of the inductively-coupled sensor system.

Based on the circuit in Figure 6, the equivalent input impedance Z_{in} at the reader coil port is given by:

$$Z_{in}(j\omega) = R_1 + j\omega L_1 + \frac{1}{j\omega C_1} + \frac{\omega^2 M^2}{R_0 + j\omega L_0 + \frac{1}{j\omega C_0}} \tag{1}$$

where L_1, R_1, and C_1 are the inductance, parasitic resistance, and parasitic capacitance of the reader coil, L_0 and R_0 are the inductance and parasitic resistance of the sensor coil. C_0 is the capacitance of the variable capacitor. M denotes the mutual inductance between reader coil and sensor coil, which is given by:

$$M = k\sqrt{L_1 L_0} \tag{2}$$

where k is the coupling coefficient with a value between 0 and ± 1. When angular frequency ω equals $1/\sqrt{L_0 C_0}$, which is the resonance frequency of the sensor circuit, the magnitude of Z_{in} is at its maximum and, meanwhile, the phase of Z_{in} will demonstrate a phase dip [23]. Thus, the resonance frequency of the sensor loop can be obtained by a frequency sweep on Z_{in}, and then picking its magnitude maximum.

Table 3 lists the physical dimensions of the passive components in the accelerometer. The square spiral inductor is used as the sensor's coil, and its inductance can be derived with an empirical equation [24]:

$$L = K_1 \mu_0 \frac{n^2 d_{avg}}{1 + K_2 \rho} \tag{3}$$

where K_1 and K_2 are empirical coefficients dependent on the coil shape. For square coils, K_1 and K_2 are 2.34 and 2.75, respectively. μ_0 is the permeability of a vacuum and n is the number of coil turns. The average diameter d_{avg} is given by $d_{avg} = (d_{in} + d_{out})/2$, and the fill ratio ρ is given by $\rho = (d_{out} - d_{in})/(d_{out} + d_{in})$, where d_{in} and d_{out} are the inner and outer diameter, respectively. The calculated inductance is 1.26 µH and the initial capacitance of the variable capacitor is estimated as 12.744 pF by the plate capacitance formula. Therefore, the calculated resonance frequency is 39.72 MHz.

Table 3. Physical dimensions of the inductor and capacitor.

Parameters	Value
Inner diameter of inductor coil	2.5 mm
Outer diameter of inductor coil	25.3 mm
Number of turns	10
Line width of coil	0.5 mm
Line spacing of coil	0.7 mm
Capacitor dimension	12 mm × 12 mm
Gap between capacitor electrodes	0.1 mm

As indicated by Equation (1), different mutual inductance M results in different maximal magnitude of Z_{in} at resonance frequency. Increasing M is the most straightforward method to increase the signal-to-noise ratio of the system. Therefore, the sensor coil and read coil should be placed close enough to maintain a detectable impedance change. The impact of distance on coupling coefficient was investigated with the electromagnetic field solver ANSYS Q3D (ANSYS Inc., Canonsburg, PA, USA). The simulated inductors (both the sensor's and reader's) are of the same size with the one we used in the accelerometer. The results, as shown in Figure 7, demonstrate that, if the distance is larger than 5 mm, the coupling coefficient is too small to be detected ($k < 0.1$). If a longer distance is desired for a specific application, increasing the inductor diameter can solve the problem, but at the expense of small size.

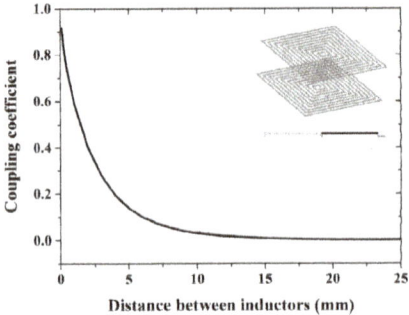

Figure 7. Effect of distance between inductors on the coupling coefficient.

3.2. Differential Capacitive Sensing

With the advantages of cancelling out the common-mode noise and high sensitivity, differential capacitive sensing is very common in accelerometers. The profile of the LTCC-based accelerometer utilizing differential capacitive sensing is shown in Figure 2b.

Figure 8 shows the evaluation board used for signal processing of the differential capacitive accelerometer. A commercially-available MS3110 (MicroSensors, Costa Mesa, CA, USA) readout chip was bonded onto the top cover. In addition, four 0306 SMT (surface mount technology) capacitors (Murata, Kyoto, Japan) were placed on the surface as filtering capacitors. Then, the differential capacitive input can be translated into the voltage output by the readout chip. From the datasheet of the MS3110, the transfer function between the differential capacitance and output voltage is given by:

$$V_{out} = \frac{2.25 \cdot 1.14 \cdot Gain \cdot \Delta C}{CF} + VREF \tag{4}$$

where ΔC is the differential capacitance. The reference voltage $VREF$ is 2.25 V, the feedback capacitor is 7.296 pF, and $Gain$ is set to 4 in the experiment. These parameters can be set with the peripheral circuits on the evaluation board.

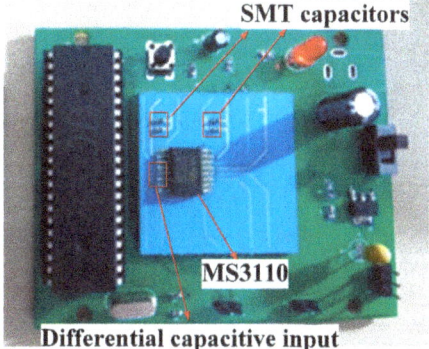

Figure 8. Differential capacitive accelerometer integrated with the signal processing circuits.

4. Performance Characterization

In this part, a static gravitational field test was performed for both the LC resonant accelerometer and the differential capacitive accelerometer to measure the acceleration sensitivity. The dynamic performance of the differential capacitive accelerometer was also evaluated using a vibration exciter. The results were obtained by averaging three repeated measurements for error reduction.

4.1. LC Resonant Accelerometer

A static gravitational field test was carried out on the LC resonant accelerometer with a dividing head, which could be easily and precisely rotated to preset angles or circular divisions with an angle error of less than 1 s. Figure 9 shows the setup of the test environment. The accelerometer was stuck to the dividing head table. The reader coil is 2 mm above the sensor, and connected to an AV3629A vector network analyzer (VNA, CETI, Shandong, China). The VNA measured the 1-port S-parameter (scattering parameter of the reader coil, which is a 1-port network) from 0.1 to 100 MHz. The input impedance is then derived by:

$$Z_{in} = Z_0 \frac{1 + S_{11}}{1 - S_{11}} \tag{5}$$

where S_{11} is the one-port S-parameter, and Z_0 is the reference impedance of the system, which is 50 Ω in this case.

The L-shaped beam described in Section 2 is used in the LC resonant accelerometer to guarantee a high sensitivity, in which case the calculate resonance frequency is 39.72 MHz and the estimated sensitivity is 598 KHz/g. The measured magnitude and phase of the input impedance with the acceleration of 1 g (the sensor inductor side faces up) is shown in Figure 10. The magnitude reaches its maximum value at 39.73 MHz, which is the resonance frequency of the sensor circuit. This corresponds well to the calculation. Figure 11 shows the measured resonance frequency vs. input acceleration using the dividing head (Tianhe Mechanical and Electrical Company, Shanghai, China). As the dividing head rotates from 0° to 180° (data was sampled once every 10°), the acceleration applied on the sensor changes from 1 g to −1 g, and the capacitance of the movable capacitor is increasing, which results in a decreasing resonance frequency. Zero offset is calculated as 40.12 MHz by averaging the outputs of accelerometer when acceleration is ±1 g. The measured sensor's sensitivity is 375 KHz/g (equivalent to 1.88 µm/g in displacement), which is smaller than the estimate, may be caused by the slight distortion of the long-beam structure as shown in Figure 5b, because the fabricated beams are not as ideal as the ones in the simulation, and deformation occurs in the center area of beams and degrades performance. The nonlinearity is caused by the measurement error, because the long cable connecting the reader coil and VNA is very sensitive. Even with careful calibration, the parasitic effect induced by the cable can be changed by slight movement during the experiment. This can be solved by designing a compact signal processing circuit into the reader, which should include functionalities of frequency sweep, demodulation, and peak value extraction.

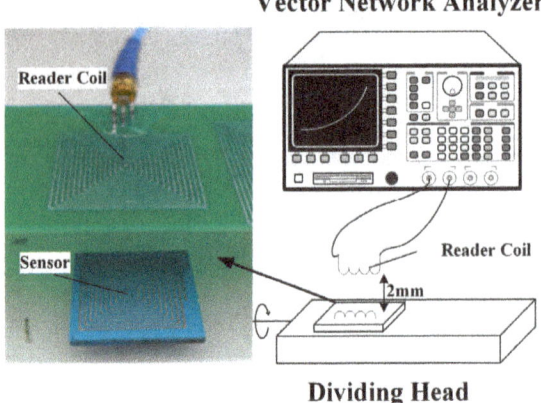

Figure 9. Dividing head test setup.

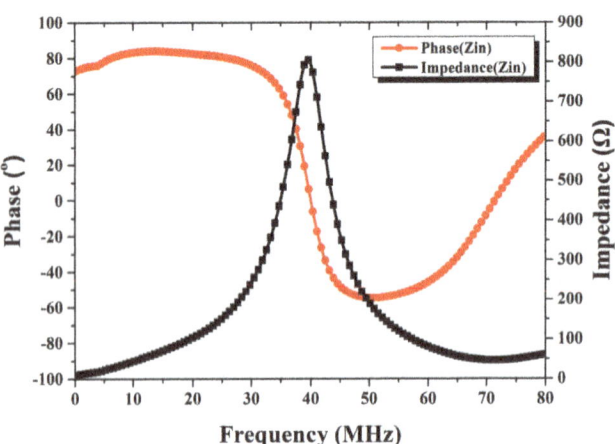

Figure 10. The impedance and phase of the input impedance with an acceleration of 1 g.

To determine the effective distance of wireless transmission, input impedance is measured at different distances of the reader coil and sensor, and the results are shown in Figure 12. As the distance increases from 4 to 8 mm, the V-shaped pattern formed by the phase curve becomes narrower and shallower, resulting in a poor signal-to-noise ratio. Similar results are also observed in impedance magnitude: the maximum value at the resonance frequency is decreasing, and the resonance wave is disappearing. The results indicate that the effective wireless transmission distance is 5 mm for our design.

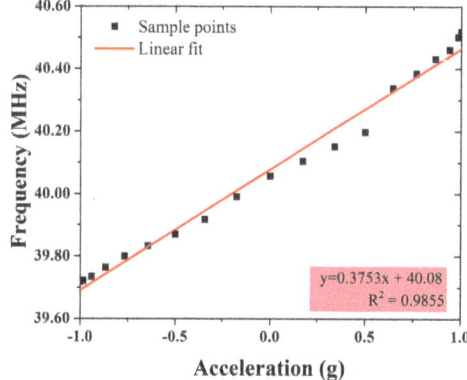

Figure 11. Resonance frequency vs. input acceleration.

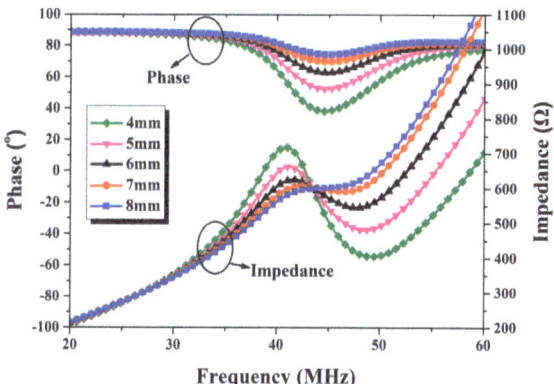

Figure 12. The impedance and phase of input impedance changes at different distances.

4.2. Differential Capacitive Accelerometer

The performances of the differential capacitive accelerometer are reported in this part. The differential capacitive accelerometer uses Z-shaped beams. First of all, the zero offset was measured. By averaging the output voltage with dividing head at 0° and 180° (acceleration input is ±1 g), the zero offset in differential capacitance was calculated as 3.20 fF by Equation (4), with the output voltage being 2.255 V.

The dividing head test was also carried out for the differential capacitive accelerometer. The input acceleration is from 1 g to −1 g and then back to 1 g as the dividing head rotates from 0° to 360°, and the results is shown in Figure 13. The measured sensitivity in the dividing head test is 30.27 mV/g, equivalent to 21.50 fF/g in differential capacitance and 0.844 µm/g in displacement, and nonlinearity is less than 1%.

Subsequently, the dynamic performance of the accelerometer was characterized with a vibration exciter (Bruel and Kjaer, Copenhagen, Denmark). Figure 14 shows the test environment. The vibration frequency and amplitude were controlled by the signal generator (Agilent Technologies, Santa Clara, CA, USA) and power amplifier (SINOCERA, Shanghai, China). An 80 Hz sinusoidal acceleration input of different amplitude was used to drive the accelerometer. As the amplitude increases from 1.41 to 10.7 g, the sensor's peak output voltage is increasing from 2.344 to 2.625 V, as shown in Figure 15.

Therefore, the measured sensitivity in the vibration test is 29.58 mV/g with the nonlinearity less than 2%, and the full-scale range is over 10 g.

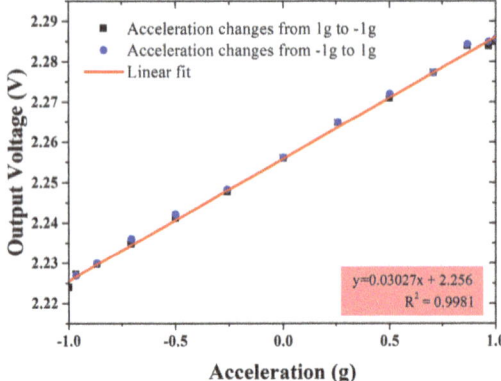

Figure 13. Dividing head test results of the differential capacitive accelerometer.

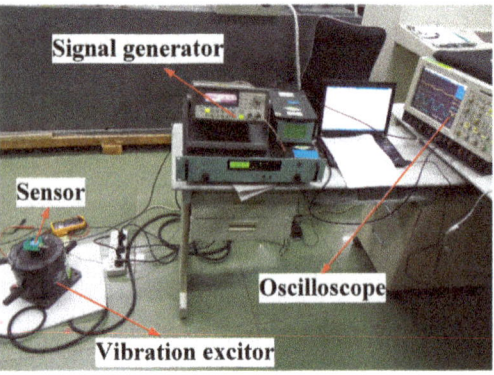

Figure 14. Test environment for the accelerometer dynamic performance.

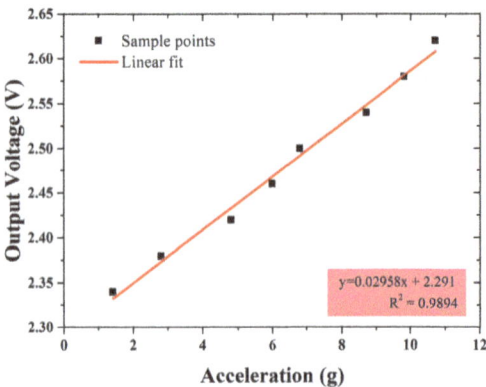

Figure 15. Vibration exciter test results of the differential capacitive accelerometer.

The performance of both the LC resonant accelerometer and the differential capacitive accelerometer is summarized in Table 4. It is noted that because the two accelerometers are based on different detection techniques, their sensitivity is in different units, and equivalent displacement sensitivity is provided in brackets for convenient comparison. Since the Z-shaped beams are stiffer than the L-shaped beams, the sensitivity of LC resonant accelerometer is 20 times higher than the differential capacitive accelerometer. However, the differential capacitive accelerometer has a larger full scale range and better characteristics with regard to nonlinearity, which benefits from the stable signal detection method. In addition, a qualitative judgment about the accuracy of the two detection techniques can be obtained. Differential capacitive sensing could cancel out the common-mode noise which still exists in coupling resonance frequency sensing. Additionally, wireless transmission without any shielding measures are more sensitive to electromagnetic noise in the environment than reliable wired interconnections, thus, differential capacitive sensing is more accurate compared with coupling resonance frequency sensing.

Table 4. Comparison of the LC resonant accelerometer and differential capacitive accelerometer.

Parameters	LC Resonant Accelerometer	Differential Capacitive Accelerometer
Type of beams	L-shaped beams	Z-shaped beams
Sensitivity	375 kHz/g (1.88 μm/g)	30.27 mV/g (0.844 μm/g)
Full scale range	1 g	10 g
Zero offset	40.12 MHz	2.255 V
Nonlinearity	Less than 6%	Less than 1%

5. Conclusions

Two versions of LTCC-based capacitive accelerometers with different detection methods and mechanical structures are developed in this paper. The optimized LTCC process is effective in fabricating the key structures, such as the heavy proof mass with thin beams embedded in a large cavity. The LC resonant accelerometer has a high sensitivity 375 kHz/g over the full scale range of 1 g, and the separated sensor part and reader circuit facilitates the application of the accelerometer in harsh environments, with wireless readout achieved as far as 5 mm. The differential capacitive accelerometer demonstrates a stable performance of sensitivity of 30.27 mV/g, with nonlinearity less than 1% over the range ±1 g, and a full scale range over 10 g. This type of accelerometer can be used for navigation in dynamic vehicles. The future work is to reduce the size of the accelerometers by fabricating capacitor electrodes and the inductor coil distributed on different layers of LTCC tapes, and to realize three-axis inertial measurement.

Acknowledgments: This work is co-funded by the National Basic Research Program of China (No. 2015CB057201), the National Natural Science Foundation of China (No. 61176102, No. 61674016 and No. U1537208), the Importation and Development of High-Caliber Talents Project of Beijing Municipal Institutions (Great Wall Scholar, No. CIT&TCD20150320), and Beijing Nova Program Interdisciplinary Studies Cooperative Projects (No. Z161100004916036).

Author Contributions: Huan Liu and Runiu Fang conceived the experiments, and performed the structure design and simulation; Yingzhan Yan, Xiaoping Tang, and Huixiang Lu helped with the process optimization; Huan Liu, Runiu Fang, and Yichuan Zhang performed the performance characterization experiments and analyzed the data; Huan Liu wrote the paper; and Min Miao and Yufeng Jin contributed creative ideas to the structure design and manuscript writing.

Conflicts of Interest: The authors declare no conflict of interest.

References

1. Ko, Y.-J.; Park, J.Y.; Ryu, J.-H.; Lee, K.-H.; Bu, J.U. A miniaturized LTCC multi-layered front-end module for dual band WLAN (802.11 a/b/g) applications. In Proceedings of the 2004 IEEE MTT-S International Microwave Symposium Digest, Fort Worth, TX, USA, 6–11 June 2004; pp. 563–566.

2.   Ponchak, G.E.; Chun, D.H.; Yook, J.G.; Katehi, L.P.B. The use of metal filled via holes for improving isolation in LTCC RF and wireless multichip packages. *IEEE Trans. Adv. Packag.* **2000**, *23*, 88–99. [CrossRef]

3.   Miao, M.; Jin, Y.F.; Fang, R.N.; Mu, F.Q.; Guo, S.C.; Zhang, X.Q.; Zhang, Y.; Hu, D.W.; Li, Z.S.; Xiang, W. Investigation of Micromachined LTCC Functional Modules for High-density 3D SIP based on LTCC Packaging Platform. In Proceedings of the IEEE 63rd Electronic Components and Technology Conference (ECTC), Las Vegas, NV, USA, 28–31 May 2014; pp. 1815–1822.

4.   Darko, E.; Thurbide, K.B.; Gerhardt, G.C.; Michienzi, J. Characterization of Low-Temperature Cofired Ceramic Tiles as Platforms for Gas Chromatographic Separations. *Anal. Chem.* **2013**, *85*, 5376–5381. [CrossRef] [PubMed]

5.   Goldbacha, M.; Axthelm, H.; Keusgen, M. LTCC-based microchips for the electrochemical detection of phenolic compounds. *Sens. Actuator B-Chem.* **2006**, *120*, 346–351. [CrossRef]

6.   Malecha, K.; Remiszewska, E.; Pijanowska, D.G. Technology and application of the LTCC-based microfluidic module for urea determination. *Microelectron. Int.* **2015**, *32*, 126–132. [CrossRef]

7.   Sadler, D.J.; Changrani, R.; Roberts, P.; Chou, C.F.; Zenhausern, F. Thermal management of BioMEMS: Temperature control for ceramic-based PCR and DNA detection devices. *IEEE Trans. Compon. Packag. Technol.* **2003**, *26*, 309–316. [CrossRef]

8.   Xu, W.; Yang, J.; Xie, G.; Wang, B.; Qu, M.; Wang, X.; Liu, X.; Tang, B. Design and Fabrication of a Slanted-Beam MEMS Accelerometer. *Micromachines* **2017**, *8*, 77. [CrossRef]

9.   Li, W.; Song, Z.; Li, X.; Che, L.; Wang, Y. A novel sandwich capacitive accelerometer with a double-sided 16-beam-mass structure. *Microelectron. Eng.* **2014**, *115*, 32–38. [CrossRef]

10.   Li, Z.; Wu, W.J.; Zheng, P.P.; Liu, J.Q.; Fan, J.; Tu, L.C. Novel Capacitive Sensing System Design of a Microelectromechanical Systems Accelerometer for Gravity Measurement Applications. *Micromachines* **2016**, *7*, 167. [CrossRef]

11.   Neubert, H.; Partsch, U.; Fleischer, D.; Gruchow, M.; Kamusella, A.; Pham, T.-Q. Thick Film Accelerometers in LTCC Technology—Design Optimization, Fabrication, and Characterization. *JMEP* **2008**, *5*, 150–155. [CrossRef]

12.   Jurkow, D.; Dabrowski, A.; Golonka, L.; Zawada, T. Preliminary Model and Technology of Piezoelectric Low Temperature Co-fired Ceramic (LTCC) Uniaxial Accelerometer. *Int. J. Appl. Ceram. Technol.* **2013**, *10*, 395–404. [CrossRef]

13.   Jurkow, D. Three axial low temperature cofired ceramic accelerometer. *Microelectron. Int.* **2013**, *30*, 125–133. [CrossRef]

14.   Hua, G.; Yufeng, J.; Min, M.; Xin, S. A novel LTCC capacitive accelerometer embedded in LTCC packaging substrate. In Proceedings of the 2011 6th IEEE International Conference on Nano/Micro Engineered and Molecular Systems (NEMS 2011), Kaohsiung, Taiwan, 20–23 February 2011; pp. 796–799.

15.   Imanaka, Y. *Multilayered Low Temperature Cofired Ceramics (LTCC) Technology*, 1st ed.; Springer: New York, NY, USA, 2005.

16.   Technical Datasheet of 951 Low Temperature Ceramic System. Available online: http://www.dupont.com/content/dam/dupont/products-and-services/electronic-and-electrical-materials/documents/prodlib/951.pdf (accessed on 20 December 2017).

17.   Birol, H.; Maeder, T.; Ryser, P. Processing of graphite-based sacrificial layer for microfabrication of low temperature co-fired ceramics (LTCC). *Sens. Actuator A Phys.* **2006**, *130*, 560–567. [CrossRef]

18.   Khoong, L.E.; Tan, Y.M.; Lam, Y.C. Overview on fabrication of three-dimensional structures in multi-layer ceramic substrate. *J. Eur. Ceram. Soc.* **2010**, *30*, 1973–1987. [CrossRef]

19.   Xiong, J.J.; Li, Y.; Hong, Y.P.; Zhang, B.Z.; Cui, T.H.; Tan, Q.L.; Zheng, S.J.; Liang, T. Wireless LTCC-based capacitive pressure sensor for harsh environment. *Sens. Actuator A Phys.* **2013**, *197*, 30–37. [CrossRef]

20.   Tan, Q.L.; Yang, M.L.; Luo, T.; Liu, W.; Li, C.; Xue, C.Y.; Liu, J.; Zhang, W.D.; Xiong, J.J. A Novel Interdigital Capacitor Pressure Sensor Based on LTCC Technology. *J. Sens.* **2014**, *431503*. [CrossRef]

21.   Sardini, E.; Serpelloni, M. High-temperature measurement system with wireless electronics for harsh environments. In Proceedings of the 2011 IEEE Sensors Applications Symposium (SAS), San Antonio, TX, USA, 22–24 February 2011; pp. 256–261.

22.   Radosavljevic, G. Wireless LTCC sensors for monitoring of pressure, temperature and moisture. *Inf. MIDEM-J. Microelectron. Electron. Compon. Mater.* **2012**, *42*, 272–281.

23. Nopper, R.; Niekrawietz, R.; Reindl, L. Wireless Readout of Passive LC Sensors. *IEEE Trans. Instrum. Meas.* **2010**, *59*, 2450–2457. [CrossRef]
24. Mohan, S.S.; Hershenson, M.D.; Boyd, S.P.; Lee, T.H. Simple accurate expressions for planar spiral inductances. *IEEE J. Solid-State Circuit* **1999**, *34*, 1419–1424. [CrossRef]

MDPI

St. Alban-Anlage 66

4052 Basel

Switzerland

Tel. +41 61 683 77 34

Fax +41 61 302 89 18

www.mdpi.com

Micromachines Editorial Office

E-mail: micromachines@mdpi.com

www.mdpi.com/journal/micromachines

Printed in June 2019
by Rotomail Italia S.p.A., Vignate (MI) - Italy